I0465619

Angel V. Oneto

Numbers, Rings, and Fields

*A Rigorous Introduction
to Numbers and
Algebraic Structures*

Printed in the United States of America

ISBN: 9781791663964

Τ his book intends to be a rigorous introduction to numbers and algebraic structures.

It is introductory because previous knowledge is not necessary for its study—although a certain mathematical maturity is necessary. Aimed at students of mathematics—which implies the use of a higher level of rigor in the treatment than that used in other careers—this does not mean that it is not useful for students of technical disciplines. We have even observed that a higher level of rigor pleasantly surprises these students, by providing deeper explanations to subjects they had learned in a more naïve way.

It will be useful both for the beginner and the more advanced student who looks forward to consolidating their knowledge. From their study, they will be able to obtain a solid foundation on numbers and on the functions most used in mathematics, which will be essential for higher studies. They will understand the ubiquity of the structures of rings and fields, characteristic of algebraic structures, that provide unity and economy of thought.

The table of contents is a good summary of the topics dealt with; here we will make an informal recount of each chapter, nonetheless.

The elementary notions about set theory are part of the basic language of mathematics, so we begin with an informal treatment of this language.

Then we axiomatically define real numbers (only the axioms that define an ordered field) and begin a formal deductive process that allows not only to become familiar with such deductive approach but also to deal with these numbers with confidence.

Next, we discuss natural numbers and induction, and their role in deriving the properties of powers and the fundamentals of combinatorics. Here, as elsewhere, the same result is proved in several ways, to emphasize the idea that "unique" proofs don't exist.

Integers and elementary divisibility lead to the study of residual arithmetic, which makes natural the, already implicitly given, definitions of rings and fields. These residual rings are a source of examples and motivation for a further study of abstract algebra.

In Chapter 6, we complete the axiomatic of real numbers and prove the essential properties of archimedianity, existence of an integral part and of roots, density of rationals, the principles of monotonous sequences, of embedded intervals, and the Cauchy criterion. We will also define the exponential function and prove its fundamental properties.

In the next chapter, we construct complex numbers from real numbers; with those, algebra reaches its fullness and roundness. Equations of degrees 2, 3 and 4 are solved, and with the help of trigonometric functions, the binomial equations are also solved.

The last chapter deals with another fundamental structure of algebra: polynomials. The study of the properties of divisibility and factorization in a general domain, in spite of raising the level of abstraction, allows us to obtain three classical Fermat theorems.

A series of appendices complement some chapters and explore new concepts.

In Appendix 1—which is a continuation of chapter 6— real numbers are built from rational numbers. This construction is the crucial step in a genetic study of numbers, that is, in the construction of natural numbers from Set Theory, then the integers, the rationals, the reals and, finally, the complex numbers.

Appendix 2 exploits the properties of the complex plane to deal with the classical constructions with straightedge and compass, proving the impossibility of trisecting the angle and duplicating the cube. On the positive side, we prove Gauss's theorem on the construction of the heptadecagon.

Appendix 3 deals with hypercomplex systems assuming a knowledge of the basic properties of linear algebra. Quaternions and octonions are constructed, and their uniqueness is shown by proving the theorems of Frobenius and Zorn. The singularity of the cross product is also proved and, as a corollary of the Euclideanity of the Hurwitz quaternions, the Four Square Theorem is obtained.

In the last appendix, Appendix 4, we fundament the exponential functions and the trigonometric functions that were already used in the chapter on complex numbers. A first course of Calculus is assumed here.

The exercises, although ordered by sections, are grouped at the end of each chapter, for practical reasons. We intended to make of these problems, genuine exercises; that is, not just exercises of first familiarization with a concept (of this class several are proposed to the reader throughout the text), nor additional, theoretical developments (except for very few of them).

A star attached to the statement of an exercise means that a solution of it will be found at the end of the book. Among these "starred exercises" there are those that can be considered the most difficult, there are others that are requisites for those and still others of which it has been wanted to present a particular solution or to make some commentary.

Two notations, already usual, have been adopted: the symbol "■" that replaces the old *"quod erat demostrandum"* and the notation "iff" to replace, in the definitions, the "if and only if."

Many people have contributed to the realization of this book, and this is the place to give them their credits and express my gratitude. Many teachings, ideas, and notes of my professor Enzo Gentile are reflected here. The encouragement of colleagues and especially of students who later became colleagues and friends has been invaluable. As the list is long, I prefer to omit it. My nephew, Leonardo Lospennato, has helped me with some language problems, with the edition and the publication. My wife, Silvia Lucarini, has interceded in my fights with the computer and has helped me in many other ways. To all of them, thank you very much!

Criticism, suggestions, and comments are welcome via email at: angeloneto@hotmail.com.

Table of contents

(Page intentionally left blank)

CHAPTER 1
SETS AND FUNCTIONS

This chapter starts with a brief treatment of formal logic, with the main purpose of precisely establish the sense in which logical connectives are used.

It is worth noting that mathematical reasoning precedes the formalization of logic, which seems to have emerged inspired by Pythagorean mathematics and had no influence in Greek mathematics, neither contemporary nor following Aristotle, author of the first systematic treatise on logic.

Additionally to that short and informal exposition of logic, we move on to set theory, and we will recall the concepts of inclusion, membership, and the elementary operations of intersection, union, difference and complement—notions which the students are already familiar with. The concepts of *relation* and *function* are also developed in finer detail, mostly for later reference.

1 - PROPOSITIONS

A mathematical theory is formally presented as a series of statements accompanied by arguments (called proofs), which try to establish the correctness of those statements.

Each statement, with its proof, is called either a theorem, a lemma, or a corollary. These words can be considered synonyms, although generally 'theorem' is reserved for a prominent result, 'lemma' for an auxiliary result prior to another, and 'corollary' for one that readily follows from a previous one. The use of such hierarchy, however, remains subjective.

They appear also definitions as well as axioms or postulates. Definitions are conventions or agreements that regard the use of certain words or symbols in a definite sense. Axioms or postulates are declarative sentences assumed true without proof, from which inferences are made to generate theorems. Such is the deductive method used universally in mathematics.

Proofs consist of inferring valid conclusions from premises considered true. The analysis of the principles of valid inference is the object of logic. It was the reflection on geometric reasoning what probably prompted the first researches on logic, stimulated as well by dialectic argumentations ("dialectic" comes from *discussion*); if a contradictory or false consequence is derived from an initial hypothesis, that hypothesis must be rejected. For example, Plato attributed to Socrates the argument that if virtue were susceptible of being taught, honest men would have instructed it to their children; it is well known, however, that Pericles, Themistocles, and Aristides failed to make their children virtuous, therefore...

Aristotle attributed to Zeno of Elea the invention of dialectics, or more precisely, the reduction to the impossible in metaphysics. Zeno could have taken it from the reduction to the absurd of Pythagorean mathematics, in particular from the proof of the irrationality of the square root of 2.

Zeno is famous for its sophisms (fallacious arguments that appear to be sound ones) which are known as Zeno's paradoxes. One of them is the paradox of Achilles and the tortoise. Achilles can never win a race against a tortoise if he gives the animal a head start, because, Zeno argued, Achilles must first reach the point where the tortoise started, and then he must reach the new point that the tortoise reached in the meantime, and so on *ad aeternum*.

Another example is the paradox of the arrow. An arrow can never reach its target, for it should first arrive at the middle of the distance between the archer and the target, then at the middle of the remaining distance, and so on—forever.

Zeno formulated these paradoxes to support the assertion of his mentor, Parmenides, that "movement is an illusion" (in opposition to those who affirmed that "all is movement").

The first systematic treatise on logic is The Organon of Aristotle. One part of it, the "Topics," was intended for the instruction of those who took part in public disputes. Indeed, the Greeks were fond to publicly discuss the most varied ubjects; an example of this inclination to controversy is the dilemma of Protagoras and Euathlus. They agreed that Protagoras would instruct Euathlus in the art of litigation, and the later would pay his master after winning his first lawsuit. Protagoras fulfilled his part of the agreement. But when Euathlus decided not to practice law, the master decided to sue. Already in court, Protagoras presents a devastating dilemma: No matter the outcome of the trial, Euathlus must pay; if the judges gave the reason to Protagoras, Euathlus should pay him (by the court's decision), whereas if they ruled in favor to Euathlus, the later would have won his first trial, being forced to honor the agreement. The replay of Euathlus is no less convincing: if Protagoras is the winner of the trial, Euathlus is not required to pay, because he would not have won his first trial; while if the favored with the decision is Euathlus, he should not pay, by the court's decision.

We will only deal with a part of logic called Propositional Calculus—specifically, with the definition of logical connectives. For a systematic treatment [20], [21], [24] or [38] can be consulted and [25] for a historical development.

A *proposition* is the meaning of a declarative sentence of which it makes sense to say that it is true (T) or false (F).

For example, "1 < 2", "2 > 1", and "one is less than two" are the same proposition; "$2 + 2 = 1$" is also a proposition; while "you must study a lot" is not a proposition: it neither enunciates anything about someone or something, nor can it be assigned a value of truth or falsity.

At this point, we are not interested in the internal structure of a proposition but only in its connections with other propositions. Such fundamental connections between propositions are: "no", "and", "or," and "implies." Since ordinary language is often imprecise or vague on the use of these connectives, we will clarify the meaning in the present context.

Negation: if p is a proposition, $\sim p$ (pronounced "not p") denotes its denial (or its contrary); so if p is "1 is less than 2", $\sim p$ is "1 *is not less than* 2" or "$1 \geq 2$." If p is true, then $\sim p$ is false, and if p is false, then $\sim p$ is true; this can be schematized in the following "truth table," which can be seen as the definition of the function $\sim$:

p	$\sim p$
T	F
F	T

Conjunction: If p and q are propositions, $p \wedge q$ (or p and q —conjunction of p, q) is the proposition that is true if, and only if, both p and q are true. Consequently, $\wedge$ is defined by:

13

p	q	$p \wedge q$
T	T	T
T	F	F
F	T	F
F	F	F

So, if p is the proposition "$1 < 2$" and q is "$2 < 3$", $p \wedge q$ is the proposition "$1 < 2$ and also $2 < 3$", which we can symbolize by "$1 < 2 < 3$".

Disjunction: If p and q are propositions, $p \vee q$ (or p or q —disjunction of p, q) is the proposition that is true if, and only if, at least one of them is true.

Its truth table is:

p	$\sim p$	$p \vee q$
T	T	T
T	F	T
F	T	T
F	F	F

Implication: If p and q are propositions, $p \Rightarrow q$ ("if p then q") is the proposition that states that, correctly reasoning, from p follows q. For example: "$-1 = 1$" $\Rightarrow$ "$0 = 2$", for starting from $-1 = 1$, adding 1 to both members, it follows that $0 = 2$. We can see that, from a false assumption, but reasoning correctly, a false statement can follow. Also, from a false assumption, a true statement can follow. For example: "$-1 = 1$" $\Rightarrow$ "$0 = 0$", since multiplying both members of $-1 = 1$ by 0, we get 0 = 0. And from a true statement, a true one may follow, of course; what cannot happen is that from a true proposition, and reasoning correctly, a false one be inferred. From these considerations we adopt the following as the truth table of $p \Rightarrow q$:

p	q	$p \Rightarrow q$
T	T	T
T	F	F
F	T	T
F	F	T

In the expression $p \Rightarrow q, p$ is called the *antecedent* and q the *consequent* of the implication, and it is also said that q is *necessary* for p and that p is *sufficient* for q.

Another connective frequently used is the *double implication*: $p \Leftrightarrow q$, which is an abbreviated way of expressing $(p \Rightarrow q) \wedge (q \Rightarrow p)$.

In ordinary language, the expression "or" can be used in an inclusive sense, i.e., in agreement with the connective $\vee$. Example: "The teachers or students of the university will receive a special discount." The advertisement does not intend to exclude teachers that are also students. On the other hand, the phrase "tonight we will have Chinese or Italian food" generally excludes the simultaneous enjoyment of both kinds of food. In such cases, the context of the sentence is enough to inform which type of "or" is used; however, there can be ambiguous cases where is best to use "either..., or"to specify the exclusive case and "and/or" to specify the inclusive case.

Also, in ordinary language, a connective is applied to a pair of propositions if they are related in some obvious sense. A sentence like: "New York is a big city, and my car is red" could generate questions about the mental health of anyone who says it; however, this is a subjective issue and sentences that sound unrelated to someone, may make perfect sense to others.

Moreover, seemingly unrelated mathematical statements as:

p: "m is a prime number of the form $2^{2^n}+1$."

q: "the regular polygon of m sides is constructible with straightedge and compass,"

are actually closely related, being $p \Rightarrow q$ a theorem proved by Gauss. Much of the beauty of a mathematical result lies in connecting objects that may seem unrelated at first glance.

In ordinary speech (but depending on the language) it happens that a double negation can be interpreted either as an affirmation (as it logically should be, since $\sim\sim p$ is equivalent to p) or as an emphatic negation. An anecdote tells us about a famous philosopher who was giving a lecture precisely on this subject; he went on to explaining how in some languages a double negation is an affirmation, but in others is a strong negation; what never happens, though, is having a double affirmation meaning negation. Then, from the back of the room, a no less famous philosopher replied with sarcasm: "yes, yes."

From given propositions, more complicated propositions can be constructed. For example: $(\sim p) \vee q$, for which a truth table can be constructed. In fact, by doing so, we are not treating $(\sim p) \vee q$ as a proposition but as a "truth function" whose truth values depend on the truth values of the "variables" p, q. The following table gather the truth tables of $p \Rightarrow q, (\sim q) \Rightarrow (\sim p)$, and $(\sim p) \vee q$:

p	q	$\sim p$	$\sim q$	$p \Rightarrow q$	$(\sim q) \Rightarrow (\sim p)$	$(\sim p) \vee q$
T	T	F	F	T	T	T
T	F	F	T	F	F	F
F	T	T	F	T	T	T
F	F	T	T	T	T	T

It can be seen that the truth values of the three propositional functions coincide. It is said that these functions are *equivalent.*

Another example of equivalent truth functions are $p \Rightarrow (q \vee r)$ and $(p \wedge \sim q) \Rightarrow r$. The corresponding truth tables are:

p	q	r	$\sim q$	$p \wedge \sim q$	$q \vee r$	$p \Rightarrow (q \vee r)$	$(p \wedge \sim q) \Rightarrow r$
T	T	T	F	F	T	T	T
T	T	F	F	F	T	T	T
T	F	T	T	T	T	T	T
T	F	F	T	T	F	F	F
F	T	T	F	F	T	T	T
F	T	F	F	F	T	T	T
F	F	T	T	F	T	T	T
F	F	F	T	F	F	T	T

Other equivalences are: $\sim (p \vee q)$ is equivalent to $(\sim p) \wedge (\sim q)$; $\sim (p \wedge q)$ to $(\sim p) \vee (\sim q)$, etc. (exercise 4).

It is clear that if two propositions are equivalent, if one of them is proved, then the other is also proved. So to prove "$ab = 0 \Rightarrow (a = 0$ or $b = 0)$" it suffices to prove "$(ab = 0$ and $a \neq 0) \Rightarrow b = 0$" and to prove "$a > 1 \Rightarrow a \neq 0$" it is enough to prove "$a = 0 \Rightarrow a \leq 1$".

Given a proposition of the form: $p \Rightarrow q$, the proposition $q \Rightarrow p$ is called its *reciprocal*, while $p \Rightarrow (\sim q)$ is called its *contrary*(of $p \Rightarrow q$) and $(\sim q) \Rightarrow (\sim p)$ its counter-reciprocal. As $p \Rightarrow q$ and its *counter-reciprocal* are, as we have seen, equivalent, it is enough to prove one in order to prove them both.

Truth functions that are true no matter the truth values of the variables are called *tautologies*, while those that are always false are called *contradictions.* For example, $p \vee \sim p$ is a tautology, and $p \wedge \sim p$ is a contradiction.

The portion of logic that we have approached so far does not embrace the classical syllogisms such as "all men are mortal, Socrates is a man, then Socrates is mortal". The logical validity of those syllogisms depends on the relationships between the subject and the predicate of each sentence that compose them. Those components assert that certain individuals or entities have definite properties: men have the property of being mortal, Socrates has the property of being a man, and they are combined to conclude that Socrates has the property of being mortal. The notion of having a particular property can be translated as belonging to a certain set—the set of elements that possess such property. The concept of set is fundamental, not only for building a theory of syllogism but as well for every field of mathematics, and the next section will be devoted to it.

2 - SETS

Voltaire used to say: "If you want to argue with me, define the terms you use." This sentence expresses an overwhelming truth: many sterile discussions arise from the use of the same word with different meanings or connotations.

To define an entity or object is to locate it within a family (the *gender*) and then to assign it a particular feature that distinguishes it from other members of the family (the *differentia*). For example, a definition of "young" may be "human being who is a few years old." Human being is the gender and being a few years old is the differentia. Now, if we wish to define *human being* we could say that it is "an animal with superior mental development"; then we define *animal* as "entity with such and such properties." Following that process, we soon arrive at a concept that is too general as to be defined by others—all we can do is to come up with synonyms. This happens with concepts like "entity", "set", "being", and many others. These words (or the concepts that they express) are said to be *primitive*.

"Set" is thus a primitive concept, as well as is "element," and from them, an axiomatic theory may be developed similarly to that of Euclid, who developed geometry from the primitive concepts "point", "straight line", and "plane." For further discussion on axiomatic set theory see [17]. Here we limit ourselves to think of a set as an aggregate or collection of objects that are the elements of the set. We write $a \in A$ ("a belongs to or is a member of A") to symbolize that a is an element of the set A. Sets are denoted by listing their elements between braces or, between braces too, by giving a property that characterizes those elements. Thus the set of natural numbers less than four is denoted by:

$$\{1, 2, 3\} \quad \text{or} \quad \{x \in \mathbb{N}/x < 4\}$$

When talking about sets, some precautions need to be taken in order to avoid paradoxes. One of such paradoxes, proposed by mathematician, philosopher, writer, and human right's defender Bertrand Russell (1872- 1970), arises when classifying the sets as "normal" and "abnormal." A set is said to be *normal* if it is *not* an element of itself. For example, the set of students in a classroom is normal because that group is not a student of the class. An *abnormal* set is a set that is also an element of itself. For example, the set of entities that are not men is an entity that is not a man. Another example: if the set of the sets that can be defined, in English, in less than 100 words, is, in turn, a set that can be defined in less than 100 words, then they are abnormal sets. Now, the paradox appears when considering the set of all normal sets (and only them): it must be either normal or abnormal. But if it were normal, it would be contained in the set of normal sets (itself), and then it would be abnormal. On the other hand, if it were abnormal, it would be not contained in the set of all normal sets (itself) and so it would be normal.

There are several versions of Russell's paradox: the barber, the mayor, the catalog (see exercise 6). One of them appears in Don Quixote, where the following embroilment is posed. There is a road that passes through a bridge that crosses a river. On one side of the bridge there is a court of law and on the other side a gallows. Every traveler who wants to cross the river is stopped at the court and asked where he is going. Then it is ruled whether he told the truth or lied. If he told the truth, he is allowed to continue. But if he told a lie, he is carried through the bridge to be hanged on the other side. A problem arises when a traveler arrives and says that he wants to cross the river just to be hanged on the other side. If he said the truth, then he should not be hanged. But if he is not hanged, he would have lied, and then he should be hanged. On the contrary, if he were lying, he must be hanged. But if he's hanged, what he said would have been the truth, and then he should not be hanged.

Others paradoxes of similar kind are: "the liar", which can be stated like "this proposition is false" (if it were true, it would be false, and if it were false, it would be true); Pinocchio's paradox, in which Pinocchio says "my nose is growing"; and Grelling's paradox, which consists in applying the adjective "heterological" to itself. A heterological adjective is one which does not comply with the qualification it establishes. For example, 'bisillabic' is heterological because that word is not bisillabic, while 'polysillabic' is not heterological.

These and others paradoxes arise by means of self-reference, i.e., by using sets that have elements defined from the prior existence of the own set. To avoid such paradoxes, it is convenient to adopt the following reasonable principle: "It is not valid to define an element of a set from the set as a whole."

Between sets, we can establish connections and a basic relation (inclusion) that we now describe.

Inclusion: A set A is said to be included in (or that it is a *subset* of) a set B if every element of A is also an element of B. That is to say, if the following implication is valid: $a \in A \Rightarrow a \in B$. In such a case we write: $A \subset B$ (A is included or contained in B). If $A \subset B$ and $B \subset A$, then A and B have the same elements and we write $A = B$. The inclusion is clearly transitive, that is, if A, B, C are sets such that $A \subset B$ and $B \subset C$, then $A \subset C$. It follows that if $\{S\}$ is the set with Socrates as its only element, and M is the set of all men, and B is the set of mortal beings, from $M \subset B$ (all men are mortal) and $\{S\} \subset M$ (Socrates is a man), it follows that $\{S\} \subset B$ (Socrates is mortal).

Another way of saying that every element of A is an element of B, is using the so-called *universal quantifier* $\forall$, meaning "for all." Instead of $A \subset B$ we can write $\forall a \in A, a \in B$. For example, $\forall n \in N, n \geq 1$ means that all natural numbers are greater than or equal to 1, or speaking in terms of sets, that the set of natural numbers is included in the set of (real) numbers greater than or equal to 1.

Intersection: If A and B are sets, $A \cap B$ (intersection of A and B) denotes the set of elements belonging to both A and B, i.e.:

$$A \cap B = \{x/x \in A \land x \in B\}$$

A and B may not have elements in common whatsoever, so it is convenient to define of a set without any elements in it: the *empty set*, noted by ∅. If we think a set as a box that contains objects (its elements), the empty set would be the empty box. To state that a set A is not empty ($A \neq ∅$), it is often used the so-called *existential quantifier*, $\exists$, which means "exists." For example, we can write $\exists a, a \in A$ (exists a such that a is an element of A).

If $A \cap B = ∅$ we say that A and B are *disjoint*.

Union: The union of sets A and B, denoted by $A \cup B$, is defined as the set of the elements belonging to at least one of A and B, i.e.:

$$A \cup B = \{x/x \in A \vee x \in B\}$$

Complement: The complement of a set A is defined as the set A' of the elements not belonging to A, i.e.:

$$A' = \{x/x \notin A\}$$

where $x \notin A$ is the negation of $x \in A$.

This definition of complement needs a clarification: to say that the complement of a set is the set of all elements not belonging to it, is too general and violates the principle that we have set in order to be safe from contradictions (that is, to avoid self-reference) because the set of all things is indeed an element of itself. When we consider the complement of a set, we refer to the elements *not* belonging to it, but belonging to a certain set of reference that we call *universe of speech*. For example, the complement of the set of all mathematicians is not the set of all the entities that are not mathematicians, but only the set of people that are not mathematicians. Here the universe of speech (or simply: "the universe") is the set of all people. Another example: the complement of the set of real numbers > 1 could be taken as the set of real numbers ≤ 1, the universe being, in this case, the set of real numbers. All the sets, treated in the same context, will be considered as subsets of one set, the universe, that will be denoted generically by the symbol U.

Difference: If A and B are sets, the difference $A - B$ is defined by:

$$A - B = \{x/x \in A, x \notin B\}$$

In other words, $A - B = A \cap B'$.

Union and intersection clearly satisfy the following properties: whatever be the sets A, B, C contained in a universe U:

1) *Associatives*: $(A \cap B) \cap C = A \cap (B \cap C)$; $(A \cup B) \cup C = A \cup (B \cup C)$

2) *Commutatives*: $A \cap B = B \cap A$; $A \cup B = B \cup A$

3) *Idempotents*: $A \cap A = A$; $A \cup A = A$

4) *Existence of neutral elements*: $A \cap U = A$; $A \cup \emptyset = A$

5) *Distributives*: $A \cap (B \cup C) = (A \cap B) \cup (A \cap C)$; $A \cup (B \cap C) = (A \cup B) \cap (A \cup C)$

6) *Morgan laws*: $(A \cap B)' = A' \cup B'$; $(A \cup B)' = A' \cap B'$

As an example, we verify the last one: $x \in (A \cup B)'$ if and only if $x \notin A \cup B$, that is $\sim (x \in A \lor x \in B) \Leftrightarrow (x \notin A \land x \notin B) \Leftrightarrow x \in A' \cap B'$.

3 - RELATIONS AND FUNCTIONS

Let A and B be sets. We define the *Cartesian product* $A \times B$ of A and B, as the set of all the ordered pairs (a, b) with $a \in A$ and $b \in B$. By an *ordered pair* we mean a pair (a, b) of elements with $a \in A$ and $b \in B$ such that $(a, b) = (c, d)$ if and only if $a = c$ and $b = d$. With greater rigor we define, for $a \in A$ and $b \in B$: $(a, b) = \{\{a\}, \{a, b\}\}$ that is the set whose elements are $\{a\}$ and $\{a, b\}$ and so we have: $(a, b) = (c, d)$ if and only if $a = c$ and b=d (exercise 8).

For example, if $A = \{a, b, c, d\}$ and $B = \{x, y, z\}$ we have:

$$A \times B = \{(a, x), (a, y), (a, z), (b, x), (b, y), (b, z), (c, x), (c, y), (c, z),$$
$$(d, x), (d, y), (d, z)\}$$

Now we will describe the concept of relation. Let us consider first some examples of ordinary language: "is a friend of" or "is a student of" are usually considered relationships between the elements of two sets of people. Let $A = \{a, b, c, d\}$ and $B = \{x, y, z\}$ be sets of people. Suppose that: a is a friend of x, y, but not of z; that b is a friend of the three x, y, z, and that c and d are not friends of anybody in B. The relation "is a friend of" among people of A and that of B, is determined by the pairs related by friendship, that is to say by the set:

$$\{(a, x), (a, y), (b, x), (b, y), (b, z)\}$$

which is a subset of the cartesian product $A \times B$.

Likewise, if we concede: a is student of y but not of x, z; b is a student of y, z, but not of x; c is a student of y, but not of x, z; and finally d is not a student of anyone in B; the relation of being a student from A to B, is determined by:

$$\{(a, y), (b, y), (b, z), (c, y)\}$$

which is also a subset of the cartesian product $A \times B$.

From these examples we see that it is reasonable to adopt the following mathematical definition: we call *relation* from a set A to a set B, to any subset of the cartesian product $A \times B$. If R is a relation from A to B, i.e., $R \subset A \times B$, we write as well "aRb" to denote $(a, b) \in R$.

Each relation R from A to B determines a relation R^{-1} from B to A, defined by:

$$R^{-1} = \{(b, a) \in B \times A / (a, b) \in R\}$$

R^{-1} is called the *inverse relation* of R.

Three kinds of relations are the most significant in mathematics: functional relations (or functions), equivalence relations, and order relations. In this section, we will only deal with functional relations.

To motivate its definition, consider the relation "is a son of" from the set of all living men and the set of all women (living or not). This relation has two remarkable characteristics: 1) Any man alive has at least a mother and 2) there cannot be two mothers of the same man. The two characteristics may be resumed saying that any man has one and only one mother. We will call functional relation to any relation with similar properties to those above.

If A and B are sets, a relation R from A to B is said a *functional relation*, *function* or *application* from A to B *iff* (if and only if) the following two conditions are verified:

1) From each $a \in A$, there exists at least $b \in B$, such that aRb.

2) aRb and aRb' ($a \in A, b, b' \in B$) $\Rightarrow$ $b = b'$.

In such case, A is said the *domain* of R and B its *codomain*.

If R is a function, since for each $a \in A$ there exists one, and only one, $b \in B$ such that aRb, it is customary to denote such b as $R(a)$. From the definition, it follows that to define a function R from A to B it is enough to give for each $a \in A$ an element $R(a) \in B$. Usually we write $R: A \rightarrow B$ to state that R is a function from A to B.

Consider the "dual" (interchanging A and B) properties of 1) and 2):

3) For each $b \in B$, there exists at least an element $a \in A$ such that aRb.

4) aRb and $a'Rb$ ($a, a' \in A, b \in B$) $\Rightarrow$ $a = a'$.

A function $R: A \rightarrow B$ that satisfy 3) is said *surjective* or *onto B*; if it fulfills 4), it is said *injective* or *one to one*, and if it satisfy 3) and 4), i.e., if it is surjective and injective, it is said *bijective* or that it is a *bijection*.

Let $R: A \to B$ be a bijective function; the properties 3) and 4), tell us that the *inverse relation* R^{-1} is a function, while 1) says that R^{-1} is surjective and 2) that R^{-1} is injective. Therefore, R^{-1} is a bijective function, called the *inverse function* of R.

Generally, the letters f, g, h, etc. are used to denote functions. If $f: A \to B$ and $g: B \to C$ are functions, the *composition* $g \circ f$, is the function from A to C defined for each $a \in A$, by $g \circ f(a) = g(f(a))$. This composition of functions is clearly associative, i.e.: $(h \circ g) \circ f = h \circ (g \circ f)$ so long as the compositions make sense, that is, if $f: A \to B, g: B \to C, h: C \to D$.

Proposition 1.1: If $f: A \to B$ is a bijective function, the inverse function f^{-1} is the unique function g: B→A such that $g \circ f = 1_A$ and $f \circ g' = 1_B$, where $1_X: X \to X$ is the *identity function* of the set X, defined by $1_X(x) = x$ whatever be $x \in X$.

Proof: As f^{-1} is defined so that $f(a) = b$ if, and only if $f^{-1}(b) = a$, it is clear that $f \circ f^{-1} = 1_B$ and that $f^{-1} \circ f = 1_A$. Moreover, if g and g' are functions that satisfy the conditions for g in the statement, from $g \circ f = 1_A$ and $f \circ g$, it follows: $g = g \circ 1_B = g \circ (f \circ g') = (g \circ f) \circ g' = 1_A \circ g' = g'$.∎

We end this chapter describing some very common notation.

If $f: A \to B$ is a function, A' is a subset of A, and B' is a subset of B, we define $f(A')$ and $f^{-1}(B')$ by:

$$f(A') = \{b \in B / \exists a \in A' \text{ with } f(a) = b\}$$
$$f^{-1}(B') = \{a \in A / f(a) \in B'\}$$

More informally, we often write the first as: $f(A') = \{f(a)/a \in A'\}$. Note that by writing $f^{-1}(B')$ it is *not* assumed that f is bijective, which is a necessary condition for the existence of the inverse function of f . In this case, f^{-1} is the inverse relation of f, which is not necessarily a function.

A special class of functions are the so-called operations. An *operation* (binary internal) on a set A is a function from $A \times A$ to A. If $*$ is an operation on A and $a, a' \in A$, it is customary to write $a * a'$ instead of $* (a, a')$. For example, the sum: $(n, m) \to n + m$; the product: $(n, m) \to n \cdot m$ and the potentiation: $(n, m) \to n^m$, are operations on the set of natural numbers. One more example: if U is a set and if $P(U)$ denotes the family of the parts (that is, the subsets) of U, then the union, intersection and difference are operations on $P(U)$.

If $f: A \rightarrow B$ is a function and A' is a subset of A, the *restriction* of f to A' is the function $f_{/A}: A' \rightarrow B$ defined by $f_{/A}(x) = f(x)$ for any $x \in A'$.

Another notation often used is the one that follows. If $f: I \rightarrow A$ is a function, putting $a_i = f(i)$, f is denoted as well by $(a_i)_{i \in I}$ or by $\{a_i\}_{i \in I} \in I\}$. This is used particularly when $I = \mathbb{N}$ is the set of natural numbers, in which case f is said a *sequence* in A and is very used as well when (being I any set) A is the set of parts of a set B, in which case f is called a family of sets of B with *indices* in I.

EXERCISES

Ex. 1: In the following propositions, "good" is anyone who always tells the truth, and "bad" is anyone who always lies.

a) A says: B is good.

B says: A is not good.

Prove that one of them tells the truth but is not good.

b) A says: B is good.

B says: A is bad.

Prove that either one of them tells the truth but is not good, or one of them lies but is not bad.

c) Suppose that each of A, B, C is good or bad.

C says: B is bad.

B says: A and C are of the same type.

Is A good, or bad?

d) Keep the assumption that each of A, B, C, is either good or bad.

A says: B and C are of the same type.

What is the answer of C to the question: "are A and B of the same type?"

Ex. 2: In a set of nine (apparently equal) coins, there is a false one that weighs less than the others. A pair of scales to compare weights is available. Show that two weighing operations are sufficient in order to find the false coin. If the set has 27 coins (or any number between 10 and 27), show that three weighings are sufficient. Generalize.

Ex 3: Three logicians receive each one hat, which are placed on their heads. They are told that the hats were selected from a set of five hats: three red and two black. Each logician can see the hats of the other two but not their own. Standing together they are asked: "Do you know the color of the hat are you wearing?" The first answers "I don't know." The second gives the same answer. The third (a blind man) gives the correct answer. How did he deduct it?

Ex. 4: Verify the following equivalences:

a) $\sim (p \vee q)$ is equivalent to $(\sim p) \wedge (\sim q)$.

b) $\sim (p \wedge q)$ to $(\sim p) \vee (\sim q)$.

c) $p \wedge (q \vee r)$ to $(p \wedge q) \vee (p \wedge r)$.

d) $p \vee (q \wedge r)$ to $(p \vee q) \wedge (p \vee r)$.

e) $p \wedge (q \wedge r)$ to $(p \wedge q) \wedge r$.

f) $p \vee (q \vee r)$ to $(p \vee q) \vee r$.

Ex. 5: Which of the following are tautologies?:

a) $p \Rightarrow (p \vee q)$.

b) $(p \wedge q) \Rightarrow p$.

c) $[p \wedge (p \Rightarrow q)] \Rightarrow q$.

d) $[(p \Rightarrow q) \wedge (\sim q)] \Rightarrow p$.

Ex. 6: Complete the statement of the following versions of Russell's paradox:

1) In a village there is only one barber. He shaves those who do not shave themselves, and only them. Being mandatory to be shaved, who shaves the barber?

2) A city is founded to be the residence of all majors that do not reside in the city they manage, and only them. Where will the major of the new city reside?

3) A library compiles a bibliographical catalog of all catalogs that do not list themselves and only those. Does the library's catalog lists itself?

Ex. 7: Which of the following relations are true?

a) $\emptyset \in \emptyset$

b) $\emptyset \subset \emptyset$

c) $\emptyset \in \{\emptyset\}$

d) $\emptyset \subset \{\emptyset\}$

Ex. 8: An ordered pair (a, b) may be defined by: $(a, b) = \{\{a\}, \{a, b\}\}$ because we have:

$$\{\{a\}, \{a, b\}\} = \{\{c\}, \{c, d\}\} \Leftrightarrow a = c, \text{and } b = d$$

Ex. 9: Let A, B, C, D be sets,

a) If $A \cap B = \emptyset$, and $A \cap C = \emptyset$, is necessarily $B \cap C = \emptyset$?

b) If the intersection of any three of them is empty, are there necessarily two of them that are disjoint?

Ex. 10: Prove or disprove:

a) $A \cap C = B \cap C \Rightarrow A = B.$

b) $A \cup C = B \cup C \Rightarrow A = B.$

c) $(A \cap C = B \cap C \text{ and } A \cup C = B \cup C) \Rightarrow A = B.$

d) $A - C = B - C \Rightarrow A = B.$

e) $A - C = B - C \text{ and } C - A = C - B \Rightarrow A = B.$

Ex. 11: Let

$$A = B_1 \cup B_2 \cup B_3 = C_1 \cup C_2 \cup C_3$$

If the C_i are pairwise disjoint, and if $B_i \subset C_i \; \forall i = 1, 2, 3$, prove that $B_i = C_i \; \forall i$.

Ex. 12: Let $f: A \to B$ and $g: B \to C$:

If f and g are bijective, then $g \circ f$ is also bijective and $(g \circ f)^{-1} = f^{-1} \circ g^{-1}$.

Ex. 13: Let $f: A \to B$.

a) If f has a right inverse (i.e., there is $g: B \to A$ such that $f \circ g = 1_B$), then f is onto g.

b) If f has a left inverse (there is $h: B \to A$ such that $h \circ f = 1_A$), then f is injective.

Ex. 14: Let $f: A \to B$.

a) f onto $\Rightarrow f$ is right cancellable (that is, $g \circ f = h \circ f \Rightarrow g = h$).

b) f injective $\Rightarrow f$ is left cancellable ($f \circ g = f \circ h \Rightarrow g = h$).

CHAPTER 2
REAL NUMBERS

Historically, the concept of real number was built in parallel with the development of a numeration system which allowed approximations of any orders of magnitude. This course is developed in Mesopotamia and, later, on other civilizations where the use of numbers, to measure and to calculate, was required.

The Pythagoreans, in contrast, were engaged in contemplation, in the love of wisdom (the terms "philosophy" (love of wisdom) and "mathematics" (knowledge, learning) are said to have been coined by Pythagoras himself) and in a higher level of rigor (tradition attributes early attempts to Thales) and they developed a philosophy with the motto: "all is number", referring to natural numbers. The arithmetic theory of proportions —a theory ascribed to Pythagoras himself— deals essentially, with rational numbers and the discovery of incommensurable segments shows its inadequacy to describe geometry. This insufficiency was overcome by the Eudoxian theory of ratios of magnitudes, a geometric theory of proportions exposed in The Elements of Euclid, which is the first rigorous theory of real numbers. For both historical and didactic reasons, in this chapter we only study the axioms of an ordered field of the real numbers, which corresponds to the arithmetic of rational numbers. We leave for a later chapter (chapter 6) the introduction to the axiom of the least upper bound, more related to irrationals and continuity.

1 - FUNDAMENTAL PROPERTIES OF REAL NUMBERS

This section presents a list of properties of real numbers. We accept these properties as true, that is, as the axioms of the theory. Any other property must be deducted from these or from some other previously proved.

At the end of section 2, it is convenient to solve exercises A and B before proceeding with the next section. These are exercises in "demonstrations," that are often difficult for the novice, because one must find a way to reach the thesis, as opposed to applying algorithms or performing verifications in which the road is drawn in advance. However, it is highly formative and should be undertaken from the beginning.

The real numbers are a set $\mathbb{R}$ with two operations, sum and product, and a relation $<$ (less than), from $\mathbb{R}$ to $\mathbb{R}$. The sum (or addition), denoted by "+" is an operation on $\mathbb{R}$, i.e., a function $+: \mathbb{R} \times \mathbb{R} \to \mathbb{R}$ and the image of the pair (a, b) by + is denoted $a + b$. In the same way, the product (or multiplication) is a function $\cdot : \mathbb{R} \times \mathbb{R} \to \mathbb{R}$ and $\cdot(a, b)$ is denoted $a \cdot b$ or ab.

We accept or postulate three sets of properties that characterize real numbers, numbered I, II and III. In what follows we state the properties of the sets I and I, leaving III for a later chapter (chapter 6).

I) S) *Properties of the sum*:

S)1)*Associative*: $(a + b) + c = a + (b + c)$ for   every $a, b, c \in \mathbb{R}$.

S)2)*Commutative*: $a + b = b + a$ for every $a, b \in R$.

S)3)*Existence of a neutral element*: There is an element $0 \in \mathbb{R}$ such that

$$a + 0 = 0 + a = a \ \forall a \in \mathbb{R}.$$

S)4)*Existence of inverse of each element*: For each $a \in \mathbb{R}$ exists $a' \in \mathbb{R}$ such that

$$a + a' = a' + a = 0.$$

P) *Properties of the product*:

P)1) *Associative*: $(ab)c = a(bc)$ for every $a, b, c \in \mathbb{R}$.

P)2) *Commutative*: $ab = ba$ for every $a, b \in \mathbb{R}$.

P)3) *Existence of a neutral element*: There is an element $1 \in \mathbb{R}$ such that $1 \neq 0$ and

$$a1 = 1a = a \ \forall a \in \mathbb{R}$$

P)4) *Existence of inverse of each nonzero element*: For each $a \in \mathbb{R}$ with $a \neq 0$, exists

$$a'' \in \mathbb{R} \text{ such that } aa'' = a''a = 1.$$

D) *Distributive property*: $a(b + c) = ab + ac$ for every $a, b, c \in \mathbb{R}$.

II) *Order properties*:

Trichotomy: For each pair of real numbers a, b, one and only one of the following is verified:

$$a < b, a = b, b < a$$

Transitivity: $a < b$ and $b < c \Rightarrow a < c$.

Compatibility with sum: $a < b \Rightarrow a + c < b + c$.

Compatibility with the product: $a < b$ and $0 < c \Rightarrow ac < bc$.

Remarks: 1) The sum is a function from $\mathbb{R} \times \mathbb{R}$ to $\mathbb{R}$, so if $(a, b) = (c, d)$, i.e., if $a = c$ and $b = d$, then

$a + b = c + d$. In the same way, being the product a function, we have $a = c$ and $b = d \Rightarrow ab = cd$. These facts, named "uniform properties" in old fashioned texts, will be used without explicit mention.

2) The equality $a = b$ means that objects a and b are the same. This is valid for any objects, not only for numbers. We will use, without explicit mention too, the properties: reflexive ($a = a$), symmetric ($a = b \Rightarrow b = a$) and transitive ($a = b$ and $b = c \Rightarrow a = c$) of equality.

3) The parallelism between the listed properties of sum and product is evident, but it is necessary to emphasize the differences: in P)3) in addition to state the existence of a neutral element, it is required that such element be not the same as the neutral element of the sum; also, in P)4) it is postulated the existence of a multiplicative inverse, but only for nonzero elements. About 0 it is not stated that it has an inverse, neither that it does not have one. Soon we will prove that the latter is the case.

4) The expression $ab + cd$ may have several interpretations according to where the parenthesis are placed: $a(b + (cd)), (ab) + (cd), a(b + c)d$ and $((ab) + c)d$, but it is an universal convention in mathematics to omit the parenthesis in the second case, i.e., to write $ab + cd$ to denote $(ab) + (cd)$. In the same way is a convention to write $ab + c$ to denote $(ab) + c$.

5) Trichotomy states two things: that at least one of the relations $a < b, a = b, b < a$ is valid and moreover that at most one of them is fulfilled.

6) The properties listed above define axiomatically a mathematical object (an ordered field); however, we haven't taken care of the independence of the axioms, that is to say, that one of them can be inferred from the others. In fact, for example, commutativity of the sum can be deducted from the rest (exercise A. 12).

In what follows we will prove some consequences derived from the set I of axioms.

Proposition 2.1:

a) $a + b = a + c \Rightarrow b = c$

b) $a \neq 0$ and $ab = ac \Rightarrow b = c$

Proof: a) By $S)4)$ exists a' such that $a' + a = 0$, then:

$$b = 0 + b = (a' + a) + b = a' + (a + b) = a' + (a + c)$$
$$= (a' + a) + c = 0 + c = c$$

where the successive equalities are justified by $S)3), S)4), S)1)$, the hypothesis $(a + b = a + c); S)1), S)4)$ and $S)3)$. Although we have said that the sum will be used without mentioning the fact that it is a function, we will, for this time, specify the passages within the previous chain of equalities in which it was used. They are:

- in the second equality $(0 = a' + a$ and $b = b \Rightarrow 0 + b = (a' + a) + b)$;
- in the fourth $(a' = a'$ and $a + b = a + c \Rightarrow a' + (a + b) = a' + (a + c))$
- in the sixth $(a' + a = 0$ and $c = c \Rightarrow (a' + a) + c = 0 + c)$.

Also, the reflexive, symmetric and transitive properties of equality have been used in several places, for example when writing: $b = 0 + b = (a' + a) + b$; we mean, of course, that $b = 0 + b$ and $0 + b = (a' + a) + b$, implies by transitivity $b = (a' + a) + b$.

b) By $P)4)$ as $a \neq 0$ exists a'' such that $a''a = 1$, then:

$$b = 1b = (a''a)b = a''(ab) = a''(ac) = (a''a)c = 1c = c$$

where the justification of the steps is the same as in a), changing S by P.∎

Proposition 2.2: $a0 = 0 \ \forall a \in \mathbb{R}$.

Proof: $a0 + 0 = a0 = a(0 + 0) = a0 + a0$, that is to say: $a0 + a0 = a0 + 0$, from where by prop. 1) a) $a0 = 0$.∎

Proposition 2.3: 0 does not have a multiplicative inverse.

Proof: If 0 had a multiplicative inverse a, it would be $a0 = 1$, but $a0 = 0$, so $0 = 1$ which contradicts $P)3)$. ∎

Property $S)3)$ asserts the existence of an element, 0, neutral for the sum, but which in principle doesn't exclude the possibility of the existence of other neutral elements. Let $0'$ a neutral element of the sum, i.e., $a + 0' = 0' + a = a \; \forall a \in R$, in particular, $0 + 0' = 0$, but from $S)3$): $0 + 0' = 0'$, therefore $0' = 0$. We conclude that 0 is then the unique neutral element of the sum.

In the same way, it is verified that 1 is the only neutral element of the product. We have proved:

Proposition 2.4: $a)$ 0 is the unique neutral element of the sum.

$b)$ 1 is the unique neutral element of the product.■

There is also uniqueness for the inverse of each element, valid for both sum and product:

Proposition 2.5:

$a)$ If $a + x = 0$ and $a + y = 0$, then $x = y$. $b)$ $ax = 1$ and $a\,y = 1$, then $x = y$.

Proof : a) From $a + x = 0$ and $a + y = 0$, it follows $a + x = a + y$ and by prop. 2.1: $x = y$. The proof of b) is similar.■

Notation: Due to the just proved uniqueness of the inverses, we can adopt the following notations:

· $-a$ denotes the unique additive inverse of a, i.e., the unique real number such that $a + (-a) = 0$.

· If $a \neq 0, a^{-1}$ denotes the unique multiplicative inverse of a, i.e., the unique real number such that $aa^{-1} = 1$.

· $b - a$ denotes $b + (-a)$.

· If $a \neq 0, \left(\frac{b}{a}\right)$ denotes ba^{-1}.

Proposition 2.6: $-(-a) = a$.

Proof: By definition we have: $(-a) + a = 0$ and $(-a) + \left(-(-a)\right) = 0$ then by prop. 2.3.: $a = -(-a)$.■

Proposition 2.7: $a(-b) = -(ab) = (-a)b$. Proof: $ab + a(-b) = a(b + (-b)) = a0 = 0$, then by 2.3 follows that $a(-b) = -(ab)$. Similarly $(-a)b = -(ab)$.■

Due to associativity, we write simply $a + b + c$ instead of $(a + b) + c$ or $a + (b + c)$, and we proceed similarly with the product. Moreover, whichever way we put brackets in $a + b + c + d$, the results are equal (for example, $(a + b) + (c + d) = a + (b + (c + d))$ by associativity of a, b and $c + d$), so we write such expression without parenthesis. Something similar happens for more summands (or factors in case of the product), but a precise formulation of this generalized associativity involves an induction process and should be postponed.

For the purposes of the next proposition and of the exercises at the end of the chapter, we define 2=1+1; 3=2+1; 4=3+1; 5=4+1 and for $a \in \mathbb{R}$ we define $a^2 = aa, a^3 = a^2a$. These definitions will be generalized later on.

Proposition 2.8: $(a + b)^2 = a^2 + 2ab + b^2$.

Proof:

$$(a + b)^2 = (a + b)(a + b) = (a + b)a + (a + b)b =$$
$$= aa + ba + ab + bb = a^2 + (1 + 1)ab + b^2 = a^2 + 2ab + b^2. \blacksquare$$

3 - SOME CONSEQUENCES OF ORDER PROPERTIES

Instead of $a < b$ we write also $b > a$ (b is greater than a). The notations $a \leq b$ and $b \geq a$ mean either $a < b$ or $a = b$.

Proposición 3.1:

a) $a < b \Rightarrow -b < -a$

b) $0 < 1$

c) $0 < a \Rightarrow 0 < a^{-1}$

d) $a < b$ and $c < d \Rightarrow a + c < b + d$

e) $a < b$ and $c < 0 \Rightarrow bc < ac$

Proof: a) From $a < b$ it follows by compatibility with the sum: $a + (-a) < b + (-a)$, or $0 < b + (-a)$ and again by compatibility with the sum: $(-b) + 0 < (-b) + b + (-a)$, so $-b < -a$.

b) According to trichotomy, one and only one of the following is valid:

$$0 < 1, 0 = 1, 1 < 0$$

then to prove the first, just exclude the others. $0 = 1$ is excluded by P)3). If it were $1 < 0$, it would result by a): $0 < -1$ (because $-0 = 0$) and from $1 < 0$ and $0 < -1$, by compatibility with the product it would result: $1(-1) < 0(-1)$, but by prop. 2.2. $0(-1)=0$ and then $-1 < 0$. Having obtained $0 < -1$ and $-1 < 0$, trichotomy is contradicted, then $0 < 1$.

c) By trichotomy one and only one of the following is valid:

$$0 < a^{-1}, 0 = a^{-1}, a^{-1} < 0$$

If $0 = a^{-1}$ then $0 = a0 = aa^{-1} = 1$ which is contradictory.

If $a^{-1} < 0$ then, as $0 < a$ by compatibility with the product, it follows $aa^{-1} < a0$, or $1 < 0$, which is absurd.

As $0 = a^{-1}$ and $a^{-1} < 0$ lead to a contradiction, we must have $0 < a^{-1}$.

d) By compatibility with the sum, from $a < b$ follows $a + c < b + c$ and from $c < d$ results $b + c < b + d$ and by transitivity: $a + c < b + d$.

e) From $c < 0$ follows by a) that $-c > 0$ and by compatibility with the product results $a(-c) < b(-c)$. By 2.5 we obtain $-(ac) < -(bc)$ and by a) and prop. 2.4.: $ac > bc$.∎

4 - GEOMETRIC INTERPRETATION AND MODULUS

It is convenient to have a graphic image of the set of real numbers. If we mark two points on a straight line, and we assign the number 0 to one point and the number 1 to the other, we assume that such correspondence extends to a bijection between the line and $\mathbb{R}$. This correspondence among the points of the line and the real numbers is usually attributed to Descartes, although it was previously clearly described by Bombelli like a correspondence between the ratios of magnitudes of Eudoxus, which satisfies the axioms of (positive) real numbers and lengths of magnitudes, which in turn can be interpreted as the points on a line.

It is assumed, as well, that in this correspondence the order is preserved, thus the notion of segment determined by two points agrees with that of the interval determined by the corresponding real numbers, where the *closed interval* $[a, b]$ determined by $a, b \in \mathbb{R}$ with $a \leq b$, is defined by:

$$[a, b] = \{x \in \mathbb{R}/a \leq x \leq b\}$$

Similarly, we define the *open* and *semi-open* intervals:

$$(a, b) = \{x \in \mathbb{R}/a < x < b\}$$
$$[a, b) = \{x \in \mathbb{R}/a \leq x < b\}$$
$$(a, b] = \{x \in \mathbb{R}/a < x \leq b\}$$

and the intervals that corresponds to half-lines:

$$[a, +\infty) = \{x \in \mathbb{R}/a \leq x\}$$
$$(-\infty, a) = \{x \in \mathbb{R}/x < a\}$$

Where ∞ (which reads "infinite") has no significance as an isolated symbol.

The idea of the distance of a point to the origin, suggests the definition of *modulus or absolute value* $|a|$ of a real number a:

$$|a| = \begin{cases} a \ if \ a \geq 0 \\ -a \ if \ a < 0 \end{cases}$$

For example, if $b < 0$, we have $|-b| = -b$ since by 3.1 $b < 0 \Rightarrow -b > 0$.

Proposition 4.1: For $a, b \in \mathbb{R}$ we have:

1) $|a| = 0 \Leftrightarrow a = 0$

2) $|a| = |-a|$

3) $|a - b| = |b - a|$

4) $|a|^2 = |a^2| = a^2$

5) $|ab| = |a||b|$

6) If $a \neq 0, |a^{-1}| = |a|^{-1}$

7) If $a \neq 0, |b/a| = ((|b|)/(|a|))$

8) $-|a| \leq a \leq |a|$

9) If $r \geq 0$ then: $|a| \leq r \Leftrightarrow -r \leq a \leq r$

10) $|a + b| \leq |a| + |b|$

11) $|a| - |b| \leq |a - b|$.

Proof: We will prove the last two, leaving the rest as exercises.

10) This can be done dividing into cases; for example, one case would be: $a \geq 0$, $b < 0$, and $a + b < 0$, and should prove that $-(a + b) \leq a - b$, or $-a \leq a$ which is evident inasmuch as $a \geq 0$. We leave as an exercise to do it this way. Doing the following is shorter: according to 8) we have,

$$-|a| \leq a \leq |a|$$
$$-|b| \leq b \leq |b|$$

and adding (proposition 3.1.d):

$$-(|a| + |b|) \leq a + b \leq |a| + |b|$$

then by 9) with $r = |a| + |b|$ we obtain $|a + b| \leq |a| + |b|$.

11) From 10) we obtain: $|a| = |(a - b) + b| \leq |a - b| + |b|$. ∎

EXERCISES

Ex. A: Prove the following properties using only the axioms of group I, except P.4, and the theorems and notations given in section 2:

1) $a = b \Leftrightarrow a - b = 0$.

2) $a + a = a \Rightarrow a = 0$.

3) $-0 = 0$.

4) $a \neq 0 \Rightarrow -a \neq 0$.

5) $-(a + b) = (-a) + (-b)$.

6) $(-a)(-b) = ab$.

7) $(b - c) = ab - ac$.

8) $(a + b)(c + d) = ac + ad + bc + bd$.

9) $a^2 - b^2 = (a + b)(a - b)$.

10) $a^3 - b^3 = (a - b)(a^2 + ab + b^2)$.

11) $a^3 + b^3 = (a + b)(a^2 - ab + b^2)$.

12) Commutativity of addition follows from the others.

Ex. B: Using P.4 additionally, show:

1) $ab = 0 \Rightarrow a = 0$ or $b = 0$.

2) $a^2 = b^2 \Rightarrow a = b$ or $a = -b$.

3) If $a \neq 0$ and $ab = ac$, then $b = c$.

4) $(-1)^{-1} = -1$.

5) If $a \neq 0$, then $a^{-1} \neq 0$ and $(a^{-1})^{-1} = a$.

6) If $a \neq 0$, then $(-a)^{-1} = -(a^{-1})$.

7) If $a \neq 0$ and $b \neq 0$, then $ab \neq 0$ y $(ab)^{-1} = a^{-1}b^{-1}$.

8) If $b \neq 0$ and $d \neq 0$, then $(a/b) = (c/d) \Leftrightarrow ad = bc$.

9) If $b \neq 0$ and $d \neq 0$, then $(a/b)(c/d) = ((ac)/(bd))$.

10) If $a \neq 0$ and $b \neq 0$, then $((a/b))^{-1} = (b/a)$.

11) If $b \neq 0$ and $c \neq 0$, then $(a/((b/c))) = ((ac)/b)$ y $(((a/b))/c) = (a/(bc))$.

12) If $b \neq 0$ then $-(a/b) = ((-a)/b) = (a/(-b))$ y $((-a)/(-b)) = (a/b)$.

13) If $b \neq 0$ and $d \neq 0$, Then $bd \neq 0$ and $(a/b) + (c/d) = ((ad + bc)/(bd))$.

Ex. C: Verify the following identities:

1) $(a + b + c)^2 = a^2 + b^2 + c^2 + 2ab + 2ac + 2bc$.

2) $(a + b)^3 = a^3 + 3a^2b + 3ab^2 + b^3$.

3) If $aa' = bb' = cc'$ and all of them are nonzero, then

$(a + b')(b + c')(c + a') = (a' + b)(b' + c)(c' + a)$.

4) $(a^2 + b^2 + c^2 + ab + ac + bc)^2 = (a + b + c)^2(a^2 + b^2 + c^2) + (ab + ac + bc)^2$.

Ex. D: Show that:

1) $a + a = 0 \Rightarrow a = 0$.

2) $a \neq 0 \Rightarrow a^2 > 0$.

3) If $a > 0$ and $b > 0$, then: $a < b \Leftrightarrow a^{-1} > b^{-1}$.

4) If $a > 0$ and $b > 0$, then: $a < b \Leftrightarrow a^2 < b^2$.

5) $a < b \Rightarrow a < ((a + b)/2) < b$.

6) $a^2 + b^2 = 0 \Rightarrow a = b = 0$.

7) It does not exist $a \in \mathbb{R}$ such that $a^2 + 1 = 0$.

8) It does not exists $a \in \mathbb{R}$ such that $a^2 + a + 1 = 0$.

9) If $a \neq 0$, then $a^2 + (1/(a^2)) \geq 2$. Moreover, the equality is valid if and only if $a = 1$ or $a = -1$.

10) If $a > 0, b > 0$ and $ab = 1$, then $a + b \geq 2$. Moreover, the equality is valid if and only if $a = b = 1$.

11) $(ab + cd)^2 \leq (a^2 + c^2)(b^2 + d^2)$.

Ex. E: Prove, or at least, design a strategy to do so, that it is impossible to prove the property of exercise C.1 ($a + a = 0 \Rightarrow a = 0$) using merely the axioms of group I, although in its statement the order is not mentioned.

Ex. F: Which of the following statements are true?

1) $a^2 = b^2 \Rightarrow a^3 = b^3$.

2) $(a + b)^2 = a^2 + b^2 \Leftrightarrow a = 0$ ó $b = 0$.

3) $a < b \Leftrightarrow a^2 < b^2$.

4) $a^2 = b^2 \Leftrightarrow |a| = |b|$.

5) $a^2 < b^2 \Rightarrow a^3 < b^3$.

6) There exists $a \in \mathbb{R}$ such that $x \leq a \, \forall x \in R$.

7) If $b \neq 0, d \neq 0$ and $b + d \neq 0$ then $(a/b) + (c/d) = ((a + c)/(b + d))$.

8) There exist $a, b \in \mathbb{R}$ such that $(1/a) + (1/b) = (1/(a + b))$.

Ex. G: Let $a, b \in \mathbb{R}$ with $a > 0$ and $b > 0$, show:

1) $(a/b) \geq 4 - ((4b)/a)$.

2) $(a/b) + (b/a) \geq 2$.

3) $((1/a) + (1/b))(a + b) \geq 4$.

4) $a + b = 1 \Rightarrow a^2 + b^2 \geq (1/2)$.

5) $a + b = 1 \Rightarrow ((1/a) - 1)((1/b) - 1) = 1$.

Ex. H: Being $a, b, c \in \mathbb{R}$ with $a > 0, b > 0$ and $c > 0$. Prove:

1) $(a + b + c)((1/a) + (1/b) + (1/c)) \geq 3^2$.

*2) $a + b + c = 1 \Rightarrow ((1/a) - 1)((1/b) - 1)((1/c) - 1) \geq 2^3$.

3) $abc = 1 \Rightarrow a + b + c \geq 3$.

Ex. I: Write each of the following subsets of $\mathbb{R}$ as an interval or union of intervals:

1) $\{x \in R/|3x + 2| > 1\}$.

2) $\{x \in R/|x - 2| < 1\}$.

3) $\{x \in R/x^2 - 4x < 5\}$.

CHAPTER 3
NATURAL NUMBERS

Despite being as ancient as civilization, natural numbers are inseparably joined to mathematical induction, which was clearly stated by Pascal only in the 17th century.

In this chapter, we first define natural numbers and state the induction principle, which is fundamental to define and then prove the elementary properties of powers, summations, finite sets, and combinatorics. We will finalize the chapter by enunciating the well-ordering principle (equivalent to the induction principle) and by proving the theorem on the division's algorithm, including its applications to positional numeration systems.

1 - DEFINITION AND BASIC PROPERTIES

Vaguely, it can be said that a natural number is a real number obtained adding "ones"; how many ones are admissible, though?

An attempt to answer this question leads to a vicious circle. So, keeping in mind that intuitive idea of what a natural number is, it will be made more precise through the concept of inductive set. However, the reader may also choose to intuitively accept the results of this section and move to the next one, where the concept of induction is treated in a more naive fashion.

A subset H of $\mathbb{R}$ is said to be *inductive* iff it has the following two properties:

    1) $1 \in H$.

    2) $h \in H \Rightarrow h + 1 \in H$.

Examples: 1) $\mathbb{R}$ is inductive because clearly satisfies the two conditions above.

    2) $\{x \in R / x \geq 1\}$ is inductive since if $h \geq 1$, then $h + 1 \geq 1 + 1 \geq 1$ as $1 \geq 0$.

3) $\{x \in \mathbb{R}/ x \geq 2\}$ is not inductive. The second condition is fulfilled but not the first.

4) $\{1, 2, 3\}$ is not inductive, since it is not true that for any $h \in \{1, 2, 3\}$ it would follow that $h + 1 \in \{1, 2, 3\}$.

Before defining the natural numbers, we will briefly digress on the intersection of an arbitrary family of sets. If $(A_i)_{i \in I}$ is a family of sets, the intersection $\bigcap_{i \in I} A_i$ of that family is defined by:

$$\bigcap_{i \in I} A_i = \{x/x \in A_i \forall i \in I\}$$

Example: For each real number $a > 0$, let $A_a = [0, a]$ (a closed interval). We assert that $\bigcap_{a \in R_{>0}} A_a = \{0\}$. For, clearly $0 \in \bigcap_{a \in R_{>0}} A_a$ inasmuch as $0 \in [0, a] \forall a \in R_{>0}$. In addition, if $x \in \bigcap_{a \in R_{>0}} A_a$ we will have $x \in A_a \forall a > 0$, or $0 \leq x \leq a \forall a > 0$, then $x = 0$ (if it were $x > 0$, it would be $x \notin A_{x/2}$).

It follows from definitions that the intersection of an arbitrary family of inductive subsets of $\mathbb{R}$ is again inductive (verify). In particular, the intersection of all inductive subsets of $\mathbb{R}$ is inductive, and we will say that a number is natural if it belongs to that intersection, which will be denoted by $\mathbb{N}$, i.e., if $(H_i)_{i \in I}$ is the family of all inductive subsets $\mathbb{R}$:

$$\mathbb{N} = \bigcap_{i \in I} H_i$$

We have then:

Proposition 1.1: $\mathbb{N}$ is inductive, and it is contained in any inductive.■

Proposition 1.2: $n \geq 1 \forall n \in \mathbb{N}$.

Proof: By an example above $H = \{x \in \mathbb{R}/x \geq 1\}$ is inductive so that $\mathbb{N} \subset H$, then if $n \in \mathbb{N}$ we will have $n \in H$, that is to say, $n \geq 1$.■

Theorem 1.3: $a, b \in \mathbb{N} \Rightarrow a + b \in \mathbb{N}$.

Proof: Let be $a, b \in \mathbb{N}$ and define:

$$H_a = \{x \in N/a + x \in N\}$$

H_a is inductive. In fact $1 \in H_a$ since $a + 1 \in \mathbb{N}$ because $a \in \mathbb{N}$ and $\mathbb{N}$ is inductive. Moreover, if $h \in H_a$, i.e., $a + h \in \mathbb{N}$, then $a + h + 1 \in \mathbb{N}$ since $\mathbb{N}$ is inductive, so that $h + 1 \in H_a$ and H_a is inductive, therefore $\mathbb{N} \subset H_a$ and as $b \in \mathbb{N}$ it follows that $b \in H_a$ or $a + b \in \mathbb{N}$.■

Theorem 1.4: $a, b \in \mathbb{N} \Rightarrow ab \in \mathbb{N}$.

Proof: Let $a, b \in \mathbb{N}$ and let $H_a{}' = \{x \in \mathbb{N}/ax \in \mathbb{N}\}$. $H_a{}'$ is inductive since $1 \in H_a{}'$ as $a \in \mathbb{N}$ and if $h \in H_a{}'$, that is $ah \in \mathbb{N}$, then $ah + a = a(h + 1) \in \mathbb{N}$ by theorem 1.3, then $h + 1 \in H_a{}'$ and $H_a{}'$ is inductive. It follows that $\mathbb{N} \subset H_a{}'$, so $b \in H_a{}'$, i.e., $ab \in \mathbb{N}$.∎

Lemma 1.5: $a \in \mathbb{N}$, $a > 1 \Rightarrow a - 1 \in \mathbb{N}$.

Proof: It is enough to verify that the set $\{1\} \cup \{x \in \mathbb{R}/x - 1 \in \mathbb{N}\}$ is inductive, and is left as an exercise.∎

Theorem 1.6: $a, b \in \mathbb{N}$ and $a > b \Rightarrow a - b \in \mathbb{N}$.

Proof: Let $a, b \in \mathbb{N}$, and define:

$$H_a = \{x \in \mathbb{R}/a > x \Rightarrow a - x \in \mathbb{N}\}$$

and check it is inductive. We have $1 \in H_a$ by the lemma above, and if $h \in H_a$, that is, if is it valid the implication:

$$a > h \Rightarrow a - h \in \mathbb{N} \quad (1)$$

we must test that $h + 1 \in H_a$, that is, the validity of the implication:

$$a > h + 1 \Rightarrow a - (h + 1) \in \mathbb{N}$$

Assume then that $a > h + 1$, so that $a > h$ and by (1): $a - h \in \mathbb{N}$. As $a - h > 1$, by the above lemma, we will have $a - h - 1 \in \mathbb{N}$, or $a - (h + 1) \in \mathbb{N}$. Since H_a inductive: $\mathbb{N} \subset H_a$ and as $b \in \mathbb{N}$, it will be $b \in H_a$, that is, is valid the implication: $a > b \Rightarrow a - b \in \mathbb{N}$.∎

Corollary 1.7: If $n \in \mathbb{N}$, it does not exist $x \in \mathbb{N}$ such that $n < x < n + 1$.

Proof: If such x existed, we would have $x - n < 1$ and, by theorem 1.6, we would have: $x - n \in \mathbb{N}$, which contradicts proposition 1.1.∎

2 - MATHEMATICAL INDUCTION

Consider, as an introduction, the following example, which is about finding the sum of the first n even natural numbers of the form $2a + 1$ with $a \in \mathbb{N}$.

Let us make a list of the values of that sum for the first values of n (in this, and in others examples, we will use the decimal system which is treated at the end of the chapter, this does not invalidate the theory since it is only used in examples):

$$1 = 1$$
$$1 + 3 = 4$$
$$1 + 3 + 5 = 9$$

$$1 + 3 + 5 + 7 = 16$$
$$1 + 3 + 5 + 7 + 9 = 25$$

From the list above, we could conjecture that the sum of the first n consecutive even natural numbers is n^2, i.e., we could conjecture the validity of the following indefinite list:

$$1 = 1^2$$
$$1 + 3 = 2^2$$
$$1 + 3 + 5 = 3^2$$
$$1 + 3 + 5 + 7 = 4^2$$
$$........$$
$$1 + 3 + ... + (2h - 1) = h^2$$
$$1 + 3 + ... + (2h - 1) + (2h + 1) = (h + 1)^2$$
$$........$$

How to prove this conjecture?. Note that adding $2h + 1$ to both members of file h, we obtain:

$$1 + 3 + ... + (2h - 1) + (2h + 1) = h^2 + 2h + 1$$

and as $h^2 + 2h + 1 = (h + 1)^2$, we obtain file $h + 1$.

But this is all that we need to prove the conjecture, since it is valid for $n = 1$, and since its validity for an arbitrary file h implies its validity for the next file $h + 1$, taking $h = 1$, we obtain its validity for the second file, then taking $h = 2$, we have the validity of the third and so on, then whatever be n∈N we have:

$$1 + 3 + ... + (2n - 1) = n^2 \quad (1)$$

This example illustrates the principle of *mathematical induction* or *complete induction* (as opposed to the incomplete induction of the empirical sciences), stated below:

Principle of mathematical induction: Let $P(n)$ be a proposition for each natural number n. If

1) $P(1)$ is true.

2) For every $h \in \mathbb{N}$, the implication $P(h) \Rightarrow P(h + 1)$ is true; then $P(n)$ is true $\forall n \in \mathbb{N}$.

In the preceding example, $P(n)$ is the proposition: "$1 + 3 + ... + (2n - 1) = n^2$".

Another, more obvious illustration of the principle, is the following: Let's put together an undefinedly long row of domino chips arranged so that each chip upon falling makes the next one fall too (condition 2 of the principle). If the first chip falls (condition 1), then the nth chip will also eventually fall, no matter which number n is.

The induction principle can be demonstrated form the results of the preceding section; consequently, it is not a principle but a theorem. In fact, let:

$$H = \{n \in \mathbb{N}/P(n) \text{ is true}\}.$$

The hypotheses of the induction principle exactly expresses that H is inductive, thus $\mathbb{N} \subset H$. That is to say that for every $n \in \mathbb{N}$ we have $n \in H$, or that $P(n)$ is true $\forall n \in N$.

Let us apply the induction principle to another example. We will prove that for each $n \in \mathbb{N}$, we have:

$$1 + 2 + \ldots + n = \left(\frac{n(n+1)}{2}\right) (2)$$

According to the induction principle, to prove this equality, it is enough to verify it for $n = 1$, which is obvious. The notation $1 + 2 + \ldots + n$ is somewhat vague, but it represents the sum of all natural numbers between 1 and n; for $n = 1$, the expression means that the sum begins and ends in 1, i.e., it has a single summand: 1. Now, to verify that for each $h \in \mathbb{N}$, its validity for $n = h$ implies its validity for $n = h + 1$, i.e., to verify the next implication:

$$1 + \ldots + h = \frac{h(h+1)}{2} \Rightarrow 1 + \ldots + h + (h+1) = \frac{h(h+2)}{2} (3)$$

Starting from $1 + \ldots + h = \frac{h(h+1)}{2}$, adding $h + 1$ to both members, we obtain:

$$1 + \ldots + h + (h+1) = \frac{h(h+1)}{2} + (h+1) = (h+1)\left(\frac{h}{2} + 1\right)$$
$$= \left(\frac{(h+1)(h+2)}{2}\right)$$

then the implication (3) is true, and we conclude that (2) is valid $\forall n \in \mathbb{N}$.

Note that the induction principle allows us to verify (2), after having guessed its validity in some way, analyzing particular cases for example.

Let's look at another way of proving (2), which can be applied, more generally, to any arithmetic progression.

Defining:

$$S = 1 + 2 + \ldots + (n-1) + n$$

we also have:

$$S = n + (n-1) + \ldots + 2 + 1$$

so $2S = (n + 1) + (n + 1) + \ldots + (n + 1) + (n + 1)$, therefore:

$$S = \frac{n(n + 1)}{2}$$

This procedure was used by Gauss as a schoolboy. The teacher, in order to train the students in addition, asked them to find the sum of the first one hundred natural numbers. All of them, except Gauss, started to add $1 + 2$, the result plus 3, and so on. Gauss instead, used the previous procedure, i.e., being S the sum to calculate:

$$S = 1 + 2 + 3 + \ldots + 98 + 99 + 100$$

then,

$$S = 100 + 99 + 98 + \ldots + 3 + 2 + 1$$

so,

$$2S = 101 + 101 + \ldots + 101 + 101 = 100 \cdot 101$$

therefore $S = 5.050$.

This and others demonstrations of his talent made Gauss—who was of humble origin—worthy of a grant (a subsidy from a nobleman) to pursue higher studies.

The relations (1) and (2) were already known by the Pythagoreans (members of the scientific-philosophical sect founded by Pythagoras) who arrived at them by graphic reasoning. To the sums of the first natural numbers, they called the triangular numbers. The following scheme shows why:

Adding the n- th triangular number to the $(n + 1)$- th, we obtain a square $(n + 1)^2$. For example if $n = 3$, it would be:

then the n-th triangular number can be obtained subtracting from the points of the square the elements of the diagonal, and dividing the result by 2:

$$\frac{(n+1)^2 - (n+1)}{2} = \frac{n(n+1)}{2}$$

obtaining again: $1 + 2 + \ldots + n = \frac{n(n+1)}{2}$.

Similarly, to the sums of the first even numbers they called quadrangular numbers, The corresponding scheme is:

Note that each square is obtained adding an even number to the former. In such a way they obtained (1).

3 - POWERS

Let a be a real number, we define $a^1 = a$ and if $h \in \mathbb{N}$: $a^{h+1} = a^h a$., Then $a^n \ \forall n \in \mathbb{N}$, since $a^2 = a^1 a = aa$; $a^3 = a^2 a = aaa$ and so on. In fact, a subtlety is presented here. In the same way, as we prove the induction principle, we should prove a principle of definition by induction, of which the above definition is a particular case. We are going to do so on the last section of the chapter, and by now we accept intuitively the validity of the definitions by induction.

Example: Analyzing particular cases, we could conjecture that $2^n > n \ \forall n \in \mathbb{N}$. Let's prove it by induction.

For the case $n = 1$ the proof is clear. Let $h \in \mathbb{N}$, assuming by inductive hypothesis that $2^h > h$, we should prove that $2^{h+1} > h + 1$. We have

$$2^{h+1} = 2^h 2 > 2h$$

since in addition $2h \geq h + 1$ (as $h \in \mathbb{N} \Rightarrow h \geq 1 \Rightarrow h + h \geq h + 1$), it follows by transitivity that $2^{h+1} > h + 1$. So $2^n > n \ \forall n \in \mathbb{N}$.

Theorem 3.1: Whatever be the real numbers a, b and the natural numbers n, m, we have:

  1) $a^m a^n = a^{m+n}$

2) $(a^m)^n = a^{mn}$

3) $(ab)^n = a^n b^n$

Proof: We will prove the first postulate, leaving the others as exercises. Proceed by induction on n, that is, prove using the induction principle the proposition $P(n)$: "Given $a \in \mathbb{R}$ and $m \in \mathbb{N}$, we have $a^m a^n = a^{m+n}$". For $n = 1$ we must verify that $a^m a^1 = a^{m+1}$, which is clear by definition of powers. It remains to verify that assuming that $h \in \mathbb{N}$ and $a^m a^h = a^{m+h}$ (inductive hypothesis), then it follows that $a^m a^{h+1} = a^{m+h+1}$. Indeed we have $a^m a^{h+1} = a^m a^h a = a^{m+h} a = a^{m+h+1}$. Then $P(n)$ is true $\forall n \in \mathbb{N}$.

Of course we could proceed by induction on m, which would be totally symmetric with what we have just done. In case 2), instead, such symmetry does not exist. As an exercise, 2) can be proved in two ways: by induction on n and by induction on m. In this last case it is convenient to prove first 3).■

Extend the definition of power to the exponent 0: if $a \in \mathbb{R}$ define $a^0 = 1$. This is simply a convention made to keep the validity of the above theorem, i.e., if we want to define a^0 in such a way so that : $a^m a^0 = a^{m+0} = a^m$ be valid whatever be $m \in \mathbb{N}$ and $a \in \mathbb{R}$, we are forced to define $a^0 = 1$ at least if $a \neq 0$. In case we have $a = 0$ we could define also $a^0 = 0^0 = 1$ or leave it without definition.

4 - FINITE SETS

"Counting" the elements of a set means to assign a natural number $(1, 2, \ldots, n)$ to each of those elements. A set is finite if its elements can be counted.

More precisely, for $n \in \mathbb{N}$ the *natural interval* I_n is defined by:

$$I_n = \{x \in \mathbb{N}/x \leq n\}$$

and it is said that a set A is *finite* iff it is the empty set or there is a bijection:

$$f: A \to I_n$$

for some $n \in \mathbb{N}$.

Proposition 4.1: If $m, n \in \mathbb{N}$ and $f: I_m \to I_n$ is an injective function, then $m \leq n$.

Proof: Proceeding by induction on n, for $n = 1$ is clear. Let $n > 1$. If $m = 1$ there is nothing to prove, assume then $m > 1$. Put $f(m) = a$ and define $g: I_n \to I_n$ as follows:

$$g(x) = \begin{cases} n \text{ if } x = a \\ a \text{ if } x = n \\ x \text{ if } x \neq a \text{ y } x \neq n \end{cases}$$

g is clearly a bijection, so $gf: I_m \to I_n$ is injective, and since $gf(m) = g(a) = n$, the restriction $gf_{/I_{m-1}}$ applies I_{m-1} on I_{n-1} and is injective, so by inductive hypothesis $m - 1 \leq n - 1$, then $m \leq n$. ∎

The above proposition, or more precisely its counter reciprocal, is often called the *box principle* or the *pigeon hole principle*, because it can be paraphrased saying that if $m > n$ and there are m objects placed in n boxes, at least one box must contain two or more objects. (If m pigeons occupy a dovecot with n nests, at least one nest must be occupied by two or more pigeons).

Corollary 4.2: Let be $m, n \in \mathbb{N}, A$ a set and $f: A \to I_n, g: A \to I_m$ bijections, then $m = n$.

Proof: As $fg^{-1}: I_m \to I_n$ and $gf^{-1}: I_n \to I_m$ are injective (moreover, they are bijective), it follows from proposition above that m≤n and n≤m. ∎

If A is a finite set, its *cardinal* # (A) is defined by:

  - # $(A) = 0$ if $A = \emptyset$.

  - # $(A) = n$ if there is a bijection $f: A \to I_n$ $(n \in \mathbb{N})$.

The preceding corollary shows that the cardinal of a finite set is uniquely determined.

If # $(A) = n$ with $n \in \mathbb{N}$, it is also said that A has n elements.

Lemma 4.3: Let A be a finite set, we have:

   1) If $a \notin A$, then $A \cup \{a\}$ is finite and # $(A \cup \{a\}) =$ # $(A) +$ 1.

   2) If $a \in A$, then $A - \{a\}$ is finite and # $(A - \{a\}) =$ # $(A) -$ 1.

   3) Every subset of A is finite.

Proof: 1) If $A = \emptyset$ is clear. Let $A \neq \emptyset$, as A is finite, there exists $n \in \mathbb{N}$ such that $(A) = n$, i. e., there exists a bijection $f: A \to I_n$. If we define $g: A \cup \{a\} \to I_{n+1}$ by :

$$g(b) = \begin{cases} f(b) \text{ if } b \neq a \\ n + 1 \text{ if } b = a \end{cases}$$

clearly, g is a bijection, so 1) is proved.

2) If A={a} then $A - \{a\} = \emptyset$ and the statement is true. Let $A \neq \{a\}$ and $f: A \to I_n$ a bijection and take $b \in A$ such that $f(b) = n$. Defining $g: A \to A$ by:

$$g(x) = \begin{cases} a \text{ if } x = b \\ b \text{ if } x = a \\ x \text{ if } x \neq a, b \end{cases}$$

then g is a bijection, so fg is also a bijection. As $fg(a) = f(b) = n$, it follows that the restriction of fg to $A - \{a\}$ is a bijection from $A - \{a\}$ onto I_{n-1}.

3) Let B be a subset of A. If $A = \emptyset$ is trivial. Let be $A \neq \emptyset$, so that exists $n \in \mathbb{N}$ such that $\#(A) = n$. Proceeding by induction on n, if $n = 1$ it must be B=$\emptyset$ or $B = A$, so B is finite. Let $n > 1$, if $B = A$ it is nothing to prove; suppose then $B \neq A$, so there exists $a \in A$ such that $a \notin B$ and we have $B \subset A - \{a\}$. As by 2) $A - \{a\}$ is finite with $n - 1$ elements, the result follows by inductive hypothesis.∎

Proposition 4.4: Let A and B be finite sets. We have,

a) If $A \cap B = \emptyset$, then $A \cup B$ is finite and $\#(A \cup B) = \#(A) + \#(B)$.

b) If $B \subset A$, then $\#(A - B) = \#(A) - \#(B)$.

c) $A \cup B$ is finite and $\#(A \cup B) = \#(A) + \#(B) - \#(A \cap B)$.

d) $A \times B$ is finite and $\#(A \times B) = \#(A) \cdot \#\#(B)$.

Proof: a) If $A = \emptyset$ or $B = \emptyset$ is clear. So let $n, m \in \mathbb{N}$ be such that $n = \#(A)$ and $m = \#(B)$. We proceed by induction on m. For $m = 1$ is clear by 1) of the lemma above. Let $m > 1$, taking $b \in B$ it results by 2) of the lemma: $\#(B - \{b\}) = m - 1$, so by inductive hypothesis: $\#(A \cup (B - \{b\})) = n + m - 1$ and as $A \cup B = (A \cup (B - \{b\}) \cup \{b\}$; the result follows from 1) of the lemma.

b) As $(A - B) \cap B = \emptyset$ we have by a): $\#(A) = \#(A - B) + \#(B)$, since $A - B$ is finite by 3) of the lemma and being $A = (A - B) \cup B$ as $B \subset A$.

c) We have $A \cup B = A \cup (B - A)$ and as this union is disjoint, it follows by a) that $A \cup B$ is finite and:

$$\#(A \cup B) = \#(A) + (\# B - A)$$

as we have in addition that $B - A = B - (A \cap B)$, it follows by b):

$$\#(B - A) = \#(B) - \#(A \cap B)$$

d) If $A = \emptyset$ or $B = \emptyset$ we have $A \times B = \emptyset$ and the statement is true. So assume $A \neq \emptyset$ and $B \neq \emptyset$ then there are $n, m \in \mathbb{N}$ with $\#(A) = n$,

$\#(B) = m$. Proceeding by induction on m, for $m = 1$ is clear (in such a case $B = \{b\}$ and if $f: A \to I_n$ is a bijection; then $g: A \times \{b\} \to I_n$ defined by $g(a, b) = f(a)$ is also a bijection). Let $m > 1$ and take $b \in B$, by inductive hypothesis we have: $(A \times (B - \{b\})) = n(m - 1)$, but as:

$$A \times B = [A \times (B - \{b\})] \cup [A \times \{b\}]$$

is a disjoint union, we have by a):

$$\#(A \times B) = n(m - 1) + n = nm. \blacksquare$$

Example: A set of n elements, has 2^n subsets.

We proceed by induction on n; if $n = 1$ is clear. Let A be a set with $n > 1$ elements and take $a \in A$. A subset of A that does not contain the element a is a subset of $A - \{a\}$ and as $A - \{a\}$ has $n - 1$ elements, by the inductive hypothesis, we have that there are exactly 2^{n-1} subsets of A that does not contain the element a. In addition, each subset of A that contains a, is the union of $\{a\}$ with a subset of $A - \{a\}$, so there are also 2^{n-1} of such subsets. Thus there are $2^{n-1} + 2^{n-1} = 2^n$ subsets of A.

Proposition 4.5: Let A be a set with m elements and B a set with n elements where $m, n \in \mathbb{N}$:

1) The number of functions from A to B is n^m.

2) The number of injective functions from A to B is 0 if $m > n$ and is $n(n - 1)\ldots(n - m + 1)$ if $m \leq n$.

Proof: left as an exercise. We suggest to do induction on m to prove 1) and induction on n to prove 2). $\blacksquare$

Both a) and d) of prop. 4.4, can be easily generalized to a union of n finite sets disjoint two to two and to a cartesian product of n finite sets respectively. The generalization of c) is more subtle and will be treated in section 8, where as a corollary the number of surjective functions from one finite set onto another will be obtained, also to be obtained, independently, in exercise 26.

5 - SUMMATORIES

Let a_i be a real number for each natural number i; we define the *summatory* $\sum_{i=1}^n a_i$ inductively by:

$$\sum_{i=1}^1 a_i = a_1 \quad ; \quad \sum_{i=1}^{h+1} a_i = \sum_{i=1}^h a_i + a_{h+1}$$

then $\sum_{i=1}^n a_i$ is defined whatever be $n \in \mathbb{N}$.

For example:

$\sum_{i=1}^{2} a_i = \sum_{i=1}^{1} a_i + a_2 = a_1 + a_2 \quad \sum_{i=1}^{3} a_i = \sum_{i=1}^{2} a_i + a_3 = a_1 + a_2 + a_3$

The lower limit of the summatory, $= 1$ above, also can be changed, for example:

$$\sum_{i=0}^{2} (i+1) = 1 + 2 + 3.$$

Proposition 5.1: If a, a_i, b_i are real numbers and $n \in \mathbb{N}$, we have:

1) $\sum_{i=1}^{n} a_i + \sum_{i=1}^{n} b_i = \sum_{i=1}^{n}(a_i + b_i)$

2) $a \sum_{i=1}^{n} a_i = \sum_{i=1}^{n} a a_i$

3) $\sum_{i=0}^{n} a_i = \sum_{i=1}^{n+1} a_i$

4) $\sum_{i=0}^{n} a_i = a_0 + \sum_{i=1}^{n} a_i$.∎

These properties can be easily proved by induction in n, which is left as an exercise. It is convenient to visualize them taking particular values of n; for example, taking $n = 3$ in 1) we have:

$$\sum_{i=1}^{3} a_i + \sum_{i=1}^{3} b_i = (a_1 + a_2 + a_3) + (b_1 + b_2 + b_3)$$

$$\sum_{i=1}^{3} (a_i + b_i) = (a_1 + b_1) + (a_2 + b_2) + (a_3 + b_3)$$

which shows that 1) is a simple consequence of the commutativity and associativity of the sum.

Example: For $x, y \in \mathbb{R}$ and $n \in \mathbb{N}$ we have:

$$x^n - y^n = (x - y) \sum_{i=1}^{n} x^{n-i} y^{i-1}$$

Indeed,

$$(x - y) \sum_{i=1}^{n} x^{n-i} y^{i-1} = \sum_{i=1}^{n} x^{n+1-i} y^{i-1} -$$

$\sum_{i=1}^{n} x^{n-i} y^{i} =$

$$= x^n + \sum_{i=2}^{n} x^{n+1-i} y^{i-1} - \sum_{i=1}^{n-1} x^{n-i} y^n$$

$$= x^n + \sum_{i=1}^{n-1} x^{n-i} y^i - \sum_{i=1}^{n-1} x^{n-i} y^i - y^n = x^n - y^n$$

Similarly, it is defined the *product* $\prod_{i=1}^{n} a_i$ by:

$$\prod_{i=1}^{1} a_i = a_1, \quad \prod_{i=1}^{h+1} a_i = \left(\prod_{i=1}^{h} a_i\right) \cdot a_{h+1}$$

6 - COMBINATORIAL NUMBERS

The *factorial* $n!$ of $n \in \mathbb{N}$, is inductively defined by:

$$1! = 1, \quad (h+1)! = (h+1)h!$$

or alternatively using the symbol of the product of the last section: $n! = \prod_{i=1}^{n} i$.

As a matter of convenience, it is also defined $0! = 1$.

For example:

$$5! = 5 \cdot 4! = 5 \cdot 4 \cdot 3! = 5 \cdot 4 \cdot 3 \cdot 2! = 5 \cdot 4 \cdot 3 \cdot 2 \cdot 1! = 5 \cdot 4 \cdot 3 \cdot 2 \cdot 1 = 120.$$

Starting from factorials, *combinatorial numbers* $\binom{n}{r}$ with $r, n \in \mathbb{N} \cup \{0\}$ and $r \leq n$, are defined by:

$$\binom{n}{r} = \frac{n!}{r!\,(n-r)!}$$

For example: $\binom{5}{3} = \frac{5}{3!2!} = \frac{5.4.3!}{3!2!} = \frac{5.4}{2} = 10.$

It follows immediately from definition that $\binom{n}{0} = \binom{n}{n} = 1$ and that $\binom{n}{r} = \binom{n}{n-r}$.

Both factorials and combinatorial numbers arise naturally in Combinatorics, i.e., in the art of counting the elements of a set. For example: how many barrels can be stacked on top of a row of n barrels? In the first row there are n barrels; in the second can be placed $n - 1$; in the next one we can place $n - 2$... and continuing like this until only 1 barrel occupies the n- th row. So the answer is the triangular number:

$$n + (n-1) + \ldots + 2 + 1 = \frac{n(n+1)}{2} = \binom{n+1}{2}$$

which is a combinatorial number.

The characteristic property of combinatorial numbers is:

Proposition 6.1: If $r, n \in \mathbb{N}$ are such that r<n, then,

$$\binom{n+1}{r} = \binom{n}{r} + \binom{n}{r-1} \quad (1)$$

Proof:

$$\binom{n}{r} + \binom{n}{r-1} = \frac{n!}{r!(n-r)!} + \frac{n!}{(r-1)!(n-r+1)!} = \frac{n!}{(r-1)!(n-r)!}\left(\frac{1}{r} + \frac{1}{n-r+1}\right) =$$

$$= \frac{n!}{(r-1)!\,r(n-r+1)}\,n+1 = \frac{(n+1)!}{r!(n-r)!} = \binom{n+1}{r}. \quad \blacksquare$$

The above proposition allows us to compute the combinatorial numbers with the following arrangement quickly:

$$\binom{1}{0} \qquad \binom{1}{1}$$

$$\binom{2}{0} \qquad \binom{2}{1} \qquad \binom{2}{1}$$

$$\binom{3}{0} \qquad \binom{3}{1} \qquad \binom{3}{2} \qquad \binom{3}{3}$$

$$\binom{4}{0} \qquad \binom{4}{1} \qquad \binom{4}{1} \qquad \binom{4}{3} \qquad \binom{4}{4}$$

where each element is the sum of the immediate two of the upper row. This arrangement is often referred to as the *arithmetic triangle* or the *Pascal Triangle*, or the *Tartaglia triangle,* although it was well known in Chinese and Indian mathematics at least since the 11th century. Let us explicitly write the arithmetic triangle up to the eighth row:

				1		1		
			1		2		1	
		1		3		3		1
	1		4		6		4		1
1		5		10		10		5		1
1	6		15		20		15		6	1
1	7	21		35		35		21	7	1
1	8	28	56		70		56	28	8	1

Looking at this arrangement, we can find some curious or interesting properties. For example, the sum of the elements of the n-th row is 2^n; or starting from either end and adding the elements of the corresponding diagonal, upon stopping at an element, that sum is the element of the following row changing the diagonal. For example, starting from the left end of the third row and stopping at 35 we have: $1 + 4 + 10 + 20 + 35 = 70$. The properties we have described, can be stated as follows:

$$\sum_{i=0}^{n}\binom{n}{i} = 2^n \quad (2)$$

$$\sum_{i=0}^{n}\binom{r+i}{i} = \binom{r+n+1}{n}$$

$$\sum_{i=1}^{n}\binom{r-1+i}{r} = \binom{r-1+1}{r} = \binom{r+n}{r+1} \quad (3)$$

they are consequences of the characteristic property of combinatorial numbers and can be proved by induction on n (ex.18).

Example: If $r, n \in \mathbb{N} \cup \{0\}$ with $r \leq n$, the number of subsets with r elements of a set A with n elements is $\binom{r}{n}$.

In fact, if $r = n$, we can assume that $r < n$. Proceeding by induction, for $n = 1$ is clear. Let $a \in A$; to count the subsets of A we will divide them in two disjoint parts: one part that contains a, and another that doesn't. The part that does *not* contain a are the subsets with r elements of $A - \{a\}$, which by inductive hypothesis are $\binom{n-1}{r}$ in number. Those which contain a are built attaching a to each subset with r-1 elements of $A - \{a\}$, so there are $\binom{n-1}{r-1}$ of them. In total, then, there are $\binom{n-1}{r} + \binom{n-1}{r-1} = \binom{n}{r}$ subsets of A with r elements. From this result we can obtain again, using the identity (2), that the number of subsets of a set of n elements es 2^n. Alternatively one also can give a combinatorial proof of (2), merging what is proved in this example and in the last example of section 4.

Example: Upon a triangular base with n balls in each side, how many balls can be stacked?

By an example above, in the base there are $\frac{n(n+1)}{2} = \binom{n+1}{2}$ balls, in the next level $\binom{n}{2}$, in which follows $\binom{n-1}{2}$ and so on. Then the number $\mathbb{N}$ of balls that can be piled on, is:

$$\mathbb{N} = \binom{n+1}{2} + \binom{n}{2} + \ldots + \binom{2}{2} = \sum_{i=1}^{n}\binom{1+i}{2}$$

53

from where, by (3): $\mathbb{N} = \binom{n+2}{3}$.

On the other side, we have:

$$\binom{n+2}{3} = \sum_{i=1}^{n}\binom{1+i}{2} = \sum_{i=1}^{n}\frac{(1+i)i}{2} = \frac{1}{2}\sum_{i=1}^{n}(1+i^2)$$

$$= \frac{1}{2}\left\{\sum_{i=1}^{n}i + \frac{1}{2}\sum_{i=1}^{n}i^2\right\}$$

so that,

$$\sum_{i=1}^{n}i^2 = 2\binom{n+2}{3} - \frac{n(n+1)}{2} = \frac{(n+2)(n+1)n}{3} - \frac{n(n+1)}{2}$$

$$= \frac{n(n+1)(2n+1)}{6}$$

Similarly, starting from $\sum_{i=1}^{n}i^2 = 2\binom{n+2}{3} - \frac{n(n+1)}{2} = \binom{n*3}{4}$, an expression for the sum of the cubes of the first natural numbers can be found; then for the fourth powers, and so on. In the next section we will show another way to obtain such relations.

Example: How many balls fit in a stack with a rectangular base of $m \cdot n$ balls?

In the base there are mn balls, in the next level $(m-1)(n-1)$, and so on, then, assuming $m \geq n$, the number of balls in the stack is:

$$\sum_{i=1}^{n}(m+1-i)(n+1-i)$$

As $(m+1-i)(n+1-i) = (m+1)(n+1) - (m+n+2)i + i^2$, we have:

$$\sum_{i=1}^{n}(m+1-i)(n+1-i)$$

$$= \sum_{i=1}^{n}(m+1)(n+1)\ (m+n+2)\sum_{i=1}^{n}i + \sum_{i=1}^{n}i^2 =$$

$$= n(m+1)(n+1) + (m+n+2)\frac{n(n+1)}{2} + \frac{n(n+1)(2n+1)}{6} =$$

$$= \frac{n(n+1)}{6}(3m - n + 1).$$

7 - BINOMIAL EXPANSION

For a, b real numbers, we have:

$$\begin{aligned}
(a + b)^1 &= a + b \\
(a + b)^2 &= a^2 + 2ab + b^2 \\
(a + b)^3 &= a^3 + 3a^2b + 3ab^2 + b^3 \\
(a + b)^4 &= a^4 + 4a^3b + 6a^2b^2 + 4ab^3 + b^4
\end{aligned}$$

We note that the coefficients are the elements of the arithmetical triangle. From the behavior of these particular cases, we can conjecture:

Theorem 7.1: (Binomial expansion) If a, b are real numbers and n is a natural number, we have:

$$(a + b)^n = \sum_{i=0}^{n} \binom{n}{i} a^{n-i} b^i$$

Proof: The first proof we give, is of combinatorial type. In the product:

$$(a + b)^n = (a + b)(a + b)\dots(a + b)$$

the coefficient of $a^{n-i}b^i$ is the number of subsets with i elements (the b) of a set with n (the b of the product) elements and that numbers is $\binom{n}{i}$.

The second proof, albeit longer, is a good exercise on manipulation of summatories. We proceed by induction on n. If $n = 1$ is clear. We must prove the validity of the implication:

$$(a + b)^h = \sum_{i=0}^{h} \binom{h}{i} a^{h-i} b^i = (a + b)^{h+1} = \sum_{i=0}^{h+1} \binom{h + 1}{i}$$

We have,

$$\begin{aligned}
(a + b)^{h+1} &= a(a + b)^h + b(a + b)^h \\
&= a \sum_{i=0}^{h} \binom{h}{i} a^{h-i} b^i + b \sum_{i=0}^{h} \binom{h}{i} a^{h-i} b^i = \\
&= \sum_{i=0}^{h} \binom{h}{i} a^{h+1-i} b^i + \sum_{i=0}^{h} \binom{h}{i} a^{h-i} b^{i+1} =
\end{aligned}$$

$$= a^{h+1} + \sum_{i=0}^{h} \binom{h}{i} a^{h+1-i} b^i$$

$$+ \sum_{i=0}^{h} \binom{h}{i} a^{h-i} b^{i+1} + b^{h+1} =$$

$$= a^{h+1} + \sum_{i=0}^{h} \binom{h}{i} a^{h+1-i} b^i + \sum_{i=0}^{h} \binom{h}{i} a^{h-i} b^{i+1}$$

$$+ b^{h+1} =$$

$$= a^{h+1} + \sum_{i=0}^{h} \binom{h+1}{i} a^{h+1-i} b^i + b^{h+1} =$$

$$\sum_{i=0}^{h+1} \binom{h+1}{i} a^{h+1-i} b^i. \blacksquare$$

The binomial expansion is known, at least, since the 11th century, thanks to Persian mathematicians Al-Karachi and Omar Khayyam. Its name is inseparably joined to Newton's, since he generalized the above expansion to a series expansion valid for any real exponent.

Example: In the previous section we saw a way to obtain the sum of the powers of the first natural numbers. We now will show another way to obtain those sums recursively. From

$$\sum_{i=1}^{n}(i+1)^k = \sum_{i=1}^{n}\sum_{j=0}^{k} \binom{k}{j} i^j = \sum_{i=1}^{n} 1 + \sum_{j=1}^{k} \binom{k}{j} \sum_{j=1}^{n} i^j + \sum_{i=1}^{n} i^k,$$

since $\sum_{i=1}^{n}(i+1)^k - \sum_{i=1}^{n} i^k = (n+1)^k - 1$ and that $\sum_{i=1}^{n} 1 = n$, it follows:

$$(n+1)^k - (n+1) = \sum_{j=1}^{k-1} \binom{k}{j} i^j \sum_{j=1}^{n} i^j \quad (*)$$

Putting up in $(*)$ $k = 2$, we have $(n+1)^2 - (n+1) = 2\sum_{i=1}^{n} i$, and we obtain again the relation:

$$\sum_{i=1}^{n} 1 = \frac{n(n+1)}{2}$$

Putting up in $(*)$ $k = 3$: $(n+1)^3 - (n+1) = 3\sum_{i=1}^{n} i^2 + 3\sum_{i=1}^{n} i$ and we obtain again:

$$\sum_{i=1}^{n} i^2 = \frac{n(n+1)(2n+1)}{6} \quad (1)$$

Putting up in $(*)$ $k = 4$: $(n+1)^4 - (n+1) = 4\sum_{i=1}^{n} i + 6\sum_{i=1}^{n} i^2 + 4\sum_{i=1}^{n} i^3$, from where we obtain:

$$\sum_{i=1}^{n} i^3 = \left(\frac{n(n+1)}{2}\right)^2 \quad (2)$$

So that we obtain an expression recursively for the sum of the k-th powers of the first natural numbers. This sums were considered in diverse epochs and civilizations. For example the relation (1) was used by Archimedes and also appears in Chinese, Indian, and even in Babylonian mathematics, where it is proved by geometric considerations. Ibn-al-Haitam, around the year 1000, in dealing with a geometric problem, finds casually the formula:

$$\sum_{i=1}^{n} i^4 = \frac{n(n+1)(2n+1)(3n^2+3n-1)}{30}$$

In his Arithmetics, Bachet proves (2) reasoning as follows:

$$1 = 1^3, \ \ 3+5 = 2^3, \ \ 7+9+11 = 3^3, \ \ 13+15+17+19 = 4^3 \ldots\ldots$$

and by adding, we arrive at (2).

Fermat and Pascal use them to derive the formula that nowadays we write:

$$\int_0^a x^n dx = \frac{a^{n+1}}{n+1}$$

Euler set forth the sum of the k powers of the first n natural numbers, as a polynomial on n of degree $k+1$, which coefficients depend on the so called Bernoulli numbers. The sums appear also in Kummer's works on Fermat's Last Theorem.

Another very usual notation is the one we describe below. If a_{ij} is a real number for each pair $i, j \in \mathbb{N} \cup \{0\}$ and if n is a natural number, the symbol:

$$\sum_{i+j=n} a_{ij}$$

denotes the sum of all the numbers a_{ij} when i, j take on all the possible values between 0 and n, but with the condition $i + j = n$. In other words:

$$\sum_{i+j=n} a_{ij} = \sum_{i=0}^{n} a_{i(n-i)}$$

With this notation, binomial expansion takes the more symmetric form:

$$(a+b)^n = \sum_{i+j=n} \frac{n!}{i!\,j!} a_i b_j$$

Similarly, $\sum_{i+j+k}^{n} a_{ijk}$ denotes the sum of the numbers a_{ijk} taking i, j, k all possible values in $\mathbb{N} \cup \{0\}$ between 0 and n, but satisfying $i + j + k = n$. More precisely:

$$\sum_{i+j+k}^{n} a_{ijk} = \sum_{i=0}^{n} \sum_{j+k=n-i} a_{ijk}$$

For example, $\sum_{i+j+k}^{n} a_{ijk} = a_{002} + a_{011} + a_{021} + a_{101} + a_{110} + a_{200}$.

Proposition 7.2: (Leibnitz formula) Let a, b, c be real numbers and n a natural number. We have:

$$(a + b + c)^n = \sum_{i+j+k=n} \frac{n!}{i!\,j!\,k!} a^i b^j c^k$$

Proof:

$$(a + b + c)^n = (a + (b + c))^n = \sum_{i=0}^{n} \frac{n!}{i!\,(n-i)!} a^i (b+c)^{n-i} =$$

$$= \sum_{i=0}^{n} \frac{n!}{i!\,(n-i)!} a^i \sum_{j+k=n} \frac{(n-i)!}{j!\,k!} b^j c^k$$

$$=$$

$$= \sum_{i=0}^{n} \sum_{j+k=n-i} \frac{n!}{i!j!k!} a^i b^j c^k =$$

$$\sum_{i+j+k=n} \frac{n!}{i!j!k!} a^i b^j c^k. \blacksquare$$

Example: Find the coefficient of x^5 in the expansion of $(x^2 + x + 1)^8$ (it is understood that similar terms are grouped). We have:

$$(x^2 + x + 1)^8 = \sum_{i+j+k=8} \frac{8!}{i!\,j!\,k!} x^{2i+j}$$

then we must group the terms such that $2i + j = 5$, i.e., the terms in which: $i = 0$ and $j = 5$; $i = 1$ and $j = 3$; $i = 2$ and $j = 1$. The coefficient we were looking for is then:

$$\frac{8!}{5!\,3!} + \frac{8!}{3!\,4!} + \frac{8!}{2!\,5!} = 504$$

The formula of prop. 7.2 can be generalized, virtually with the same proof, to obtain:

$$(a_1 + a_2 + \ldots + a_m)^n = \sum_{i_1+i_2+\cdots+i_m} \frac{n!}{i_1!\,i_2!\ldots i_m!} a_1^{i_1} a_2^{i_2} \ldots a_m^{i_m}$$

In prop.4.4 c) we proved that for finite sets A and B, we have:

$$\# (A \cup B) = \# (A) + \# (B) - \# (A \cap B) \ (*)$$

This section is devoted to generalize that result for any finite family of finite sets.

First consider three finite sets A, B, C. Using repeteadly $(*)$ we have:

$$\# (A \cup B \cup C) = \# (A \cup B) + \# (C) - \# ((A \cup B) \cap C) =$$
$$= \# (A \cup B) + \# (C) - \# ((A \cap C) \cup (B \cap C)) =$$
$$= \# (A) + \# (B) - \# (A \cap B) + \# (C) - \# (A \cap C) - \# (B \cap C) +$$
$$\# (A \cap B \cap C)$$

Doing the same with four sets, we arrive at the following generalization:

Theorem 8.1 (Principle of Inclusion-Exclusion): Let $A_1, \ldots, A_n$ be finite sets. For each $k = 1, \ldots, n$,

$$S_k = \sum_{\{i_1, \ldots, i_k\} \subset I_n} \# \left(A_1 \cap \ldots \cap A_{i_k} \right)$$

where the sum extends over all the subsets with k elements of $i_k = \{1, \ldots, n\}$ (there are then $\binom{n}{k}$ summands). With this notation, we have:

$$\# (A_1 \cup \ldots \cup A_n) = \sum_{k=1}^{n} \sum (-1)^{k-1} S_k \ (1)$$

Proof: Proceeding by induction on n, for $n = 1$ is clear. Assume $n > 1$, by $(*)$ results:

$$\# (A_1 \cup \ldots \cup A_n) = \# (A_1 \cup \ldots \cup A_{n-1}) + \# (A_n) - \# ((A_1 \cap) \cup \ldots \cup (A_{n-1} \cap A_n)) \ (2)$$

By inductive hypothesis, we have:

$$\# (A_1 \cup \ldots \cup A_n) = \# (A_1) + \ldots + \# (A_n) +$$
$$\sum_{k=2}^{n-1} (-1)^{k-1} \sum_{\{i_1, \ldots, i_k\} \subset I_{n-1}} \# \left(A_1 \cap \ldots \cap A_{i_k} \right) \ (3)$$

and also:

$$-\# ((A_1 \cap A_n) \cup \ldots \cup (A_{n-1} \cup A_n)) =$$

$$= \sum_{K=1}^{n-2} (-1)^k \sum_{\{i_1, \ldots, i_k\} \subset I_{n-1}} \# \left(A_{i_1} \cap \ldots \cap A_{i_k} \cap A_n \right)$$
$$+ (-1)^{n-1} \# \left(A_1 \cap \ldots \cap A_{i_k} \right) =$$

$$= \sum_{k=2}^{n-1}(-1)^{k-1}\sum_{\{i_1,...,i_k\}\subset I_{n-1}} \# \left(A_{i_1} \cap...\cap A_{i_{k-1}} \cap A_n\right) +$$
$$(-1)^{n-1} \# \left(A_1 \cap A_n\right) (4)$$

and since that:

$$\sum_{\{i_1,...,i_k\}\subset I_{n-1}} \# \left(A_{i_1} \cap...\cap A_{i_k}\right)$$

$$+ \sum_{\{i_1,...,i_k\}\subset I_{n-1}} \# \left(A_{i_1} \cap...\cap A_{i_{k-1}} \cap A_n\right) =$$

$$= \sum_{\{i_1,...,i_k\}\subset I_n} \# \left(A_{i_1} \cap...\cap A_{i_k}\right)$$

from (2), (3), and (4), it follows (1).∎

Corollary 8.2: The number of surjective functions from a set A with m elements onto a set B with n elements, where $m, n \in \mathbb{N}$ and $m \geq n$ is given by:

$$n^m - \sum_{K=1}^{n}(-1)^{k-1}\binom{n}{k}(n-k)^m = \sum_{K=0}^{n}(-1)^{k-1}\binom{n}{k}(n-k)^m$$

Proof: Let F be the set of all functions from A to B. Put $B = \{b_1,...,b_n\}$ and for each $i = 1,...,n$, put:

$$A_i = \{f \in F / b_i \notin Im(f)\}$$

it is clear that A_i is the set of all functions from A to $B - \{b\_\{i\}\}$, so (prop. 4.5.1) $\# (A_i) = (n-1)^m$. Furthermore, if $i \neq j$, $A_i \cap A_j$ is the set of all functions from A to $B - \{b_i, b_j\}$, so then $\# (A_i \cap A_j) = (n-2)^m$ and, in general, if $\{i_1,...,i_k\} \subset I_n$ we have: $\# (A_{i_1} \cap...\cap A_{i_k}) = (n-k)^m$.

By the principle of inclusion-exclusion, results:

$$\# \left(A_1 \cup ... \cup A_n\right) = \sum_{k=1}^{n}(-1)^{k-1}\sum_{\{i^1,...,i_k\}\subset I_n} \# \left(A_{i^1} \cap ... \cap A_{i_k}\right)$$

$$=$$

$$= \sum_{k=1}^{n}(-1)^{k-1}\binom{n}{k}(n-k)^m$$

and since $A_1 \cup...\cup A_n$ is the subset of F composed by the functions that are not surjective, the searched number of surjective functions from A onto B is given by:

$$\# (F) - \# \left(A_1 \cup...\cup A_n\right) = n^m - \sum_{k=1}^{n}(-1)^{k-1}\binom{n}{k}(n-k)^m =$$
$$\sum_{k=0}^{n}(-1)^k \binom{n}{k}(n-k)^m.\blacksquare$$

Example: If A and B are two finite sets with the same number of elements n, any surjective function from A to B must necessarily be bijective (exercise 17) and since there are $n!$ bijections from A to B, by the above corollary we obtain the useful identity (see exercise 22, Chap. 5 for an application):

$$n! = \sum_{k=0}^{n} (-1)^k \binom{n}{k} (n-k)^n.$$

9 - WELL ORDERING

For a subset A of $\mathbb{R}$, it is said that it has a *first element* or a *smallest element* iff exists $m \in \mathbb{R}$ satisfying the following two conditions:

   a) $m \in A$.

   b) $a \in A \Rightarrow m \leq a$

That is, a smallest element of A is an element of A that is less than any other element of A.

Examples: 1) $A = \{\frac{1}{n}/n \in \mathbb{N}\}$ does not have a smallest element, since if it had such element m, it should satisfy a) $m \in A$, i.e., $m = \frac{1}{r}$ for some $r \in \mathbb{N}$, and b) $m \leq \frac{1}{n} \forall n \in \mathbb{N}$, i.e., $\frac{1}{r} \leq \frac{1}{n} \forall n \in \mathbb{N}$ and we simply take $n = r + 1$ to obtain a contradiction.

2) In contrast $\{0\} \cup \{\frac{1}{n}/n \in \mathbb{N}\}$ has first element, because 0 clearly fulfills a) and b).

A subset A of $\mathbb{R}$ is said to be *well ordered* iff every non-empty subset of A has a first element.

Examples: 1) In example 2) above, $\{0\} \cup \{\frac{1}{n}/n \in \mathbb{N}\}$ has a first element, but it is not well ordered, because, by the example, the subset $\{\frac{1}{n}/n \in \mathbb{N}\}$ is non-void and it does not have first element.

   2) $\emptyset$ is well ordered since it does not have non-void subsets.

Theorem 9.1 (*Well ordering principle*): $\mathbb{N}$ is well ordered.

Proof: Let A be a non-empty subset of $\mathbb{N}$. If $1 \in A$, then 1 is clearly the first element of A. Let's assume $1 \notin A$ and write:

$$B = \{b \in \mathbb{N} / b < a \forall a \in A\}$$

As $1 \notin A$ and $1 \leq n \forall n \in \mathbb{N}$, it follows that $1 \in B$. But B can not be inductive (because if it were, it would be $B = \mathbb{N}$ and taking $a \in A$ ($A \neq \emptyset$)would result in $a > n \forall n \in \mathbb{N}$ which is absurd since a$\in$N); then there must exist b$\in$B such that $b + 1 \notin B$. We assert that $b + 1$ is smallest element of A. In fact, as $b \in B$ we have $b < a \forall a \in$

A, so by corollary 1.7 results that $b + 1 \leq a$ whatever be $a \in A$, and as if it were $b + 1 < a \forall a \in A$ it should be $b + 1 \in B$, it must be $b + 1 = a$ for some $a \in A$, i. e. $b + 1 \in A$.∎

From the well-ordering principle we can deduct the following useful version of the induction principle:

Theorem 9.2: (Second induction principle) Let be $P(n)$ a proposition for each $n \in \mathbb{N}$. If

    1) $P(1)$ is true.

    2) For each $k \in \mathbb{N}$ with $k > 1$, the veracity of $P(h)$ for every $h \in \mathbb{N}$ such that $h < k$, implies that of $P(k)$.

Then $P(n)$ is true $\forall n \in \mathbb{N}$.

Proof: Put $A = \{n \in \mathbb{N} / P(n)$ is false$\}$. If A were non-empty, it would have a smallest element m. By condition 1) it must be $m \neq 1$, so $m > 1$ and $\forall h \in \mathbb{N}$ such that $h < m, P(h)$ would be true, by the choice of m, from where, by condition 2) $P(m)$ should be true. Then it must be $A = \emptyset$ and $P(n)$ true $\forall n \in \mathbb{N}$.∎

To appreciate the usefulness of the second induction principle, look at the following:

Example: Consider the Fibonacci's sequence $a_1, a_2, \ldots, an, \ldots$ defined by the conditions:

$$a_1 = 1, \qquad a_2 = 2, \qquad a_{n+2} = a_n + a_{n+1}$$

that is the sequence: $1, 2, 3, 5, 8, 13, \ldots$. This sequence was used by Leonardo of Pisa (1175-1250), called Fibonacci ("son of Bonaccio"), to find the number of pairs of rabbits after a certain time, beginning with a single fertile couple, knowing that rabbits take a month to be fertile and the gestation time is also one month. At the end of the first month there is one pair, at the end of the second, three, after the third, five...If instead of a fertile couple, the process begins with a newborn one, another 1 must be added at the beginning of the sequence. Testing values of n, it can be guessed that whatever be $n \in \mathbb{N}$, we have:

$$a_n < \left(\frac{7}{4}\right)^n$$

For $n = 1$ that is obvious. We apply the second induction principle.

Let $k > 1$ and assume $a_h < \left(\frac{7}{4}\right)^h \forall h < k$. Then we have to verify that $a_k < \left(\frac{7}{4}\right)^k$. If $k =$

2, then it is obvious. Assume $k \geq 3$, then $k - 1, k - 2 \in \mathbb{N}$ and $a_k = a_{k-2} + a_{k-1}$. By inductive hypothesis we have:

$$a_{k-2} < \left(\frac{7}{4}\right)^{k-2} \quad and \quad a_{k-1} < \left(\frac{7}{4}\right)^{k-1}$$

then,

$$a_k < \left(\frac{7}{4}\right)^{k} + \left(\frac{7}{4}\right)^{k-1} = \left(\frac{7}{4}\right)^{k-2}\left(1 + \frac{7}{4}\right) = \left(\frac{7}{4}\right)^{k-2}\frac{11}{4}$$

and as $\frac{11}{4} < \frac{49}{16}$, it follows that $a_k < \left(\frac{7}{4}\right)^{k}$.

Fibonacci published the problem of the rabbits in his "Liber Abaci," a book that had a deep influence in the introduction and diffusion of Hindu Arabic numerals, i.e., decimal system, in Europe, in substitution of the roman system.

The Fibonacci's sequence forms a pattern that repeatedly appears in nature. For example, in the genealogy of bees. A female bee comes from a fertilized egg and a male bee from an unfertilized one; in other words, a female bee has a father and a mother, but a male bee has only a mother. Consider the genealogic tree of a female bee upwards: in the first level there is one element (the female bee) in the next level up there are two bees (the parents), in the following there are three (two parents of the mother and one for the father). Following in that way we recognize the pattern of Fibonacci's sequence. If we start with a male, 1 must be added at the beginning, obtaining the other version of the sequence. The same pattern is founded in the branching process of some trees and algae and in many other natural phenomena [13].

The Fibonacci sequence is related to the so-called Golden Ratio $((1+\sqrt{5})/2)$ (exercise 6, e)). It is used, with its closely related Lucas sequence, in primality testing.

10 - ENTIRE DIVISION

The entire division process, learned in school, can be formalized as follows:

Theorem 10.1 (of the Entire Division or Division Algorithm): Given $a \in \mathbb{N} \cup \{0\}$ and b$\in$N, there exist $q, r \in \mathbb{N} \cup \{0\}$ such that:

$$a = bq + r \text{ and } r < b$$

Furthermore, such q (quotient) and r (residue) are unique.

Proof: (Existence) Let

$$A = \{a - bx / x \in \mathbb{N} \cup \{0\}\} \cap (\mathbb{N} \cup \{0\})$$

As $\mathbb{N}$ is well ordered it is clear that $\mathbb{N} \cup \{0\}$ is too, and since $A \neq \emptyset$ because $a \in A$, A has a smallest element r. Then $r \in \mathbb{N} \cup \{0\}$ and there exist $q \in \mathbb{N} \cup \{0\}$ such that $r = a - bq$. If it were $r \geq b$, it should be $r - b = a - b(q + 1) \in A$, contradicting the minimality of r.

Alternatively, using the second principle of induction, we can proceed as follows. If $a < b$, just take $q = 0$ and $r = a$; if $a = b$, just take $q = 1$ and $r = 0$. Let then $a > b$, so $a - b \in \mathbb{N}$ and $a - b < a$, by inductive hypothesis we can suppose the result valid for para a-b, that is there exist $q', r \in \mathbb{N} \cup \{0\}$ such that:

$$a - b = bq' + r \text{ and } r < b$$

then $a = b(q' + 1) + r$, and putting $q = q' + 1$, it is also valid for a.

(Uniqueness) Assume we have: $q, q', r, r' \in \mathbb{N} \cup \{0\}$ such that:

$$a = bq + r = bq' + r' \, r, r' < b$$

and that we have (say) $r' < r$. Then $0 \leq r - r' = b(q' - q)$, from what follows $q' - q \geq 0$. If it were $q' - q > 0$, it should be $q' - q \geq 1$ and then $r - r' = b(q' - q) \geq b$ what is impossible inasmuch as $0 \leq r, r' < b$. It follows that $q' = q$ and so $r = r'$.∎

11 - NUMBER-SYSTEMS

Humankind, through history, has used several systems of numeration, that is, means of writing numbers in a way suitable for making calculations. In western culture, the most known systems are the Roman and the decimal or Hindu-Arabic. Mayans used a system of base 20 and Babylonians one with base 60. These last ones (including the decimal) are positional systems, more appropriate to deal with big numbers and to compute than, for example, the Roman. So we will deal only with positional or place-value systems like the decimal system.

The first positional system was that of Babylon and the election of 60 as a base seems to be the result of the unification of units of diverse measurement systems (or diverse number-systems) in a unit multiple of them (see [32]), although that explanation is controversial (for the points in dispute see [23]). Traces of this system remains in our use of measuring time and angles.

In the Hindu-Arabic system, the election of the number ten as a base, has an anthropomorphic reason: we have ten fingers in our hands, but, of course, number ten has not a particular mathematical property that makes it more preferable than others to be a base of a number-system, despite the opinion of the Pythagoreans who considered ten as a "perfect" number.

The following theorem is the basis of positional numeration systems.

Theorem 11.1 (b-adic expansion, or expansion in base b): Selected a natural number b (the *base* of the system) with $b \geq 2$, each natural number a can be expressed in the form:

$$a = a_n b^n + a_{n-1} b^{n-1} + \ldots + a_1 b + a_0$$

where $n \in \mathbb{N} \cup \{0\}$ and the a_i (the *digits* of the system) are natural numbers or zero such that $a_n \neq 0$ and $a_i < b$ $\forall i = 0, \ldots, n$.

Moreover, such expression is unique, that is, if we also have:

$$a = c_m b^m + c_{m-1} b^{m-1} + \ldots + c_1 b + c_0$$

with $m \in \mathbb{N} \cup \{0\}$ and $c_j \in \mathbb{N} \cup \{0\}$ in such a way that $c_m \neq 0$ and $c_j < b$ $\forall j = 0, \ldots, m$; then $m = n$ and $a_i = c_i$ $\forall i = 0, \ldots, n$.

Proof:

(Existence): If $a \leq b$, the proof is trivial. Let be $a > b$, by the Theorem of Entire Division, there exist $q, a_0 \in \mathbb{N} \cup \{0\}$ such that:

$$a = bq + a_0 \text{ and } a_0 < b$$

Proceeding by induction, as $q < a$ (because if it were $q \geq a$ then we should have $a - a_0 = bq > a$, which is a contradiction because $a_0 \geq 0$), and $q \in \mathbb{N}$ (since $a > b$), we can assume by inductive hypothesis, the result valid for q, i.e., we have:

$$q = a_n b^{n-1} + \ldots + a_1, \ a_n \neq 0, \ a_i < b$$

with $n - 1 \in \mathbb{N} \cup \{0\}$, then $a = a_n b^n + \ldots + a_1 b + a_0$ and the statement is true.

(Uniqueness): From

$$a = a_n b^n + \ldots + a_1 b + a_0 = c_m b^m + \ldots + c_1 b + c_0$$

it follows (due to the uniqueness of quotient a residue) that

$$a_0 = c_0 \text{ and } a_n b^{n-1} + \ldots + a_1 b^{m-1} + \ldots + a_1 = c_m b^{m-1} + \ldots + c_1$$

and by inductive hypothesis, we have $n - 1 = m - 1, a_1 = b_1, \ldots, a_n = b_n$. ∎

For example, taking $b = 6$ and $a = 296$, the proof of the theorem tells us how to expand a in base b by successive divisions:

$$296 = 49 \cdot 6 + 2 = (8 \cdot 6 + 1) \cdot 6 + 2 = 8 \cdot 6^2 + 1 \cdot 6 + 2$$
$$= 1 \cdot 6^3 + 2 \cdot 6^2 + 1 \cdot 6 + 2$$

Let us clarify that in this example, as well as in some previous ones, we have freely used the decimal system; based on this section, this is not a *petitio principii* since the decimal system has been used only in examples but not in the theoretical development.

Corollary 11.2: With the notations of the theorem, if $a \leq b^m - 1$ with $m \in \mathbb{N}$, then in the b − adic expansion of a, we have $n < m$.

Proof: Let $a = a_n b^n + \ldots + a_0$ be an expansion of a in base b, if it were $n \geq m$ as $a_n \neq 0$, it should be, $a \geq a_n b^n \geq b^n \geq b^m$, which contradicts the hypothesis.∎

In what follows we will see some examples where the theorem presented above is applied.

Example 1: There are four cards. On one of them, with the heading "1" has all odd numbers between 1 and 15 written on it. A second card, with the heading "2", contains the numbers 2, 3, 6, 7, 10, 11, 14, and 15. The third card, with heading "4", contains the numbers 4, 5, 6, 7, 12, 13, 14 and 15. The last card, with heading "8" has the numbers from 8 to 15 (included). Ask someone to choose a number between 1 and 15 and ask him to give you the cards where appears the chosen number. Add the heading of these, and it will result in that number. Why does it work?

The above theorem provides an explanation: the numbers on the cards are located according to their binary expansion (base 2). For example:

$$13 = 1 \cdot 2^3 + 1 \cdot 2^2 + 1 = 8 + 4 + 1$$

and 13 appears on the cards with headings 8, 4, and 1.

Example 2: (Bachet's problem of the weights) This is about finding, in an optimal way, the value of the weights that allow weighing objects (with a weight equal to some integer number) up to (say) 63 kilograms, on a weighing balance scale. One of the dishes of the scale is for the weights, and the other dish is for the objects.

Clearly, we must have a weight of value 1. Then to weigh 2Kg we can use a weight of value 1 or 2; more efficient is using a 2 Kg weight because with it we also can weigh 3 Kg., following such reasoning, we see that the optimal selection is taking values that are powers of 2. Taking $b = 2$ and $m = 5$ from corollary 11.2, we see that using weighs of values 1, 2, 4, 8, 16, and 32 we can weigh any object up to $2^5 - 1 =$

63 Kg. For example, take an object of 39 Kg. Developing 39 in base 2 yields:

$$39 = 1 \cdot 2^5 + 1 \cdot 2^2 + 1 \cdot 2 + 1$$

and for 39 Kg it is enough to use weights of 1, 2, 4, and 32 Kg. In this problem, clearly, 63 can be replaced by any number of the form $2^m - 1$.

A variation of this problem consists in allowing the weighs to occupy both dishes. In the original problem posed by Bachet such was the case; the requirement is to weigh objects up to 40 Kg. We assert that it is enough to take weighs with values: 1, 3, 9 and 27. For an object of 17 Kg, by example, we develop $40 - 17 = 23$ on base 3, so that:

$$23 = 2 \cdot 3^2 + 3 + 2$$

and subtracting this from $40 = 1 + 3 + 3^2 + 3^3$, we obtain:

$$17 = 3^3 - 3^2 - 1$$

so we locate the object and the weights with value 1 and 9 on one pan, and the weigh with value 27 on the other.

In general to weigh an object of x Kg with $x \in \mathbb{N}$ and $x \leq \frac{3^m-1}{2} = y$, weighs of values $1, 3, 3^2, \ldots, 3^{m-1}$ are sufficient. In fact, expanding $y - x$ on base 3:

$$y - x = a_0 + a_1 3 + \ldots + a_{m-1} 3^{m-1}$$

and as $y = 1 + 3 + \ldots + 3^{m-1}$, we obtain:

$$x = b_0 + b_1 3 + \ldots + b_{m-1} 3^{m-1}$$

where $b_i = 1 - a_i$ so $b_i = -1, 0$ or 1. Then, pulling apart negative and positive b_i we can write:

$$x + \sum c_i 3^i = \sum d_i 3^i$$

with $c_i, d_i \geq 0$. Then on one pan we put the object and the weighs of values such that $c_i = 1$ and in the other the weighs such that $d_i = 1$.

The proof that the election of that values of the weighs is optimal is somewhat more involved, and we refer the interested reader to [18].

Example 3: (Multiplication by duplication) Russian peasants (and before them ancient Egypts) have a method to multiply two numbers, by multiplying and dividing only by 2. We exemplify it with the numbers 39 and 425. Schematically:

39	425
19	850
9	1.700
4	3.400
2	6.800
1	13.600
	16.575

In the column on the left are placed, beginning with 39, the successive quotients of division by 2 without considering the remainders, and in the column on the right, each number is obtained, starting with 425, by doubling the preceding one. Then the numbers of this column are added without considering those corresponding to even numbers of the left column.

The correction of the method can be explained expanding 39 on base 2:

$$39 \cdot 425 = (2^5 + 2^2 + 2 + 1) \cdot 425$$
$$= 2^5 \cdot 425 + 2^2 \cdot 425 + 2 \cdot 425 + 425 =$$
$$= 13.600 + 1.700 + 850 + 425 = 16.575$$

The same method, with a slight variation, was used in ancient Egypt. While the right column remains as above, the Egyptians instead of dividing by 2, in the left column they placed the successive powers of two: $1, 2, 4, 8, 16, 32$, and then select the numbers which compose 39: $1, 2, 4, 32$, adding the corresponding numbers of the right column.

A positional or place-value numbering system of base b, consists of set of digits or symbols that represents 0 and the natural numbers $< b$ (we will use for convenience the same symbols as in the decimal system even if $b > 10$) and in writting each natural number according with its expansion on base b using the notation:

$$a_n b^n + a_{n-1} b^{n-1} + \ldots + a_1 b + a_0 = (a_n a_{n-1} \ldots a_1 a_0)_b$$

with a_i digits.

For example, take $0, 1, 2 = 1 + 1, 3 = 2 + 1, 4 = 3 + 1, b = 4 + 1$, in such a way that $0, 1, 2, 3, 4$ are the digits of the system, then:

$$(2103)_b = 2 \cdot b^3 + 1 \cdot b^2 + 0 \cdot b + 3$$
$$(10)_b = b$$

To perform the elementary operations in a number-system, first we build the summation and multiplication of the digits. In the case we are modeling, the tables are:

+	0	1	2	3	4
0	0	1	2	3	4
1	1	2	3	4	10
2	2	3	4	10	11
3	3	4	10	11	12
4	4	10	11	12	13

.	0	1	2	3	4
0	0	0	0	0	0
1	0	1	2	3	4
2	0	2	4	11	13
3	0	3	11	14	22
4	0	4	13	22	31

and have been constructed according to the following examples:

$$3 + 3 = 3 + 2 + 1 = 4 + 1 + 1 = b + 1 = (11)_b$$
$$2 \cdot 2 = 2(1 + 1) = 2 + 2 = 2 + 1 + 1 = 3 + 1 = 4$$
$$3 \cdot 2 = 3(1 + 1) = 3 + 3 = (11)_b$$
$$3 \cdot 3 = 3(2 + 1) = 3 \cdot 2 + 3 = (11)_b + 3 = b + 1 + 3$$
$$= b + 4 = (11)_b$$

To add two numbers in this system, for example $(243)_b + (44)_b$, proceed like this:

$$(243)_b + (44)_b = 2b^2 + 4b + 3 + 4b + 4 =$$
$$= 2b^2 + (4 + 4)b + (3 + 4)$$
$$= 2b^2 + (13)_b b$$
$$+ (12)\_\{b\} =$$
$$= 2b^2 + (b + 3)b + b + 2$$
$$= 3b^2 + 4b + 2$$
$$= (243)_b$$

Schematically one can proceed "carrying," just like in school, as follows:

$$1\ 1$$
$$243$$
$$44$$
$$342$$

Let us exemplify the product:

$$(44)_b \cdot (243)_b = (4b + 4)(2b^2 + 4b + 3)$$
$$= 4b(2b^2 + 4b + 3) + 4(2b^2 + 4b + 3) =$$
$$= (4 \cdot 2b^3 + 4 \cdot 4b^2 + 4 \cdot 3b) + (4 \cdot 2b^2 + 4$$
$$\cdot 4b + 4 \cdot 3) =$$
$$= (13)_b b^3 + (31)_b b^2 + (22)_b b + (13)_b b^2$$
$$+ (31)_b b + (22)_b =$$

69

$$= b^4 + (3+3)b^3 + (1+2)b^2 + 2b + b^3 + (3 + 3)b^2 + (1+2)b + 2 =$$
$$= b^4 + (11)_b b^3 + 3b^2 + 2b + b^3 + (11)_b b^2 + 3b + 2 =$$
$$= (21320)_b + (2132)_b = (24002)_b$$

Then we can multiply according to the following arrangement:

243

44

2132

2132

24002

In a similar manner, it can be justified the schemes of subtracting, dividing, and so on.

Binary number-system (base two) is the simplest; since it has only two digits, no effort is needed to memorize the summation and multiplication tables. In contrast, it has the practical inconvenience that in order to write a relatively small number, several digits are needed; for example, the number that in decimal system is written 89, in the binary system is denoted as 1011001. It has, however the advantage that in a switch 1 can be represented by current flow and 0 by its interruption, so it is used in calculators and computers.

Example: We will analyze the so-called game of Nim, by applying the binary system. Starting from an arbitrary number of rows with an arbitrary number of matches each, two players A and B, alternately withdraw matches with the conditions that each player, in turn, must remove at least one match, and may remove as many as he wishes, but from a single row (at his choice, being able to choose distinct rows in each turn). The winner is the player who removes the last match. We will show an strategy to win in this game with the condition of choosing the start. In order to be specific we take an example, where at the right of each row, we write in the binary system, the numbers of matches in that row:

||||||	110
|||||	101
|||||||||	1001
||||||||||	1010
|||	100

pipp

With the numbers in binary arranged in a natural way, under each column with zeros and ones we write the parity of ones in that column, p if there is an even number of ones and i if there is an odd number of them. The strategy consist in leaving the adversary all columns with p, so then if A may choose the start, since in the previous array not all columns have parity p, he choose to start off, and remove, for example, the last row (also he could remove 4 matches of the second row, etc.) leaving B the situation:

$$
\begin{array}{ll}
|\,|\,|\,|\,|\,| & 110 \\
|\,|\,|\,|\,| & 101 \\
|\,|\,|\,|\,|\,|\,|\,|\,| & 1001 \\
|\,|\,|\,|\,|\,|\,|\,|\,|\,| & 1010 \\
& pppp
\end{array}
$$

It is B's turn, who must remove at least a match from one, and only one row, which results in modifying one, and only one, of the numbers in binary at the right. But modifying a number involves changing some of its digits, which implies changing the parity of the corresponding column. It is therefore impossible to B not to modify the parity of at least one column. Assume he removes 8 matches from the last row, so he left for A the following situation:

$$
\begin{array}{ll}
|\,|\,|\,|\,|\,| & 110 \\
|\,|\,|\,|\,| & 101 \\
|\,|\,|\,|\,|\,|\,|\,|\,| & 1001 \\
|\,| & 10 \\
& ippp
\end{array}
$$

A is then forced to play in the third row changing 1001 by 1, that is removing 8 matches (in this case the winning play is unique):

$$
\begin{array}{ll}
|\,|\,|\,|\,|\,| & 110 \\
|\,|\,|\,|\,| & 101 \\
| & 1 \\
|\,| & 10 \\
& ppp
\end{array}
$$

Continuing like this, it will remain only two rows, for example the situation: :

$$
\begin{array}{ll}
|\,|\,|\,|\,| & 101 \\
|\,| & 10 \\
& iii
\end{array}
$$

It must be the turn of A since A leaves to B parity p in all columns. Here the winning play is unique too: to remove 3 matches from the firtst row, leaving to B:

||	10
||	10

and it is already clear that A should be the winner.

To conclude this section, we will comment on certain curiosities. The number 6174, expressed in the decimal system, has an interesting property, if one chooses any number of four digits, not all the same, orders the digits from highest to lowest and from lowest to highest, subtracts the results and reiterates the process, arrives in few steps to the number 6.174. For example, starting from 5949, we have:

9.954		5.553	9.981	8.820
	8.532			
4.599		3.555	1.899	288
	2.358			
5.535		1.998	8.082	8.532
	6.174			

What happens if we change the number of digits or the base of the system?. The answers are left to the reader.

Another number with a remarkable property is 1729, also expressed in decimal system. The distinguished English mathematician G.H. Hardy tells that on visiting on his sickbed to S. Ramanujan, an extraordinaire Hindu self-taught mathematician, told him that he had arrived there in a car with license plate 1729 and that he (Hardy) did not see any striking property in that number. Ramanujan immediately answered him that 1729 is the smallest natural number which is the sum of two cubes in two different ways:

$$1.729 = 10^3 + 9^3 = 12^3 + 1^3$$

It is interesting to note that this result was known, three centuries earlier, by P. Fermat, a self-taught and brilliant French mathematician.

Note that there is a marked difference between the above-mentioned properties of the numbers 6174 and 1729; the first is a characteristic of the decimal representation of the number, whereas the one of 1729 is independent of its representation in any number-ing system and is one of the first examples in the study of the representation of integers by cubic forms.

12- DEFINITIONS BY INDUCTION

This section is dedicated to justifying the definitions by induction and can be omitted without prejudice of understanding what follows.

Theorem 12.1: Let A be a non-empty set, a an element of A and for each $n \in N$, $g_n: A \to A$ a function. There exists one, and only one, function $f: N \to A$ such that:

a) $f(1) = a$,

b) $f(n + 1) = g_n(f(n)) \forall n \in N$.

Proof: Consider the family F of all relations R from N to A that satisfy:

a) $(1, a) \in R$,

b) $(n, x) \in R \Rightarrow (n + 1, g_n(x)) \in R$.

If R is a function, these conditions are the same as the statement of the theorem. Note that F is not empty as, for example, $N \times A$ is a relation belonging to F. If we put:

$$f = \bigcap_{R \in F} R$$

that is f is the intersection of all the relations belonging to F, we easily verify that $f \in F$, i. e., f is a relation from N to A which fulfills a) and b).

To prove its existence, it is enough to verify that f is a function, and for this it is enough to prove that the set:

$H = \{n \in N/\text{there is one, and only one, } x \in A \text{ such that } (n, x) \in f\}$

is the set of all natural numbers, and for that it is sufficient to check that H is inductive:

1) $1 \in H$.

In fact, $(1, a) \in f$ by a). If it were $(1, b) \in f$ with $b \neq a$, the relation: $f - \{(1, b)\}$ should belong to F, contradicting the election of f.

2) $n \in H \Rightarrow n + 1 \in H$.

Indeed, if $n \in H$ there is one and only one x∈A, such that (n, x)∈f. As f satisfies b), it follows that $(n + 1, g_n(x)) \in f$. If it were $(n +$

$1, y) \in f$ with $y \neq g_n(x)$; taking into account the relation: $f - \{(n + 1, y)\}$ there would be a contradiction with the election of f.

The existence has been proven. Let us show the uniqueness.

Let $f': N \rightarrow A$ be a function verifying a) and b). Consider

$$K = \{n \in N / f(n) = f'(n)\}$$

It is easily verified that K is inductive. Then $K = N$ and, therefore, $f = f'$. ∎

Example 1: Given $a \in R$, let $g: R \rightarrow R$ be defined by $g(x) = xa$. Taking $g_n = g \forall n \in N$ in the above theorem, there exists only one function $f: N \rightarrow R$ such that:

$$f(1) = a, \qquad f(n + 1) = f(n)a$$

Putting $f(n) = a^n$, which translates as:

$$a^1 = a, \qquad a^{n+1} = a^n a$$

Example 2: For each $i \in N$ let $a\_\{i\}$ be a real number. Defining, for each $n \in N, g_n: R \rightarrow R$ by $g_n(x) = x + a_{n+1}$, there exists, by the above theorem, one and only one function $f: N \rightarrow R$ such that:

$$f(1) = a_1, \qquad f(n + 1) = f(n) + a_{n+1}$$

or, putting $f(n) = \sum_{i=1}^{n} a_i$:

$$\sum_{i=1}^{1} a_i = a_1, \qquad \sum_{i=1}^{n+1} a_i = \sum_{i=1}^{n} a_i + a_{n+1}$$

Example 3: Let $g_n: R \rightarrow R$ such that $g_n(x) = x(n + 1)$, there exists a single $f: N \rightarrow R$ such that:

$f(1) = 1$, $f(n + 1) = f(n)(n + 1)$ and putting $f(n) = n!$ results:

$$1! = 1, \qquad (n + 1)! = n! (n + 1)$$

EXERCISES

Ex.1: Prove the following versions of the induction principle:

a) Let be $n_0 \in \mathbb{N}$ and $P(n)$ a proposition for each natural $n \geq n_0$. If

  1) $P(n_0)$ is true.

  2) Being $h \in \mathbb{N}$ with $h \geq n_0$ the following implication is true: $P(h) \Rightarrow P(h + 1)$.

Then $P(n)$ is true $\forall n \in \mathbb{N}$ with $n \geq n_0$. (Hint. : define $Q(n)$ by $Q(n) = P(n + n_0 - 1)$).

b) Let $r \in \mathbb{N}$ and $P(n)$ a proposition for each natural number n such that $n \leq r$. If

1) $P(1)$ is true.

2) $h \in N, h \leq r - 1, P(h) \Rightarrow P(h + 1)$.

Then $P(n)$ is true $\forall n \leq r$.

Ex. 2: Justify the following rule of Fibonacci (s.13) for, given $a \in$ N, finding $b \in$ N such that $a^2 + b^2$ be a square:

If a is odd, take $b^2 = 1 + 3 + 5 + \ldots + (a^2 - 2)$.

If a is even, take $b^2 = 1 + 3 + 5 + \ldots + \left(\frac{a^2}{2} - 1\right)$.

Ex. 3: Prove inductively:

a) $1 \cdot 2 + 2 \cdot 3 + \ldots + n(n + 1) = \frac{1}{3}n(n + 1)(n + 2)$.

b) $\quad\quad 0 \cdot 1 \cdot 2 + 1 \cdot 2 \cdot 3 + \ldots + (n - 1)n(n + 1) = \frac{1}{4}(n^2 + n)(n^2 + n - 2)$.

c) $1^2 + 3^2 + \ldots + (2n - 1)^2 = \frac{1}{3}n(4n^2 - 1)$.

d) $1 + q + q^2 + \ldots + q^n = \frac{1 - q^{n+1}}{1 - q}$ with $q \neq 1$.

e) $\frac{1}{1 \cdot 2} + \frac{1}{2.3} + \ldots + \frac{1}{n.(n+1)} = \frac{n}{n+1}$.

f) $\frac{1}{n+1} + \frac{1}{n+2} + \ldots + \frac{1}{n+(n+1)} \leq \frac{5}{6}$.

Ex. 4: The following inequalities are valid:

a) $2^n > n \; \forall n \in$ N.

b) $n^2 \geq 2n + 1 \; \forall n \geq 3$.

c) $2^n \geq n^2 \; \forall n \geq 4$.

Ex. 5: a) Let $a, b \in$ R, $a \geq 0$ and $a \leq b$, then $a^n \leq b^n \; \forall n \in$ N.

b) If $a \in$ R, $a \geq -1$, then $(1 + a)^n \geq 1 + na \; \forall n \in$ N.

c) If $a > 0$ and $n \in$ N with $n \geq 2$, then: $(1 + a)^n \geq 1 + na + \frac{n(n-1)}{2}a^2$.

Ex. 6: Consider the Fibonacci's sequence:

$$1, 1, 2, 3, 5, 8, \ldots, a_n, a_{n+1}, a_{n+2}, \ldots$$

where each term, starting from the third, is the sum of the two previous ones. Prove the following relations $\forall n \in$ N:

a) $a_{n+1}^2 - a_n a_{n+2} = (-1)^{n+1}$

b) $a_2 + a_4 + \ldots + a_{2n} = a_{2n+1} - 1$

c) $a_1 + a_3 + \ldots + a_{2n-1} = a_{2n} - 1$

d) $a_1 - a_2 + \ldots + a_{2n-1} - a_{2n} = -a_{2n-1}$

e) $a_n = \frac{1}{\sqrt{5}}\left[\left(\frac{1+\sqrt{5}}{2}\right)^n - \left(\frac{1-\sqrt{5}}{2}\right)^n\right]$

Ex. 7: A finite set has a greatest element, i.e., an element of the set $\geq$ than any other.

Ex. 8: Find an error in the following argument. We will "prove" the following proposition $P(n)$: "If in a set of n boys there is at least one with blue eyes, then all of them have blue eyes".

$P(1)$ is obvious. Assume $P(h)$ true and prove $P(h+1)$. Let A be a set of h+1 boys with at least one with blue eyes. Pick $a \in A$ with blue eyes. As $h \in \mathbb{N}$, $h+1 \geq 2$ we can take $b \in A$ such that $b \neq a$. $A - \{b\}$ is then a set of h boys with at least one $(a \in A - \{b\})$ with blue eyes and by inductive hypothesis all them boys of $A - \{b\}$ have blue eyes. Taking $c \in A$ such that $c \neq a$ and $c \neq b$, and applying again the inductive hypothesis, it follows that all the boys of $A - \{c\}$ have blue eyes; in particular, as $b \in A - \{c\}$, b has blue eyes and since all of the boys of $A - \{b\}$ have them of that color, the same is also true for all the boys of A.

Ex. 9: Prove the following dual of prop. 4.1. If $f: I_m \to I_n$ is surjective, then $m \geq n$.

Ex. 10: If A and B are non-empty finite sets and if there exists a bijection from A to B, then A and B have the same cardinal.

Ex. 11: Let $P(n)$ be the proposition:

$$\sum_{i=1}^{n} i = \frac{1}{2}\left(n + \frac{1}{2}\right)^2$$

Prove that $P(h) \Rightarrow P(h+1)$ and that $P(n)$ is false whatever be $n \in \mathbb{N}$.

Ex. 12: Evaluate:

a) $\sum_{i=17}^{54} i$, b) $\sum_{i=17}^{54} i^2$

Ex. 13: Prove:

a) $\sum_{i=1}^{n} i^4 = \frac{1}{30}n(n+1)(2n+1)(3n^2 + 3n - 1)$

b) $\sum_{i=1}^{n} i^5 + \sum_{i=1}^{n} i^7 = \frac{n^4(n+1)^4}{8}$

Ex. 14: Let $a_i, b_i \in \mathbb{R}$. Prove inductively the following inequality (Cauchy's inequality):

$$\left(\sum_{i=1}^{n} a_i b_i\right)^2 \leq \left(\sum_{i=1}^{n} a_i\right)^2 \left(\sum_{i=1}^{n} b_i\right)^2 \quad \forall n \in \mathbb{N}$$

Ex. 15: Let $n \in \mathbb{N}$, $a_i \in \mathbb{R}$ with $a_i > 0$ ($i = 1, \ldots, n$). Prove the following generalizations of exercise H, chapter2:

a) $\left(\sum_{i=1}^{n} a_i \right) \left(\sum_{i=1}^{n} \frac{1}{a_i} \right) \geq n^2$

* b) $\sum_{i=1}^{n} a_i = 1$ and $n \geq 2 \Rightarrow \prod_{i=1}^{n} \frac{1}{a_i} \geq 2^{3(n-2)}$

* c) $\prod_{i=1}^{n} a_i = 1 \Rightarrow \sum_{i=1}^{n} a_i \geq n$.

Ex. 16: Find $n \in \mathbb{N}$ such that $3\binom{n}{4} = 5\binom{n-1}{5}$

Ex. 17: Let $f: A \to B$ with A, B non-empty finite sets of the same cardinal. Prove:

$$f \text{ is injective} \Leftrightarrow f \text{ is surjective.}$$

and conclude that the number of bijections from A onto B es $n!$.

Ex. 18: Being $m, n \in \mathbb{N}$, prove

$$\sum_{i=0}^{n} \binom{m+i}{i} = \binom{m+n+1}{n},$$

$$\sum_{i=0}^{n} \binom{m+i-1}{m} = \binom{m+n}{m+1}$$

Ex. 19: By a combinatorial argument, show: ($n \geq r \geq 2$):

$$\binom{n+2}{r} = \binom{n}{r} + 2\binom{n}{r-1} + \binom{n}{r-2}$$

Ex. 20: If $r \leq m$ and $r \leq n$ with $r, m, n \in \mathbb{N} \cup \{0\}$, prove:

$$\sum_{k=0}^{r} \binom{m}{k} \binom{n}{r-k} = \binom{m+n}{r}$$

a) by induction on m or on n.

b) by a combinatorial argument.

Derive the following useful identity:

$$\sum_{k=0}^{n} \binom{n}{k}^2 = \binom{2n}{n}$$

Ex. 21: How many parallelograms are formed when intersecting a set of 6 parallel lines, with another set of 4 parallel lines, but oblique to the first ones? (R.: 90).

Ex. 22: For $n \in \mathbb{N}$ prove:

a) $n > 1 \Rightarrow 2 < \left(1 + \frac{1}{n} \right)^n$

b) $n > 1 \Rightarrow n! < \left(\frac{n+1}{2} \right)^n$

c) $n > 2 \Rightarrow 2^{n-1} < n!$

d) $n > 1 \Rightarrow \left(1 + \frac{1}{n}\right)^n < \sum_{i=0}^{n} \frac{1}{i!}$

*e) $\left(1 + \frac{1}{n}\right)^n < 3$

f) $n \geq 3 \Rightarrow (n+1)^n < n^{n+1}$

*g) $r, s \in \mathbb{N}, r > s \geq 3 \Rightarrow r^s < s^r$

Ex. 23: Find without calculations:

$$\binom{8}{0} + \binom{8}{2} + \binom{8}{4} + \binom{8}{6} + \binom{8}{8}$$

Ex. 24: For any $n \in \mathbb{N}$, is valid:

$$\binom{n}{0} + \binom{n}{2} + \binom{n}{4} + \ldots = \binom{n}{1} + \binom{n}{3} + \binom{n}{5}$$

Ex. 25: a) If a_{ij} are real numbers and $n \in \mathbb{N}$, we have:

$$\sum_{j=0}^{n} \sum_{i=0}^{j} a_{ij} = \sum_{i=0}^{n} \sum_{k=0}^{n-i} a_{i(i+j)}$$

b) If $a_n = \sum_{i=0}^{n} \binom{n}{i} b_i$, then:

$$(-1)^n b_n = \sum_{j=0}^{n} \binom{n}{j} (-1)^j a_j$$

Ex. 26: Let $m, n \in \mathbb{N}$ with $m \geq n$ and let $\sigma_m(n)$ be the number of surjective functions from a set of m elements onto a set of n elements. Setting $\sigma_m(0) = 0$, prove:

$$\sum_{i=0}^{n} \binom{n}{i} \sigma_m(i) = n^m$$

and using the preceding exercise, conclude that:

$$\sigma_m(n) = \sum_{k=0}^{n} (-1)^k \binom{n}{k} (n-k)^m$$

Ex. 27: a) Every finite set is well ordered.

b) the union of two well-ordered sets is well ordered.

Ex. 28: Consider the sequence defined by the conditions:

$$a_1 = 3, \qquad a_2 = 7, \qquad a_n = 3a_{n-1} - 2a_{n-2} \text{ if } n > 2$$

Prove: $a_n = 2^{n+1} - 1$.

Ex. 29 : a) Let $f: \mathbb{N} \to \mathbb{N}$ be the function defined by:

$$f(n) = \begin{cases} \dfrac{n}{2} & \text{if } n \text{ is even} \\ 3n + 1 & \text{if } n = 4k + 1 \\ 3n - 1 & \text{if } n = 4k - 1 \end{cases}$$

prove that by iteration it reaches 1 (for example: $7 \to 20 \to 10 \to 5 \to 16 \to 8 \to 4 \to 2 \to 1$).

c) (Open problem) Analyze the same as in b) but with the function defined by:

$$f(n) = \begin{cases} \dfrac{n}{2} & \text{if } n \text{ is even} \\ 3n + 1 & \text{if } n \text{ is odd} \end{cases}$$

Ex. 30 : a) Construct tables of summation and multiplication in the numeration systems of bases $b = 5$ and $b = 12..$

b) Using a) calculate $(332)_b^3$ in both systems.

c) Idem that in b) but trough decimal system.

Ex. 31: a) Find a base b such that $(301)_b$ be a square.

b) 1234321 is a square in any system with base ≥ 5.

Ex. 32: a) Find the digits a, b, c such that $(abc)_7 = (bca)_9$.

b) In which base b: $(79)_{10} = (142)_b$?.

CHAPTER 4
INTEGERS

In this chapter we study the notions related to elementary divisibility which, most of them, already appear in Euclid's Elements: prime numbers, greatest common divisor, Euclid algorithm, prime factorization (the fundamental theorem of arithmetic is not explicitly stated in the Elements, the first explicit statement and proof of the theorem was made by Gauss in 1801). It is likely that the Greeks have obtained these results while carrying on their research on perfect numbers. As applications, the even perfect numbers and the Pythagorean triples are determined, and the linear Diophantine equation is solved.

Although the Greeks, and in fact nobody in Europe until modern times, did not work with negative numbers, it is convenient, for future generalizations to polynomials and other domains, to develop those concepts in the more general setting of integers and so we begin defining them.

1 - DEFINITION OF INTEGERS AND SOME CONSEQUENCES

We call *integer, whole number* or *integral number* to any real number that is either natural, or zero, or the opposite (additive inverse) of a natural. In other words, if we define:

$$\mathbb{N}' = \{x \in \mathbb{R} \, / \, \exists n \in N \text{ with } x = -n\},$$

that is $\mathbb{N}'$ is the set of the opposites of the natural numbers, then we define the set $\mathbb{Z}$ (from the German Zahl=number) of integers, by:

$$\mathbb{Z} = \mathbb{N} \cup \{0\} \cup \mathbb{N}'$$

Proposition 1.1: Given $a, b \in \mathbb{Z}$, we have,

a) $-a \in \mathbb{Z}$.

b) $|a| \in \mathbb{N}$ or $a = 0$.

c) $a + b \in \mathbb{Z}$.

d) $ab \in \mathbb{Z}$.

e) If $m \in \mathbb{Z}$, there is no integer c such that $m < c < m + 1$.

f) (Division algorithm) If $b \neq 0$, there exist $q, r \in \mathbb{Z}$ such that:

$$a = bq + r \text{ and } 0 \leq r < |b|$$

Furthermore, such q and r are unique.

Proof: We will prove the existence in f), leaving the rest as exercises. We have proved (9.1 Chap. 3) the existence if $a, b \in \mathbb{N}$ (with $q, r \in \mathbb{N} \cup \{0\}$). It remains to verify the following cases:

1) $a \in \mathbb{N}$ and $b \in \mathbb{N}'$: as $- b \in \mathbb{N}$, by the case already proved, there exist $q', r \in \mathbb{N} \cup \{0\}$ such that $a = (-b)q' + r$ with $0 \leq r < -b = |b|$ so it is enough to take $q = -q'$.

2) $a \in \mathbb{N}'$ and $b \in \mathbb{N}$: as $- a \in \mathbb{N}$, there exist $q', r' \in \mathbb{N} \cup \{0\}$ such that $- a = bq' + r'$ and $0 \leq r' < b$, then if $r' = 0$ it is clear, while if $r' > 0$, we have $a = b(-q' - 1) + (b - r')$ with $0 < b - r' < b$, and it suffices to take $q = -q' - 1, r = b - r'$.

3) $a \in \mathbb{N}'$ and $b \in \mathbb{N}'$: we have $- a, -b \in \mathbb{N}$, then there exist $q', r' \in \mathbb{N}$ such that $- a = (-b)q' + r'$ and $0 \leq r' < -b = |b|$. If $r' = 0$ is clear, so assume $r' > 0$ and then $a = b(-q' + 1) + (-b - r')$ with $0 < -b - r' < -b$ and it is enough to take: $q = -q' + 1, r = -b - r'$.

4) $a = 0$: this case is obvious. ∎

2 - POWERS WITH INTEGRAL EXPONENT

In section 3 of the preceding chapter, we have defined a^n for $a \in \mathbb{R}$ and $n \in \mathbb{N} \cup \{0\}$. If $a \neq 0$, we can extend the definition to integral exponents putting, if $r \in \mathbb{N}', r = -n$ con $n \in \mathbb{N}$:

$$a^r = (a^{-1})^n$$

and the properties shown for powers with natural or zero exponents, are extended to integral exponents:

Proposition 2.1: Let $a, b \in \mathbb{R}$ non zero and r, s integers. We have:

1) $a^r a^s = a^{r+s}$

2) $(a^r)^s = a^{rs}$

3) $(ab)^r = a^r a^s$

4) If $a > 1$, then: $r < s \Leftrightarrow a^r < a^s$

If $0 < a < 1$, then: $r < s \Leftrightarrow a^s < a^r$.

Proof: We will prove1) and 4), leaving 2) and 3) as exercises. 1) The case: $r, s \in \mathbb{N}' \cup \{0\}$ have been proved in 3.1, Chap. 3. Let $r \in \mathbb{N} \cup \{0\}, s \in \mathbb{N}'$, so $s = -n$ with $n \in \mathbb{N}$.

If $r \geq n$, then $r = m + n$ with $m \in \mathbb{N} \cup \{0\}$, so,

$$a^r a^s = a^{(m+n)}(a^{-1})^n = a^m a^n (a^{-1})^n = a^m (aa^{-1})^n = a^m = a^{r-s}$$
$$= a^{r+s}$$

If $r < n$, then $n = r + t$ with $t \in \mathbb{N}$, so,

$$a^r a^s = a^r (a^{-1})^n = a^r (a^{-1})^{r+t} = a^r (a^{-1})^r (a^{-1})^t =$$
$$= (aa^{-1})^r (a^{-1})^t = (a^{-1})^t = a^{-t} = a^{r-t} = a^{r+t}$$

The case $r \in \mathbb{N}', s \in \mathbb{N} \cup \{0\}$, is the same as the above.

It remains to consider the case: $r, s \in \mathbb{N}'$. Then $r = -n, s = -m$ with $n, m \in \mathbb{N}$, so,

$$a^r a^s = (a^{-1})^n (a^{-1})^m = (a^{-1})^{n+m} = a^{r+t}$$

4)   Let $a > 1$,   if $r < s$,   put $s = r + t$ with $t \in \mathbb{N}$, as $a > 1$ we have $a^t > 1$, so $a^s = a^{r+t} = a^r a^t > a^r$. Conversely, if $a^r < a^s$ we must have $r < s$, for if $r \geq s$ we would have $a^r \geq a^s$. Finally, in case: $0 < a < 1$, we have $a^{-1} > 1$, and by the case just proved, we have: $r < s \Leftrightarrow (a^{-1})^r < (a^{-1})^s \Leftrightarrow a^s < a^r$. ∎

3 - DIVISIBILITY

Let $a, b \in \mathbb{Z}$. It is said that a *divides* b, (or b is   a   *multiple* of a, or a es *divisor* of b) and it is written $a \mid b$, iff there exists $c \in \mathbb{Z}$ such that $b = ac$.

Note that if $a \neq 0$ this is equivalent to say that (b/a) is an integer.

Examples: 1) 2   does   not   divide   3   (2 ∤ 3) since, if it does, then there would exist $c \in \mathbb{Z}$ such that $3 = 2c$, then $c = \frac{3}{2}$ and as $1 < \frac{3}{2} < 2$ this contradicts 1.1. e).

2) $a \mid 0 \; \forall a \in \mathbb{Z}.$, since we have $0 = a0$.

3) $0 \mid a \Rightarrow a = 0$, as $a = 0c \Rightarrow a = 0$.

4) $\pm 1 \mid a$ and $\pm a \mid a \; \forall a \in \mathbb{Z}$, since $a = 1a = (-1)(-a)$.

Proposition 3.1: Let a, b, c be integers,

   1) $a \mid b$ and $b \mid c \Rightarrow a \mid c$.

   2) $a \mid b$ and $a \mid c \Rightarrow a \mid ha + kb$ whichever be $h, k \in \mathbb{Z}$.

   3) $a \mid b \Rightarrow b = 0$ or $|a| \leq |b|$.

   4) $a \mid b$ and $b \mid a \Rightarrow a = \pm b$.

   5) $a \mid 1 \Rightarrow a = \pm 1$.

Proof: 1) From $b = au$ and $\;\; c = bv$ with $u, v \in \mathbb{Z}$ it follows $uv \in \mathbb{Z}$ and $c = auv$, then $a \mid c$.

2) If $b = au$ and $c = av$ with $u, v \in \mathbb{Z}$ then whatever be $h, k \in \mathbb{Z}$, we have $hu + kv \in \mathbb{Z}$ and $hb + kc = a(hu + kv)$, so then $a \mid hb + kc$.

3) From $b = ad \; (d \in \mathbb{Z})$ and $b \neq 0$ follows $d \neq 0$, then $|d| \geq 1$ and so $|a| \leq |a||d| = |b|$.

4) From $a \mid b$ and $b \mid a$ follows $a = 0 \Leftrightarrow b = 0$ in which case $a = \pm b$. Assume then $a \neq 0$ and $b \neq 0$. From $a \mid b$ and $b \neq 0$ follows by 3), $|a| \leq |b|$, and from $b \mid a$ and $a \neq 0$, follows $|b| \leq |a|$. Then $|a| = |b|$ and $a = \pm b$.

5) From $a \mid 1$, as $1 \mid a$ it follows by 4): $a = \pm 1.\blacksquare$

4 - PRIME NUMBERS

As we have seen in example 4), any integer a is divisible by ± 1 and by $\pm a$. An integer is said to be *prime* ó *irreducible* iff it is positive and has exactly four divisors or, in other words, an integer p is said prime iff $p > 1$ and verifies:

$$a \mid p \Rightarrow a = \pm 1 \; or \; a = \pm p$$

Examples: 1) 2 is prime since $a \mid 2 \Rightarrow |a| \leq 2$ by 4) of prop. 3.1. So by 1.1. e., we have $|a| = 1$ or $|a| = 2$, That is $a = \pm 1$ or $a = \pm 2$.

2) 3 is prime, since if $a \mid 3$, then $|a| = 1, 2,$ or 3, but $|a| = 2 \Rightarrow 2 \mid 3$ and we have seen that is not possible, so $a = \pm 1$ or $a = \pm 3$.

3) 4 is not prime because $2 \mid 4$.

Proposition 4.1: Set $a \in \mathbb{N}$ with $a > 1$. a is not prime if, and only if, there exist $c, d \in \mathbb{N}$ such that $a = cd$ and $1 < c, d < a$. (such an a is said to be *composite*)

Proof: If a is not prime, it has a divisor e with $e \neq \pm1$ and $e \neq \pm a$, $a = ef$ ($f \in \mathbb{Z}$), and taking $c = |e|, d = |f|$, The implication for the right follows. The reciprocal is immediate.■

Corollary 4.2: If a is a natural number $a > 1$ and a is not prime, then there are a divisor c of a such that $c > 1$ and $c^2 \leq a$.

Proof: By the above proposition, we have $a = cd$ with $c, d \in \mathbb{N}$ and $1 < c, d < a$. Changing notation, if it were neccesary, we can assume $c \leq d$, then $c^2 \leq cd = a$.■

This corollary allows saving work when building a table of primes. Exemplify it building a table of primes up to 120.

1	7	13	19	25	31	37	43	49	55	61	67	73	79	85	91	97	103	109	115
2	8	14	20	26	32	38	44	50	56	62	68	74	80	86	92	98	104	110	116
3	9	15	21	27	33	39	45	51	57	63	69	75	81	87	93	99	105	111	117
4	10	16	22	28	34	40	46	52	58	64	70	76	82	88	94	100	106	112	118
5	11	17	23	29	35	41	47	53	59	65	71	77	83	89	95	101	107	113	119
6	12	18	24	30	36	42	48	54	60	66	72	78	84	90	96	102	108	114	120

Written the numbers from 1 to 120, we cross out 1, which is not prime; then we cross out the multiples of 2 (except 2 itself, a prime) which are all composites by prop. 4.1. The next number which was not crossed out (3), must be prime, because otherwise by the corollary, it would be multiple of one strictly between 1 and 3, that is of 2, but the multiples of 2 was already crossed out. Then we cross out the multiples of 3 (except itself) which are composites by the above proposition. Again the next number which was not crossed out (5) must be prime, because otherwise by the corollary, it would be divisible by 2, 3 or 4, but all multiples of these were crossed out. The next one that has not been crossed out (7) must be prime. Finally, we cross out the multiples of 7. We assert that the unmarked numbers are prime. Indeed, by the above corollary, if a number $a \leq$

120 is composite, it has a divisor c such that $c > 1$ and $c^2 \leq a$, then $c < 11$ ($c \geq 11 \Rightarrow c^2 \geq$ 121) and as all multiples of the numbers < 11 were already crossed out, a, as any composite, was crossed out.

By this method (with several refinements), called *Sieve of Eratosthenes*, tables of primes up to several million have been constructed.

From the examination of a table, it follows that the distribution of primes is, on the small, extremely irregular. For example, there are prime pairs that differ by 2 in any range; $2111, 2113$; $3557, 3559$; $4481, 4483$; $8819, 8821$; $10007, 10009$; are some of such pairs (it is a conjecture that there is an infinitude of them). In contrast, there are chains of arbitrary length of consecutive composite numbers: between $n! + 2$ and $n! + n$ ($n \geq 2$) all the integers are, clearly, composite, making up a chain of length $n - 1$.

However, in the long run, there is a certain regularity. Examining a table of primes, Legendre conjectured the *Prime Number Theorem*, which asserts that:

$$\pi(n) \sim \frac{n}{\log n}$$

where $\pi(n)$ is the number of primes $\leq n$, log the natural logarithm and, if f and g are functions with real values, $f(n) \sim g(n)$ means that the quotient $\frac{f(n)}{g(n)}$ tends to 1 when $n \to \infty$. Note that $f(n) \sim g(n)$ does not assert that the difference $f(n) - g(n)$ tends to 0; for example, we have $n^2 + n \sim n^2$ since $\frac{n^2 + n}{n^2} = 1 + \frac{1}{n} \to 1$ when $n \to \infty$.

The Prime Number Theorem was conjectured, independently, by Gauss and by Chebyshev in the form:

$$\pi(n) \sim \int_2^n \frac{dx}{\log x}$$

which is equivalent to the given above.

Essential tools to deal with the Prime Number Theorem were developed by Riemann, and it was first proved by Hadamard and La Vallee Poussin, independently, in the same year (1896).

Another observation from the tables is that between two consecutive squares there is always a prime, that is, for any $n \in \mathbb{N}$ exists p prime such that $n^2 < p < (n + 1)^2$, however, until now, nobody could prove or disprove this. If it were true, it would have as a consequence Bertrand's Postulate, which asserts that for any $m >$

1, there exists a prime between m and $2m$. This was verified by Bertrand for $m < 3000000$ and then proved by Chebishev in 1850 $\forall m > 1$. Let us show how to derive Bertrand Postulate (which, as we said, actually is a theorem) from the conjecture above. Given $m > 1$, if $m \leq 25$ it can be verified from the above table of primes. Let $m > 25$ and take $n \in \mathbb{N}$ such that $(n-1)^2 < m \leq n^2$ and pick a prime p such that $n^2 < p < (n+1)^2$. As $25 < m \leq n^2$, we have n≥6 and so $(n-1)^2 \geq 4n$, from which follows that m>4n and then:

$$m \leq n^2 < p < (n+1)^2 = (n-1)^2 + 4n < 2m$$

Another famous conjecture is the so-called Goldbach's conjecture, raised by Goldbach in a letter to Euler:

a) Any integer > 5 is a sum of three primes, which is equivalent to:

b) Every even number > 2 is a sum of two primes.

Let us show the equivalence:

a)⇒b): If n is even > 2, then $n + 2 > 5$, and by a) $n + 2$ is a sum of three primes, but at least one of them must be even ($n + 2$ is even), i.e., $= 2$, then n is the sum of the two remaining ones.

b)⇒a): Let $n > 5$ integer. If n is evn, by b) we have: $n - 2 = p + q$ con p, q primos. If is n oldd, we have by b) $n - 3 = p + q$ con p, q primos. In both cases, n is a sum of three primes.

Although significant progress has been made, Goldbach's conjecture is still an open problem.

5 - GREATEST COMMON DIVISOR

We will call *greatest common divisor* (g.c.d.) of two integers a and b, to any integer d that satisfies:

1) $d \mid a$ and $d \mid b$ (i. e., d is a common divisor of a and b).

2) $c \mid a$ and $c \mid b \Rightarrow c \mid d$ (any common divisor of a and b is a divisor of d).

3) $d \geq 0$.

Proposition 5.1: (Uniqueness) If d and d' are the greatest common divisors of a and b, then $d = d'$.

Proof: As $d' \mid a$ and $d' \mid b$ from 2) results, $' \mid d$. By symmetry $d \mid d'$, then $d = \pm d'$, but as both are positive $d = d'$

We will denote by (a, b) to the unique greatest common divisor of a and b, which, as we will see bellow, always exists.

Examples: 1) $(0, a) = |a|$, since $|a|$ is a common divisor of 0 and a, and if $c \mid a$ and $c \mid 0$, then $c \mid |a|$. Therefore $(0, a) = |a|$.

2) If $x = yw + z$ with $x, y, w, z \in \mathbb{Z}$, then $(x, y) = (y, z)$. In fact, putting $d = (x, y)$ and $d' = (y, z)$, from $d \mid x$, $d \mid y$ and $x = yw + z$, follows $d \mid z$ so $d \mid d'$. By symmetry $d' \mid d$ and, so $d = d'$.

Theorem 5.2: (Existence) Any two integers a and b, have a greatest common divisor d. Moreover, there exist integers s, t such that $d = sa + tb$.

Proof: If a=0, the result is valid by example 1) above. Assume then a≠0 and define:

$$A = \{ha + kb / h, k \in \mathbb{Z}\} \cap \mathbb{N}$$

$A \neq \emptyset$, because, for example, $a^2 = aa + 0b \in A$, then by well ordering, A posses a smallest element d. As $d \in A$ we have $d = sa + tb$ for certain integers s, t, and from this it follows that $c \mid a$ and $c \mid b \Rightarrow c \mid d$. It is also clear that $d \geq 0$ (in fact $d \in \mathbb{N}$). It remains to verify that $d \mid a$ and $d \mid b$.

Let us see that $d \mid a$. By the division algorithm, there area $q, r \in \mathbb{Z}$ such that $a = dq + r$ with $0 \leq r < d$, then:

$$r = a - dq = a - (sa + tb)q = a(1 - sq) + btq$$

And from this follows that if it were $r \neq 0$, it would be $r \in A$ which is contradictory since $r <$
d and d is the smallest element of A. So $r = 0$ and $d \mid a$. Similarly $d \mid b$.■

Another way of proving the existence of the g.c.d., that at the same time, is a practical way to compute and to write it as a "linear combination" $sa + tb$ of a and b, is the so-called *Euclid's algorithm*, described bellow.

If $b = 0$, we already know (a, b) by example1 above. Le be $b \neq 0$, by the division algorithm and by example 2, we obtain:

$$a = bq_0 + r_0, r_0 < |b|, (a, b) = (b, r_0)$$

If $r_0 = 0$ then $(a, b) = (b, 0) = |b|$. If $r_0 \neq 0$ we can repeat the process to obtain:

$$b = r_0 q_1 + r_1, r_1 < r_0, (b, r_0) = (r_0, r_1)$$

$$\dots\dots\dots\dots\dots\dots\dots\dots\dots\dots\dots\dots\dots\dots\dots\dots\dots$$

$$r_{n-2} = r_{n-1} q_n + r_n, r_n < r_{n-1}, (r_{n-2}, r_{n-1}) = (r_{n-1}, r_n)$$
$$r_{n-1} = r_n q_{n+1} + 0, (r_{n-1}, r_n) = (r_n, 0)$$

It is plain that, being the remainders positive integers smaller and smaller each time, in a finite number of steps we must reach zero remainder, and according to the right column, we have,

$$(a, b) = (r_0, r_1) = \dots = (r_{n-2}, r_{n-1}) = (r_{n-1}, r_n) = (r_n, 0) = r_n$$

i. e., (a, b) is the last nonzero remainder in this process of succesive divisions.

Example: Let $a = 308, b = 52$. We will find $d = (a, b)$ and write it as a linear combination $sa + tb$ of a and b.

By Euclid's algorithm, we have:

$$308 = 52 \cdot 5 + 48 \ (308, 52) = (52, 48)$$
$$52 = 48 \cdot 1 + 4 \ (52, 48) = (48, 4)$$
$$48 = 4 \cdot 12 + 0 \ (48, 4) = (4, 0) = 4$$

then $d = 4$. Moreover:

$$4 = 52 - 48 = 52 - (308 - 52 \cdot 5) = 52 \cdot 6 - 308$$

and taking $s = -1, t = 6$ we have $d = sa + tb$.

Note that the expression of $d = (a, b)$ as a linear combination of a and b is not unique, in fact if $d = sa + tb$ with $s, t \in \mathbb{Z}$, we have $d = (s + kb)a + (t - ka)b$, for any $k \in \mathbb{Z}$. Note also that if we have $e = sa + tb$ for certain integers s, t, it does not necessarily follows that $e = (a, b)$ even if $e \geq 0$, since, while it is true that $c \mid a$ and $c \mid b \Rightarrow c \mid e$, it is not necesarily true that $e \mid a$ and $e \mid b$. This is necesarily valid only if $e = 1$, i.e.if $1 = sa + tb$ with $s, t \in \mathbb{Z}$, then $1 = (a, b)$.

Two integers are said to be *coprime* or *prime each other* iff their greatest common divisor is 1.

Proposition 5.3: If $a \mid bc$ and a, b are coprime, then $a \mid c$.

Proof: As $(a, b) = 1$, we have $1 = sa + tb$ with $s, t \in \mathbb{Z}$, then $c = sac + tbc$, and from the hypothesis, $a \mid bc$, and prop. 3.1, results $a \mid c$.■

Corollary 5.4: If a and b are coprime and $a \mid c$ and $b \mid c$, then $ab \mid c$.

Proof: There are integers u, v such that $c = au = bv$ then $b \mid au$ and as $(b, a) = 1$, from the above proposition follows $b \mid u$, so $u = bw$ with $w \in \mathbb{Z}$ then $c = abw$ and $ab \mid c$.■

Proposition 5.5: Let p be an integer > 1. Then p is prime if, and only if, the following is true:

$$p \mid ab \Leftrightarrow p \mid a \text{ or } p \mid b$$

Proof: Let p be a prime such that $p \mid ab$. If $p \mid a$ there is nothing to prove. If $p \nmid a$ (does not divide a) we have $(p, a) = 1$ (since if $d = (p, a)$, as $d \mid p$ and $d \geq 0$ it must be $d = 1$ or $d = p$, but in this last case it would result $p \mid a$) and by the above proposition, $p \mid b$.

Reciprocally, let $p > 1$ be an integer with the mentioned property and let $a \in \mathbb{Z}$ such that $a \mid p$, then $p = ab$ with $b \in \mathbb{Z}$, so $p \mid ab$, then $p \mid a$ or $p \mid b$. If $p \mid a$ we have $a = \pm p$, while if $p \mid b$ we have: $b = pc, p = apc$, and $ac = 1$, from where $a = \pm 1$.■

Example: If $i \in \mathbb{Z}$ with $0 < i < p$ where p is prime, then:

$$p \mid \binom{p}{i}$$

Note in the first place that combinatorial are natural numbers, that follows easily by induction from their fundamental property (prop.6.1, Chap. 3); so the statement makes sense. Put $\binom{p}{i} = a \in \mathbb{N}$ then: $p(p - 1)\ldots(p - i + 1) = ai(i - 1)\ldots 1$. As p does not divide any of the numbers $i, i - 1, \ldots, 1$, because all of them are $< p$, by the preceding proposition follows that $p \mid a$.

6 - UNIQUE FACTORIZATION

Theorem 6.1 (on the Unique Factorization or Fundamental Theorem of Arithmetic):

1) Given a>1 integer, there are primes p_1,...,p_n such that a=p_1...p_n.

2) If $p_1 \ldots p_n = q_1 \ldots q_m$ with p_i, q_j primes, then $n = m$ and, except for a reordering of the q_j, we have $p_i = q_i \; \forall i = 1, \ldots, n$.

In other words, any integer > 1 is a product of primes in an essentially unique way.

Proof: 1) If 1) were false, it would exist a smallest natural $a > 1$, that would not be a product of primes. Such a would not be prime (if not it would be a product of primes with only one factor) and by prop. 4.1 it would be $a = bc$ with $1 < b, c < a$, but by the election of a, both b and c would be the product of primes and, as consequence, $a = bc$ also.

2)   From $p_1 \ldots p_n = q_1 \ldots$ and prop. 5.5, it follows that $p_1 \mid q_j$ for some $j = 1, \ldots, m$. Reordering the q_j, if necessary, we can assume that $j = 1$, that is $p_1 \mid q_1$, but as q_1 is prime, it must be $p_1 = q_1$, then $p_2 \ldots p_n = q_2 \ldots q_m$ Proceeding by induction we obtain $n - 1 = m - 1, p_2 = q_2, \ldots, p_n = q_n$}.∎

Corollary 6.2: Any integer > 1 is divisible by some prime.∎

Grouping repeated factors, we conclude that any integer $a > 1$ is expressed in a unique manner like:
$$a = p_1^{\alpha_1} \ldots p_r^{\alpha_r}$$
with the p_i distinct primes and the α_i natural numbers.

It follows from the above theorem that if $c \mid a$, it must be:
$$c = p_1^{\beta_1} \ldots p_r^{\beta_r}$$
with $\beta i \in \mathbb{N} \cup \{0\}$ and $\beta i \leq \alpha_i \; \forall i$ and reciprocally.

Also,   it   follows   that if
$p^1, \ldots, p_r$ are the common primes, that appear on both factorizations of a and b, in such a way that:
$$a = p_1^{\alpha_1} \ldots p_r^{\alpha_r} q_1^{\beta_1} \ldots q_r^{\beta_r} \quad b = p_1^{\gamma_1} \ldots p_r^{\gamma_r} h_t^{\delta_1} \ldots h_t^{\delta_t}$$
then
$$(a, b) = p_1^{\varepsilon_1} \ldots p_r^{\varepsilon r}$$
where $\varepsilon_i = min(\alpha_i, \beta_i)$.

Example: They do not exist natural numbers a, b such that

$$a^2 = 2b^2$$

since factoring a and b as a product of primes, in the first member 2 would appear with an even exponent and in the second with an odd exponent.

7 - INFINITUDE OF PRIMES

Several proofs of the infinitude of primes are known. Here we will treat only two, one that is essentially the given by Euclid in his "Elements" and one due to G. Pólya. In [8] or [24] for example, other proofs can be consulted.

Theorem 7.1: (Euclid) There is an infinitude of primes.

Proof: Assume by contradiction that $p_1, \ldots, p_n$ were all the primes, and set:

$$a = p_1 \ldots p_n + 1$$

As $a > 1$, by the above theorem, a must be divisible by some prime p, but p must be one of the $p_1, \ldots, p_n$, then $p \mid 1$, a contradiction.∎

Another idea to prove the infinitude of primes consists in finding an indefinite sequence: $a_1, a_2, \ldots, a_n, \ldots$ of integers > 1 and pairwise coprime, since, if such sequence is available and if p_n is a prime dividing a_n (that exists by the unique factorization theorem), in the sequence $p_1, \ldots, p_n, \ldots$ can not be repeated primes ($p_i = p_j \Rightarrow a_i$ and a_j are not coprime).

The proof of Pólya consists in taking the sequence of *Fermat's numbers*: $F_n = 2^{2^n} + 1$ (a^{b^c} denotes $a^{(b^c)}$ and not $\left(a^b\right)^c = a^{bc}$). This numbers were considered by Fermat, who asserts that all of them were prime ([12] or [35]), which is false.

The first Fermat's numbers are:

$$F_0 = 3, F_1 = 5, F_2 = 17, F_3 = 257, F_4 = 65.537$$

and all of them are prime, although, as Euler noted, F_5 is not prime. Moreover, it has not been found any other prime Fermat number.

Let us prove that any two Fermat numbers are coprime. Indeed, we have:

$$F_{n+1} = F_0 F_1 \ldots F_n + 2 \; (*)$$

because

$$F_{n+2} - 2 = 2^{2^{n+2}} - 1 = \left(2^{2^{n+2}}\right)^2 - 1 = \left(2^{2^{n+2}} - 1\right)\left(2^{2^{n+2}} + 1\right)$$
$$= (F_{n+1} - 2)F_{n+1}\}$$

from which $(*)$ follows readily by induction. From $(*)$ it follows that the greatest common divisor d of two distinct Fermat numbers must divide 2, and as the Fermat numbers are odd, it must be $d = 1$.

Another sequence with the property that any two of its terms are coprime is the one defined recursively by :

$$a_{n+1} = a_n^2 - 2$$

with, say, $a_1 = 3$. This sequence was used by Lucas to obtain primality criteria. To verify that its terms are pairwise coprime, it is convenient to wait until the properties of congruences are available (Chap. 5).

8 - PERFECT NUMBERS

It is likely that the basic theorems of arithmetic presented in the previous sections, most of them already known by the Greeks, were obtained in their study of the so-called *perfect numbers*. The investigation on this numbers, has given rise as well, to Fermat's Little Theorem (5.2, Chap. 5), and perhaps to the Quadratic Reciprocity Law [37] and has given rise to the, possibly, older open problem in mathematics: the existence of odd perfect numbers.

If a is a natural number, the sum of its *proper divisors* (i.e., of the positive divisors of a other than a) may be greater, lesser or equal to a. In the first case, it is said that a es *abundant*, in the second that it is *deficient* and in the third that it is *perfect*. For example: 12 is abundant, since $1 + 2 + 4 + 3 + 6 = 16 > 12$.; 8 is deficient since $1 + 2 + 4 = 7 < 8$ and $6 = 1 + 2 + 3$ and $28 = 1 + 2 + 4 + 7 + 14$ are perfect.

The first perfect numbers: 6; 28; 496; 8128 were known by the Greeks. Cataldi at the beginning of the 17th century, constructed extensive tables of primes to study perfect numbers, corrected errors of other authors and gave, correctly, the fifth: 33550336 (previously known) and the sixth: 8589869056.

If $\sigma(a)$ denotes the sum of the positive divisors of a (a included), it can be said that a is perfect if:

$$\sigma(a) = 2a$$

Theorem 8.1: (Euclid) If $n \in \mathbb{N}$ and $2^n - 1$ is prime, then $a = 2^{n-1}(2^n - 1)$ is perfect.

Proof: Put $p = 2^n - 1$ which is prime by hypothesis. As $a = 2^{n-1}p$, by the Fundamental Theorem of Arithmetic follows that the positive divisors of a are: $1, 2, ..., 2^{n-1}$ and $p, 2p, ..., 2^{n-1}p$, from where:

$$\sigma(a) = 1 + 2 + ... + 2^{n-1} + p(1 + 2 + ... + 2^{n-1}) = (1+p)(2^n - 1) = 2^n p = 2a. \blacksquare$$

All perfect numbers mentioned above (and, in fact, all known hitherto) are as described in the preceding theorem. Moreover, as Euler proved in the 18th century, all even perfect numbers are that way.

Lemma 8.2: If a and b are coprime natural numbers, then:

$$\sigma(ab) = \sigma(a)\sigma(b)$$

Proof: If a and b are coprime it follows, by the Fundamental Theorem of Arithmetic, that each divisor of ab is obtained by multiplication of a divisor of a by one of b. $\blacksquare$

Theorem 8.3: (Euler) If a is an even perfect number, then exists $n \in \mathbb{N}$ such that $a = 2^{n-1}(2^n - 1)$ with $2^n - 1$ prime.

Proof: Let a be an even perfect number and put $a=2^{n-1}p$ where $n \geq 2$ and p is odd (this can be done by the Fundamental Theorem). As a is perfect, by the above lemma we obtain:

$$2^n p = 2a = \sigma(a) = \sigma(2^{n-1})\sigma(p)$$

and as $\sigma(2^{n-1}) = 1 + 2 + ... + 2^{n-1} = 2^n - 1$, we have $2^n p = (2^n - 1)\sigma(p)$, or:

$$\sigma(p) = p + \frac{p}{2^n - 1} \quad (*)$$

From $(*)$ follows that $\frac{p}{2^n-1}$ is an integer (it is equal to $\sigma(p) - p$) so $\frac{p}{2^n-1}$ is a positive divisor of p. Then $(*)$ expresses that the sum

94

of all positive divisors of p, is the sum of two of them: p and $p/(2^n - 1)$, hence p must be prime and $p/(2^n - 1) = 1$.∎

According to the theorems of Euclid and Euler, the even perfect numbers are determined by the primes of the form $2^n - 1$. It is easy to check (exercise 18) that if $2^n - 1$ is prime, n itself must be prime. The numbers of the form $2^p - 1$ with p prime, are called *Mersenne numbers* for they having been studied by P.M. Mersenne, active correspondent of the European mathematicians of his time (17th century). Currently, 49 Mersenne primes (Mersenne numbers that are prime) are known, the first being those corresponding to $p = 2, 3, 5, 7, 13, 17, 19, 31$.   While   for $p = 11$ and $p = 23$ the   corresponding Mersenne numbers are composite. The largest known prime until now is the Mersenne number corresponding to $p = 77232917$. It is conjectured that there are infinite Mersenne primes which, if true, would imply, of course, the existence of infinite perfect numbers.

With regard to odd perfect numbers, it is an open problem to determine their existence. It is known that if there exists an odd perfect number, in its prime factorization, must appear at least 10 distinct primes and that it must be greater than 10^{1500}.

9 - LINEAR DIOPHANTINE EQUATION

Diophantus of Alexandria (around 250 A.D.), in his "Aritmetics," studied problems on "indeterminate analysis", that is, problems on finding integral solutions of polynomial equations in several variables with integral coefficients, now called *Diophantine equations*. Actually, he was satisfied with finding one solution, and he admitted rational solutions too. The most simple Diophantine equation is the linear one, not treated by Diophantus, but extensively studied in Indian mathematics, for example by Aryabhata (5th century), Brahmagupta (7th century), Bháscara (12th century), in relation with astronomical problems.

The linear Diophantine equation in two variables is the following:

$$ax + by = c \quad (1)$$

where a, b, c are given integers and the problem is to find all the integers (if any) x, y that satisfies it.

Proposition 9.1: There exist integers x, y that satisfies (1) if, and only if, $(a, b) \mid c$.

Proof: If there exists a solution x, y in integers of (1), we will have $ax + by = c$ and $\qquad$ if $d = (a, b)$, $\qquad\qquad$ as $d \mid a$ and $d \mid b$, we must have $d \mid c$.

Reciprocally, $\qquad\qquad\qquad\qquad$ let $d \mid c$ with $d = (a, b)$, so $c = de$ for som integer e. There are integers s, t such that $d = sa + tb$, hence:

$$c = a(se) + b(te)$$

and se, te is a solution of (1).∎

The above proof gives us a way to find a solution, if it exists. The next proposition tells us how to find all solutions from one of them.

Proposition 9.2: Let x_0, y_0 be a solution of (1). Any solution x, y is obtained from the relations:

$$x = x_0 + \frac{b}{d}t \quad y = y_0 - \frac{a}{d}t$$

where $d = (a, b)$ and t is an arbitrary integer.

Proof: Note that we can assume $a \neq 0$ and $b \neq 0$ for if not, the equation (1) would be trivial. So $d \neq 0$. If x, y is a solution of (1), then $c = ax + by = ax_0 + by_0$ and so $a(x - x_0) = b(y_0 - y)$. Putting $a = da', b = db'$, we obtain $a'(x - x_0) = b'(y_0 - y)$ and as $a'b' = 1$ it follows that $b' \mid x - x_0$, That is, $\exists t \in \mathbb{Z}$ such that $x - x_0 = b't$ and hence, $y_0 - y = a't$, i.e.:

$$x = x_0 + \frac{b}{d}t \quad y = y_0 - \frac{a}{d}t. \blacksquare$$

Example: (proposed in the Arithmetic of Bachet) Find all solutions in positive integers of the system:

$$x + y + z = 41 \ (2)$$
$$12x + 9y + z = 120 \ (3)$$

Subtracting (2) from (3), we have:

$$11x + 8y = 79 \ (4)$$

The g.c.d. of 11 and 8 is $1 = 3 \cdot 11 - 4 \cdot 8$, then $x_0 = 3 \cdot 79 = 237$, $y_0 = (-4) \cdot 79 = -316$ is a solution of (4). According to the above proposition, any solution x, y of (4), must comply:

$$x = 237 + 8t, y = -316 - 11t$$

for some $t \in \mathbb{Z}$. But as $x \geq 0$, it must be $237 + 8t \geq 0$, or $t \geq -\frac{237}{8} > -30$ and as $y \geq 0$, it must be $-316 - 11t \geq 0$, or $t \leq -\frac{316}{11} < -28$, hence $t = -29, x = 5, y = 3$ and (from (1) or (2)): $z = 33$.

10 – PYTHAGOREAN TRIPLES

The triples (x, y, z) of positive integers that satisfy:

$$x^2 + y^2 = z^2 \ (1)$$

are called *Pythagorean triples* because, according to Pythagoras theorem, they determine the rectangle triangles with integral sides. Actually, the Pythagoreans thought that, by conveniently choosing a unit measure, any triangle would have integral sides. The discovery that this belief leads to a contradiction marked a crisis in Greek mathematics.

The Pythagoreans (members and disciples of the mystical-philosophical brotherhood founded by Pythagoras) found some solutions of the Diophantine equation (1). Cuneiform written tablets have been found in Mesopotamia containing tables of Pythagorean triples, indicating that a method to find them were available more than a thousand years before Pythagoras. In Euclid's Elements there is a general method to determine primitive Pythagorean triples.

In order to determine the Pythagorean triples, it is enough to determine the *primitive* ones, that is, those in which x, y, z has no common factors. If (x, y, z) is a primitive pythagorean triple, then x, y, z are coprime in pairs (because any common factor of two of them, must be by (1), a factor of the third). In particular, one of them must be even and the other two odd. Moreover, z can not be even, for if not we would have $z = 2z', x = 2x' + 1, y = 2y' + 1$ and replacing this values in (1) we reach 4|2 which is an absurd. By symmetry we can assume $x = 2x'$ is even, so then y, z are odd, and we have:

$$x'^2 = \frac{z+y}{2} \frac{z-y}{2} \ (2)$$

As y, z are odd, $\frac{z+y}{2}$ and $\frac{z-y}{2}$ are integers and, furthermore, they are coprime, since if $d = \frac{z+y}{2}, \frac{z-y}{2}$, from $d \mid \frac{z+y}{2}$ and $d \mid \frac{z-y}{2}$ it follows $d \mid \frac{z+y}{2} + \frac{z-y}{2} = z$ and $d \mid \frac{z+y}{2} - \frac{z-y}{2} = y$ and as y and z are coprime: $d = 1$. Hence by (2) and the Fundamental Theorem of Arithmetic $\frac{z+y}{2}$ and $\frac{z-y}{2}$ must be both squares, i.e., there exist $u, v \in \mathbb{N}$ such that:

$$\frac{z+y}{2} = u^2, \frac{z-y}{2} = v^2 \text{ (3)}$$

and from (2) follows $x' = uv$, or $x = 2uv$. In addition from (3) it follows $z = u^2 + v^2, y = u^2 - v^2$. Also, it is clear that, as $d = 1$, u and v must be coprime. We have proved the implication ($\Rightarrow$) of the following:

Theorem 10.1: (x, y, z) is a primitive pythagorean triple with x even $\Leftrightarrow$ there exist $u, v \in \mathbb{N}$, coprime and of distinct parity with $u > v$, such that:

$$x = 2uv \; ; \; y = u^2 - v^2 \; ; \; z = u^2 + v^2$$

Proof: ($\Leftarrow$) By direct replacement, it is checked that $x^2 + y^2 = z^2$. If p were a common prime factor of $2uv, u^2 - v^2, u^2 + v^2$, as $p \mid 2uv$ it would be $p \mid 2, p \mid u$ or $p \mid v$; but $p \nmid 2$ since $p \mid u^2 - v^2$ which is odd, being u, v of distinct parity. In addition, from $p \mid u$, as $p \mid u^2 - v^2$, it follows that $p \mid v$, a contradiction as u and v are coprime. Similarly, from $p \mid v$ would follows $p \mid u$. Hence x, y, z do not have prime factors in common and (x, y, z) is a primitive pythagorean triple.■

Assigning values to u and v, we obtain the primitive Pythagorean triples. Let us make a small table:

u	v	$x = 2uv$	$y = u^2 - v^2$	$z = u^2 + v^2$
2	1	4	3	5
3	2	12	5	13
4	1	8	15	17
4	3	24	7	25
5	2	20	21	29
5	4	40	9	41

6	1	12	35	37
6	5	60	11	61
7	2	28	45	53

Note that in the last column, they appear the first primes of the form $4n + 1$: $5, 13, 17, 29, 37, 41, 53, 61$ (which are then a sum of two squares). The observation of such a table, led Fermat and his contemporaries to the discovery of the so-called "Two square theorem", enounced first by A. Girard and surely proved by Fermat (although the first published proof is from Euler), which states that any prime of the form $4n + 1$ is a sum of two squares.

On the other way, the Pythagorean triples, which Fermat studied from the Bachet version of Diophantus Arithmetic, led him to consider, more generally, the Diophantine equation:

$$x^n + y^n = z^n \ (*)$$

Fermat claimed, writing in a margin of his copy of the aforementioned text, that he has found a "truly marvelous" proof of the theorem which now bears the name of "Fermat's Last Theorem", which states that the Diophantine equation (∗) does not have solution in natural numbers if $n \geq 3$, but, he continued, the margin at which he wrote was too narrow to contain it. The marginal note of Fermat was a private observation, published only after his dead. In his correspondence with his contemporaries, Fermat only poses the cases $n = 3$ and $n = 4$.

The history of Fermat's Last Theorem, very rich and interesting, can be found in the masterpieces [12] and [35]. In a very succinct way it can be summarized as follows: Euler proves the case $n = 3$ although with some gaps, Legendre, Dirichlet and later Gauss the case $n = 5$, Lamé proves it for $n = 7$. Then Kummer comes into the scene, bringing new and more profound methods, stablishing the foundations of Algebraic Number Theory, exhaustively studying the cyclotomic fields, being able to prove the theorem for all exponents that are regular primes (characterized by not dividing the numerators of Bernoulli numbers) and also for some irregular primes. Then comes a period of deepening of Kummer's ideas and application of Class Field Theory, until arriving at present days. In the theory of Elliptic Curves it was conjectured for some time the so-called Taniyama-Shimura conjecture. P. Ribet proves that Fermat's Last Theorem follows as a consequence of this conjecture. This stimulated A. Wiles to the investigation of the conjecture, achieving its proof in 1995 after several years of work.

Next, we will prove the case $n = 4$, the easier, which is an immediate consequence of the following:

Proposition 10.2: There are no natural numbers x, y, z, such that $x^4 + y^4 = z^2$.

Proof: If there exist such triples, it must be one x, y, z, with smallest z. The minimality of z implies that each two of x, y, z are coprime, hence by the above theorem, there exist $u, v \in \mathbb{N}$, coprime, of distinct parity, with $u > v$ such that:

$$x^2 = 2uv \quad y^2 = u^2 - v^2 \quad z = u^2 + v^2$$

From the second of these: $y^2 + v^2 = u^2$, it follows that u must be odd and v even, and applying again the previous theorem, there are natural numbers s, t coprime, such that:

$$v = 2st \quad y = s^2 - t^2 \quad u = s^2 + t^2$$

but from $x^2 = 2uv$ it follows that $2v$ and u must be squares and as $2v = 4st$, it follows that s and t must be squares as well, hence there exist a, b, c natural, such that $s = a^2, t = b^2, u = c^2$ and by replacing them in $u = s^2 + t^2$, allow to obtain:

$$a^4 + b^4 = c^2$$

and as $c \leq u < z$, the minimality of z is contradicted.∎

EXERCISES

Ex. 1: For any $n \in \mathbb{N}$ we have,

 a) $4^n - 1$ is divisible by 3.

 b) $3^{2n+1} + 2^{n+2}$ is divisible by 7.

 c) $3^{2n+2} - 8n - 9$ is divisible by 64.

 d) $7^{2n} + 16n - 1$ is divisible by 64.

 e) $2^{2n+1} - 9n^2 + 3n - 2$ is divisible by 54.

 f) $3 \cdot 5^{2n+1} + 2^{3n+1}$ is divisible by 17.

 *g) $(3 + \sqrt{5})^n + (3 - \sqrt{5})^n$ is natural and divisible by 2^n.

Ex 2: a) 421 is prime.

 b) Find all primes < 200.

Ex 3: Using the Prime number theorem: approximately how many primes of one hundred digits are there?. Compare the result with the number of primes of at most one hundred digits.

Ex. 4: Assuming Bertrand's postulate, prove that if $p_1, p_2, \ldots, p_r, \ldots$ are the primes in ascending order, then:

$$p_{n+1} \leq p_1 + p_2 + \ldots + p_n \ \forall n \in \mathbb{N}$$

Ex. 5: Let $a, b \in \mathbb{Z}$ and $n \in \mathbb{N}$.

   a) $a^n - b^n$ is divisible by $a - b$.

   b) If n is odd, $a^n + b^n$ is divisible by $a + b$.

   c) If n is even, $a^n - b^n$ is divisible by $a + b$.

Ex. 6: Check Goldbach's conjecture for the number 150.

Ex. 7: Compute (a, b) and write it in the form $sa + tb$ in the following cases:

   a) $a = 240, b = 48$ b) $a = -120, b = 45$

Ex. 8: For $a, b, c \in \mathbb{Z}$ prove that:

   a) $(a, b) = d \Rightarrow (\frac{a}{d}, \frac{b}{d}) = 1$.

   b) $a(b, c) = (ab, ac)$.

   c) $(a, (b, c)) = ((a, b), c)$.

Ex. 9: If p is prime:

   a) $p > 3 \Rightarrow 24 | p^2 - 1$.

   b) $p > 5 \Rightarrow 240 | p^4 - 1$.

Ex. 10: If a and b are coprime integers, then:

   a) $(a, a + b) = 1$

   b) $(a + b, ab) = 1$

   c) $(a + b, a - b) = 1$ or 2

   d) $(a + b, a^2 - ab + b^2) = 1$ or 3.

Ex. 11: Let $a, b \in \mathbb{Z}, m, n \in \mathbb{N}$:

   a) $m | n \Rightarrow a^m - b^m | a^n - b^n$.

   *b) If $a > 1$, then: $a^m - 1 | a^n - 1 \Leftrightarrow m | n$.

   *c) If $d = (n, m)$ then $(a^n - 1, a^m - 1) = a^d - 1$.

Ex. 12: ¿True or false?:

   a) $(a, b) = d \Rightarrow (a, cb) = cd$.

   b) $a | bc$ and $a \nmid b \Rightarrow a | c$.

   c) $d = (a, b) \Rightarrow d^3 = (a^3, b^3)$.

Ex. 13: The least common multiple of two integers a, b is any integer m such that:

    1) $a \mid m$ $b \mid m$.

    2) $a \mid c$ $b \mid c \Rightarrow m \mid c$.

    3) $m \geq 0$.

Prove:

    a) If there exists a *least common multiple*, it is unique.

    b) If d is the greatest common divisor of a and b, then $\frac{ab}{d}$ is a least common multiple of a and b.

*Ex. 14: For any $n \in \mathbb{N}$, $\frac{1}{n+1}\binom{2n}{n}$ is an integer.

Ex. 15: If $(a, b) = 1$ and ab is a square, then a is a square.

Ex. 16: Prove the following assertion of E. Lucas: The problem of finding the integral solutions of:

$$1 + x + x^2 + x^3 = y^2$$

is equivalent to the solution of the system:

$$1 + x = 2u^2, 1 + x^2 = 2v^2, y = 2uv$$

Ex 17: The only numbers that are the product of its proper divisors are those of the form p^3 and pq with p, q distinct primes.

Ex. 18: a) Which are the possible values of (a^2, b^3) if $(a, b) = 4$?.

    b) and those of (a^2, b^2) if $(a, b) = p^3$ with p prime?.

Ex. 19: Let $a, n \in \mathbb{N}$.

    a) If $a \geq 2$ and $a^n + 1$ is prime, then a is even and $n = 2^m$ with $m \in \mathbb{N} \cup \{0\}$.

    b) If $n \geq 2$ and $a^n - 1$ is prime, then $a = 2$ and n is prime.

Ex. 20: There are no nonzero integers x, y such that $x^2 = 10y^2$.

Ex. 21: a) For any $n \in \mathbb{N}$, there are allways a prime p such that $n \leq p \leq n! + 1$.

    b) There exist infinite primes of the form $4n + 3$ (imitating Euclid's proof, if $p_1, \ldots, p_n$ were all the primes of that form except for 3, prove that $4p_1 \ldots p_n + 3$ is divisible by a prime of that form. Note that the similar argument, changing 3 by 1 does not work to prove the infinitude of primes of the form $4n + 1$).

    c) There are infinite primes of the form $6n + 5$.

Ex. 22: Two positive integers are amicable if each is the sum of the proper divisors of the other. The Pythagorean known the pair

220, 284. In the 9th century, Arab mathematician Thâbit ben Korrah noted that if $h = 3 \cdot 2^n - 1, k = 3 \cdot 2^{n-1} - 1$ and $l = 9 \cdot 2^{n+1} - 1$ are all prime with $n > 1$, then $2^n hk$ and $2^n l$ are amicable. Prove it.

Ex 23: An even perfect number cannot be of the form p^r nor of the form $p^r q^s$ where p, q are distinct primes and r, s natural numbers.

CHAPTER 5
RINGS OF RESIDUES

With modular arithmetic an infinite number of examples of rings and fields are available, making natural the introduction of these algebraic structures. The consideration of the invertible elements in the rings of residues leads to the theorems of Fermat-Euler and Wilson. The first-degree equation, some linear systems (Chinese remainder theorem), and the second-degree equation, with the statement of the Quadratic Reciprocity Law, are studied in these rings of classes of remainders.

1 - CONGRUENCES

Let $m \in \mathbb{N}$. If $a, b \in \mathbb{Z}$, we say that a is *congruent* to b to modulus m and we write:

$$a \equiv b \pmod{m}$$

iff m is a divisor of $a - b$.

Usually, a new notation is introduced, to condense into a few symbols a rather long expression, the notation of congruence does not meet that rule since it is even shorter to write $m \mid a - b$ than $a \equiv b \pmod{m}$. However, this notation, introduced by Gauss, is suggestive and useful as shown by the following:

Proposition   1.1:   Let $m \in N$. Whichever be the integers a, b, c, d, the following properties are valid:

1) (Reflexive) $a \equiv a \pmod{m}$

2) (Symmetric) $a \equiv b \pmod{m} \Rightarrow b \equiv a \pmod{m}$

3)   (Transitive) $a \equiv b \pmod{m}$ and $b \equiv c \pmod{m} \Rightarrow a \equiv c \pmod{m}$

4) (Compatibility with the sum) $a \equiv c \pmod{m}$ and $b \equiv d \pmod{m} \Rightarrow a + b \equiv c + d \pmod{m}$

5) (Compatibility with the product) $a \equiv c(modm)$ and $b \equiv d(modm) \Rightarrow ab \equiv cd(modm)$

6) If $n \in \mathbb{N}$ and $a \equiv b(modm)$, then $a^n \equiv b^n(modm)$

7) (Cancellation) $ac \equiv bc(modm)$ and $(c,m) = 1 \Rightarrow a \equiv b(modm)$

8) $a \equiv b(modm) \Leftrightarrow$ a and b have the same remainder when divided by m.

Proof: We will prove the last four, leaving the others as exercises.

5) If $a \equiv c(modm)$ and $b \equiv d(modm)$, then $a = c + mq, b = d + mq'$ for some integers q, q'. Hence $ab = cd + m(qd + cq' + mqq')$, so $m \mid ab - cd$, or $ab \equiv cd(modm)$.

6) Follows from 5) by induction on n.

7) $ac \equiv bc(modm)$ means $m \mid (a - b)c$, then as $(c,m) = 1$ by prop. 5.3, Chap. 4, results $m \mid a - b$, or $a \equiv b(modm)$.

8) Let r be the remainder when dividing a in by m and r' the one when dividing b by m, in such a way that:
$$a = mq + r, 0 \le r < m, b = mq' + r', 0 \le r' < m$$
and, say, $r \ge r'$. Hence $a - b = m(q - q') + r - r'$ with $0 \le r - r' < m$, then,
$$a \equiv b(modm) \Leftrightarrow m \mid a - b \Leftrightarrow m \mid r - r' \Leftrightarrow r - r' = 0. \blacksquare$$

Note that the cancellation property is not valid without the restriction $(c,m) = 1$, for example, we have $1 \cdot 2 \equiv 3 \cdot 2(mod4)$, but it is not true that $1 \equiv 3(mod4)$.

To appreciate the usefulness of the concept of congruence, we will see, in what follows, some illustrative examples.

Example: Find the remainder in the division of 7^{4510} by 15. We have $7^2 \equiv 4(mod15)$, then:
$$7^4 \equiv 16 \equiv 1(mod15)$$
As $4510 = 4 \cdot 1127 + 2$, results:
$$7^{4510} = (7^4)^{1127} \cdot 7^2 \equiv 7^2 \equiv 4(mod15)$$
so 4 is the remainder searched.

Example: We have proved in 6, Chap. 4, that two distinct Fermat numbers are coprime. Let us prove it in another way. Let $F_n = 2^{2^n} + 1$ and $F_m = 2^{2^m} + 1$ with $n > m$, and let d be a common divisor of F_n and F_m, then: $F_n \equiv 0, (mod\ d)$ and $F_m \equiv 0(mod\ d)$, or $2^{2^n} \equiv -1(mod\ d)$ and $2^{2^m} \equiv -1(mod\ d)$. Putting $n = m + r$, we have:

$$2^{2^n} = \left(2^{2^m}\right)^{2^r} \equiv (-1)^{2^r} = 1(mod\ d)$$

but as $2^{2^m} \equiv -1(mod\ d)$, results in $1 \equiv -1(mod\ d)$, that is $d \mid 2$ and as the Fermat numbers are even, we must have $d = \pm 1$ and $(F_n, F_m) = 1$.

Example: Let us show that the Lucas sequence, defined in 6, Chap. 4, i.e., the sequence defined by:

$$a_{n+1} = a_n^2 - 2$$

has its terms pairwise coprime if we start with a_1 odd. Let a_m, a_n be two terms of that sequence, with $n > m$, and let d be a common divisor of a_m and a_n. Then:

$$a_{m+1} = a_m^2 - 2 \equiv -2(mod\ d);$$
$$a_{m+2} = a_{m+1}^2 - 2 \equiv 4 - 2 = 2(mod\ d);$$
$$a_{m+3} = a_{m+2}^2 - 2 \equiv 2(mod\ d),$$

and so on. Hence $a_n \equiv \pm 2(mod\ d)$, but as $d \mid a_n$, it must be $d \mid 2$ and, being a_1 odd, all the terms of the sequence are odd, hence $d = \pm 1$ and $(a_n, a_m) = 1$.

Example: (Divisibility criteria) The widely known divisibility criteria by 3 and by 9, in decimal system: a number is divisible by 3 (by 9) if, and only if, the sum of its digits is divisible by 3 (by 9), can be justified as follows:

Take a number writing in decimal system: $a = a_n 10^n + \cdots + a_0$ with the a_i dígits. As $10 \equiv 1(mod\ 3)$ (an also modulus 9), we have $10^2 \equiv 1(mod\ 3)$; $10^3 \equiv 1(mod\ 3)$; and in general, $10^r \equiv 1(mod 3)$ $\forall r \in \mathbb{N}$, then $a_r 10^r \equiv a_r(mod\ 3)$ $\forall r$ such that $0 \leq r \leq n$; and adding:

$$a \equiv a_n + a_{n-1} + \ldots + a_0(mod 3)$$

hence $a \equiv 0(mod 3) \Leftrightarrow \equiv a\_n + a\_(n - 1) + \ldots + a_0(mod 3)$ or, a is divisible by 3 if, and only if, the sum of its digits is divisible by 3 (similarly by 9).

In the same manner, divisibility criteria by other numbers and in any numeration system can be obtained.

Let us find a criterion of divisibility by 11 in the decimal system. We have:

$$10 \equiv -1(mod\,11);$$
$$10^2 \equiv 1(mod\,11);$$
$$10^3 \equiv -1(mod\,11);$$

etc., so if $a = a_n 10^n + \cdots + a^0$, then:

$$a \equiv a_0 - a_1 + a_2 - \ldots \ldots (mod\,11)$$

hence a number is divisible by 11 if, and only if, the alternating sum of its digits is divisible by 11.

2 - EQUIVALENCE RELATIONS

Let A be a set. An *equivalence relation* in A is a relation $\sim$ from A to A, such that whatever be $a, b, c \in A$:

1) $a \sim a$.

2) $a \sim b \Rightarrow b \sim a$.

3) $a \sim b$ and $b \sim c \Rightarrow a \sim c$.

So the congruences modulus m, are equivalence relations in $\mathbb{Z}$; the parallelism between lines in a plane, is an equivalence relation in the set of all lines in the plane, provided we consider a line parallel to itself; having the same parents is an equivalence relation between people.

If $\sim$ is an equivalence relation in A, we define for each $a \in A$, the *class* of a by $\sim$ as the following subset of A:

$$\bar{a} = \{b \in A / b \sim a\}$$

For example, if the equivalence relation is the congruence modulus 2, we have:

$$\bar{0} = \{\ldots, -4, -2, 0, 2, 4, \ldots\}$$
$$\bar{1} = \{\ldots, -3, -1, 1, 3, 5, \ldots\}$$
$$\bar{2} = \{\ldots, -4, -2, 0, 2, 4, \ldots\}$$

we have then only two classes: odd numbers and even numbers.

If the relation is the parallelism, the class of a line is the set of lines which have the same direction that the given line.

If the relation is that of having the same parents, the class of a person is the set of his or her brothers and sisters (he or she included).

108

If the relation, in the set of the days, is to differ in a multiple of seven days, the class of $16 - 07 - 2017$ is the set of Sundays.

If in a family of finite sets, the relation is the existence of a bijection, (if there exist a bijection between two sets we said that they are coordinable), the class of a set is the family of sets having the same number of elements as the given one.

Usually in mathematics, the equivalence relations are used to define mathematical objects, for example, to define the direction of a line we could say that it is the class of that line; that is, the set of the lines that are parallel to it. Or to define the cardinal of a set, we could say that it is the family of all sets coordinable with it. Extending this beyond the realm of mathematics, we could define "Sunday" as the class of July 16th, 2017, or "family" (in a restricted sense) of a person by the class of that person by the obvious equivalence relation above.

Returning to the general case, we have:

Proposition 2.1: If $\sim$ is an equivalence relation in a set A and $a, b \in A$, then,

$$a \sim b \Leftrightarrow \bar{a} = \bar{b}$$

Proof: ($\Rightarrow$) Let $c \in \bar{a}$, that is $c \sim a$. As $a \sim b$ it follows $c \sim b$, so $c \in \bar{b}$. We have proved: $\bar{a} \subset \bar{b}$. By symmetry results $\bar{b} \subset \bar{a}$.

($\Leftarrow$) As $a \in \bar{a}$ (since $a \sim a$) and $\bar{a} = \bar{b}$, we have $a \in \bar{b}$, then $a \sim b$.■

Corollary 2.2: With the same notations as in 2.1, two distinct classes are disjoint.

Proof: $a \cap b \neq \emptyset \Rightarrow \exists c \in a \cap b$, then $c \sim a$ and $c \sim b$ hence $a \sim b$, and by the above proposition results $a = b$.■

Given an equivalence relation $\sim$ in A, the *quotient set* of A by the relation $\sim$ is the set of all equivalence classes and it is denoted by $\frac{A}{\sim}$. That is:

$$\frac{A}{\sim} = \{\bar{a}/a \in A\}$$

where $\bar{a}$ is the class of $\bar{a}$ by $\sim$.

Thus, in the case of parallelism, the quotient set is the set of directions in the plane. In the case of the days, the quotient set has seven elements, and can be thought as the week.

In the case of the congruences modulus 2, the quotient set $(Z/(\equiv (2)))$ have two classes 0 and 1, i.e., that of the even and that of the odd numbers. Since the sum of two even numbers is even, the sum of an even and an odd number is odd, etc., we can symbolize those relations by the following tables of sum and product:

+	$\bar{0}$	$\bar{1}$
$\bar{0}$	$\bar{0}$	$\bar{1}$
$\bar{1}$	$\bar{1}$	$\bar{0}$

$\cdot$	$\bar{0}$	$\bar{1}$
$\bar{0}$	$\bar{0}$	$\bar{0}$
$\bar{1}$	$\bar{0}$	$\bar{1}$

where, for example, $\bar{1} + \bar{1} = \bar{0}$ means that the sum of two odd numbers is even.

It is routine then to verify that $\frac{\mathbb{Z}}{\sim}$ with the operations defined by the tables above, satisfies the properties S.1, ..., D (that is all of the properties of group I) stated for the real numbers, and so that quotient set becomes a field, as it will be defined in the next section.

Returning again to the general case, let $\sim$ be an equivalence relation in   a   set A. Moreover, assume A finite and non — empty. If $\overline{a_1}, \ldots, \overline{a_r}$ are the distinct classes, we have:

$$A = \bigcup_{i=1}^{r} \overline{a_i}$$

According to the above corollary, the classes are pairwise disjoint, then:

$$\#(A) = \sum_{i=1}^{r} \#(\overline{a_i})$$

In case that all classes have the same number m of elements, we have:

$$\#(A) = rm$$

thus the next proposition was proved:

Proposition 2.3: If $\sim$ is an equivalence relation in a finite, non-empty, set A such that all clases have the same number m of elements, then:

$$\# \left(\frac{A}{\sim} \right) = \frac{\#(A)}{m}. \blacksquare$$

Example: How many numbers can be formed permuting the ciphers of 1234454?

Putting sub-indices to the repeated digit, we can count, in the first place, the number of 7-uples formed reordering the seven objects: $1, 2, 3, 4_1, 5, 4_2, 4_3$. That number is 7! Each of those 7 $-$ uples determines a number, for example, the 7 $-$ uple $4_2 5 1 4_3 2 4_1 3$ determines the number 4514243. If we say that two of those 7- uples are equivalent iff they determine the same number, we have clearly an equivalence relation and each equivalence class has 3! elements. For example, the class of the above 7- uple have the following elements: $4_2 5 1 4_3 2 4_1 3$; $\qquad\qquad\qquad 4_2 5 1 4_1 2 4_3 3$; $4^1 5 1 4^2 2 4^3 3$; $4^1 5 1 4^3 2 4^2 3$; $4_3 5 1 4_1 2 4_2 3$ and $4_3 5 1 4_2 2 4_1 3$. Hence by prop. 2.3, when permuting the ciphers of 1234454 we obtain $\frac{7!}{3!}$ numbers.

3 - RINGS AND FIELDS

In this section, we define two algebraic structures: rings and fields, notions widely distributed throughout mathematics. The concept of field was implicit in the works of Abel and Galois as, in modern words, the field generated by the coefficients of a polynomial equation and the field obtained adjoining its roots. The concept of "ring" comes from two sources, in the progressive abstraction process along the 19th century, the rings of polynomials and the rings, called maximal orders by Dedekind, of algebraic integers in number fields. The name ring appears as "Zahlring," referring to rings of algebraic integers, with Hilbert at the end of the century. The ubiquity of the concept is notable: it appears in the study of polynomials, algebraic integers as we said, but also in series, matrices, functions, quaternions, octonions, Boolean structures, modular arithmetic, etc.

We will call *ring* to any triple $(A, +, \cdot)$ formed by a set A and two operations (called sum and product), $+ : A \times A \rightarrow A : (a, b) \mapsto a + b$ and $\cdot : A \times A \rightarrow A : (a, b) \mapsto ab$, such that the following properties are satisfied:

Properties of the sum:

S.1- (Associative) $(a + b) + c = a + (b + c)$ for any $a, b, c \in A$.

S.2- (Commutative) $a + b = b + a$ for any $a, b \in A$.

S.3- (Existence of neutral element) There exists an element in A, denoted by 0, such that $a + 0 = a \; \forall a \in A$.

S.4- (Existence of additive inverse) For each $a \in A$, there exists $a' \in A$ such that $a + a' = 0$.

Properties of the product:

P.1- (Associative) $(ab)c = a(bc)$ for any $a, b, c \in A$.

P.2- (Commutative) $ab = ba$ for any $a, b \in A$.

P.3- (Existence of neutral element) There exists an element in A, denoted by 1, such that $1 \neq 0$ and $a1 = a \; \forall a \in A$.

Distributive: D- $a(b + c) = ab + ac$ for any $a, b, c \in A$.

It is convenient to clarify, that the definition of ring just given is somewhat restrictive. Usually, it is called ring a set with operations that match the enounced properties of the sum, but without requiring the commutativity of the product (in which case the distributive property becomes two: one on the left and one on the right); quaternions and matrices, for example, are not commutative. Also, it is appropriate, in some cases, to leave out the existence of a neutral element of the product and even the associativity of the product; octonions, for example, are not associative. We will use the word ring only in the more restricted sense mentioned above, unless we specify something else.

In what follows we refer to a ring $(A, +, \cdot)$, simply as the ring A without explicit mention of the operations.

If in a ring A it is valid:

P.4- (Existence of multiplicative inverse) For each $a \in A$ such that $a \neq 0$, there exists $a'' \in A$ such that $a''a = 1$,

then it is said that A is a *field*.

If in a ring A is valid the property:

$$ab = 0 \Rightarrow a = 0 \text{ ó } b = 0$$

then it is said that A is an *integral domain* or, simply, a *domain*.

The properties stated at the beginning of chapter 2, which do not depend on P.4. nor of the order properties, are valid in any ring, since for its proofs only were used the formal properties defining a ring and not the nature of the elements.

In particular, in any ring A the opposite of $a \in A$ is unique and is denoted $-a$; the notation: $b - a = b + (-a)$ is adopted too; it is valid $a0 = 0$ for any $a \in A$; it is defined as above a^n for $n \in \mathbb{N} \cup \{0\}$; also it is defined na for $n \in \mathbb{N}$ and $a \in A$ inductively by $1a = a$ and $(h + 1)a = ha + a$ and with this notations the binomial expansion is valid (this is not so in a noncommutative ring).

In the same way, if A is a field there are valid, among other, the properties of exercise B of chapter 2; the multiplicative inverse of $a \neq 0$ is unique and is denoted a^{-1}, ba^{-1} is also written $\frac{b}{a}$, etc.

Proposition 3.1: Any field is a domain.

Proof: If $ab = 0$ and $a \neq 0$, by P.4 there exists a'' such that $a''a = 1$, hence:

$$0 = a''(ab) = (a''a)b = 1b = b. \blacksquare$$

Before proving that the reciprocal is false, i.e., that there are domains that are not fields, it is advisable to set another definition.

If A is a ring, a *subring* of A is a subset B of A such that with the operations restricted, B is a ring. Then in order to verify that a subset B of A is a subring of A, it is not necessary to verify the associative properties (of both the sum and the product), nor the commutative ones, nor the distributive, since they are valid in any subset of A, but only the following:

    a) $a, b \in B \Rightarrow a + b \in B$

    b) $a, b \in B \Rightarrow ab \in B$

    c) $0 \in B$

    d) $a \in B \Rightarrow -a \in B$

    e) B has a neutral element of the product.

The first two, express that the restrictions of the operations to B, are operations in B. Regarding the third, at first sight the neutral element of the sum in B, could not be the same as the one of A, but if 0 were a neutral element of the sum in B, taking $b \in B$ we would have $b + 0 = b + 0$, from where, adding $-b$, results in

0=0. Note that a similar reasoning is not valid for the neutral element of the product, because a multiplicative inverse is not available, and, in fact, there are examples of subrings in which its neutral element does not coincide with that of the ring. Finally, regarding the fourth, note that if $a' \in B$ is an opposite or additive inverse of a, necessarily we have $a' = -a$.

It follows from 1.1, Chap. 4 that $\mathbb{Z}$ is a subring of $\mathbb{R}$ with the same neutral element of the product (since $1 \in \mathbb{Z}$) and, it is also a domain, as the property $ab = 0 \Rightarrow a = 0$ or $b = 0$ is valid in any subset of a field. As $\mathbb{Z}$ is not a field, for to be a field it would be necessary that the multiplicative inverse of each nonzero integer, would belong to $\mathbb{Z}$; hence the reciprocal of the above proposition is false.

In a similar way, a *subfield* of a field K, is a subset F of K such that with the restriction of the operations, it is a field. Then F must be a subfield of K, but now, as a multiplicative inverse of each nonzero element is available, we can conclude, as in the additive case, that the neutral element of F must be the same as that of K and that if b is the multiplicative inverse in F of $a \neq 0$, it must be $b = a^{-1}$. Hence, F is a *subfield* of K if, and only if, it verify the properties a, b, c, d of subrings and, in addition:

e') $1 \in F$

f) $a \in F$ and $a \neq 0 \Rightarrow a^{-1} \in F$.

The examples of rings that we have until now are, the field $\mathbb{R}$ of real numbers and the domain $\mathbb{Z}$ of integers. We conclude this section with another example of a ring.

Example: Let $X \neq \emptyset$ be a set. In $P(X)$, the set of subsets or parts of X, we define a product as the intersection and a sum as the symmetric difference $\triangle$:

$$A \triangle B = (A - B) \cup (B - A) = (A \cup B) - (A \cap B)$$

then it is straightforward to verify that $P(X)$ with the above operations is a ring, with X as the neutral element of $\cap$ and $\emptyset$ as the neutral element of $\triangle$. Furthermore, $-A = A^{\wedge'}$ (the complement of A)and $P(X)$is a field or an integral domain only in case X is a set with a single element (verify). In addition, $P(X)$ is a boolean ring, that is to say, every element A verifies $A^2 = A$, i.e.: $A \cap A = A$.

By prop. 1.1., the congruence modulus m is an equivalence relation in $\mathbb{Z}$, so we can form the quotient set $\dfrac{\mathbb{Z}}{\equiv(m)}$, briefly denoted $\mathbb{Z}_m$. In this quotient set, we define operations:

$$\oplus: \mathbb{Z}_m \times \mathbb{Z}_m \to \mathbb{Z}_m \quad \odot: \mathbb{Z}_m \times \mathbb{Z}_m \to \mathbb{Z}_m$$

as follows:

$$a \oplus b = a + b \quad a \odot b = ab \quad (1)$$

These tell us that the remainder, or residue, in the division by m of the sum (product) of two integers, is the remainder of the sum (product) of the remainders. These are, effectively, operations in $\mathbb{Z}_m$, as they do not depend on the election of the representatives of the classes, but only on the classes themselves.

The fact that $\oplus$ is a function is expressed by:

$$\bar{a} = \bar{c} \text{ and } \bar{b} = \bar{d} \Rightarrow \bar{a} \oplus \bar{b} = \bar{c} \oplus \bar{d} \quad (2)$$

but, since by prop.2.1: $\bar{a} = \bar{b}$ is equivalent to $a \equiv b(mod\ m)$, (2) is translated as:

$$a \equiv c(mod m) \text{ and } b \equiv d(mod m) \Rightarrow a + b \equiv c + d(mod m)$$

which is nothing more than compatibility with the sum (prop. 1.1.). Similarly, $\odot$ is well defined, i.e., it is a function.

Proposition 4.1: $(\mathbb{Z}_m, \oplus, \odot)$ is a ring (the *ring of residues modulo m*).

Proof: We have already seen that $\oplus$ and $\odot$ are functions. Let us check that $\oplus$ is associative:

$$(a \oplus b) \oplus c = a + b \oplus c = (a + b) + c = a + (b + c)$$
$$= a \oplus b + c = a \oplus (b \oplus c)$$

where we have used the definition of $\oplus$ and the associativity of the sum in $\mathbb{Z}$. Similarly, the commutativity of $\oplus$, the associativity and commutativity of $\odot$, and the distributivity are proved. Also, $\bar{0}$ is the neutral element of $\oplus$, $\bar{1}$ that of $\odot$ and $-\bar{a}$ is the opposite of $\bar{a}$.∎

In what follows, we will write simply $+$ and $\cdot$ to denote $\oplus$ and $\odot$ respectively, as there can be no confusion with the sum and product of integers, since according to the elements to which they are applied (integers or classes of integers), it will be clear to which sum or product we refer.

Proposition 4.2: $\mathbb{Z}_m = \{\bar{0}, \bar{1}, \overline{m-1}\}$

Proof: Let $a \in \mathbb{Z}_m$ with $a \in \mathbb{Z}$. By the Division Algorithm, there are integers q, r such that: $a = mq + r, 0 \leq r < m$, so $\bar{a} = \bar{r}$. Moreover, the elements $\bar{0}, \bar{1}, \ldots, \overline{m-1}$ are all distinct, since if $\bar{r} = \bar{s}$ with $0 \leq s \leq r \leq m - 1$, then $r \equiv s (mod\ m)$, i.e., $m \mid r - s$, hence $r - s = 0$ or $m \leq r - s$, but as $r - s \leq m - 1$, results $r = s$. $\mathbb{Z}_m$ has, then, exactly m elements.∎

Any set of m integers $\{a_1, \ldots, a_{m-1}\}$ such that all classes in $\mathbb{Z}_m$ are represented, i.e., such that:
$$\{\overline{a_1}, \ldots, \overline{a_m}\} = \mathbb{Z}_m$$
is often called a *complete system of representatives mod m*. So then, by the above proposition, $\{\bar{0}, \bar{1}, \ldots, \overline{m-1}\}$ is a complete system of representatives mod m.

$\mathbb{Z}_m$ does not have to be a field, not even a domain. For example in $\mathbb{Z}_4 = \{\bar{0}, \bar{1}, \bar{2}, \bar{3}\}$ we have $\bar{2} \cdot \bar{2} = \bar{4} = \bar{0}$ (since $4 \equiv 0 (mod\ 4)$) and $\bar{2} \neq \bar{0}$, for if not we would have $4 \mid 2$ which is absurd. However in $\mathbb{Z}_m$ as in any ring, it is interesting to know the invertible elements.

An element a of a ring A is said to be *invertible* or an *unity*, iff there exists $b \in A$ such that $ab = 1$. We denote by $U(A)$ to the set of invertible elements of A. If A is a field, we have by definition, $U(A) = A - \{0\}$. According to 3.1 Chap.4, $U(\mathbb{Z}) = \{1, -1\}$. The invertible elements of $\mathbb{Z}_m$ are characterized by:

Theorem 4.3: $a \in U(\mathbb{Z}_m) \Leftrightarrow (a, m) = 1$.

Proof: $a \in U(\mathbb{Z}_m) \Leftrightarrow \exists\ b \in \mathbb{Z}_m$ such that $a \cdot b = ab = 1 \Leftrightarrow$ exist $b, q \in \mathbb{Z}$ such that $ab = 1 + mq \Leftrightarrow (a, m) = 1$.∎

For example, the invertible elements of $\mathbb{Z}_{12}$ are: $\bar{1}, \bar{5}, \bar{7}$ and $\overline{11}$. In particular, $\mathbb{Z}_{12}$ is not a field. The following theorem characterizes those $\mathbb{Z}_m$ which are fields and which are domains.

Theorem 4.4: The following conditions on $m \in \mathbb{N}, m \geq 2$, are equivalent:

1) $\mathbb{Z}_m$ is a field.

2) $\mathbb{Z}_m$ is a domain.

3) m es prime.

Proof: 1) $\Rightarrow$ 2) since any field is a domain (prop.3.1).

2) $\Rightarrow$ 3) for if m is not prime, then by 4.1, Chap. 4, there exist $a, b \in \mathbb{Z}$ such that $m = ab$ with $1 < a, b < m$, then $\bar{a} \cdot \bar{b} = \bar{m} = \bar{0}$ with $\bar{a} \neq \bar{0}$ (if not we would have $m \mid a$ and, therefore, $m \leq a$) and $\bar{b} \neq \bar{0}$, hence $\mathbb{Z}_m$ would not be a domain.

3) $\Rightarrow$ 1) Let $\bar{a} \neq \bar{0}$, i.e., $m \nmid a$, from where, being m prime, follows $(a, m) = 1$, and by the above theorem a is invertible.∎

If $\phi(m)$ denotes the number of invertible elements of $\mathbb{Z}_m$, we will have by theorem 4.3.:

$$\phi(m) = \# \left(U(\mathbb{Z}_m)\right) = \# \{a \in N / (a, m) = 1 \text{ and } a < m\}$$

where $\#(A)$ is the cardinal or number of elements of A.

The function $\phi: \mathbb{N} \to \mathbb{N}$ defined by $m \mapsto \phi(m)$ is called *Euler's totient function* or *Euler's phi function*.

For example, $\phi(12) = 4$ since as we saw above Z_{12} has four invertible elements: $\bar{1}, \bar{5}, \bar{7}$ and $\overline{11}$. We will now show how to compute $\phi(m)$ without having to found the invertible elements of $\mathbb{Z}_m$.

Let us see first, how to compute $\phi(m)$ when m is a power of a prime. Let $m = p^\alpha$ with p prime and $\alpha \in \mathbb{N}$. To compute $\phi(p^\alpha)$, from the p^α natural numbers $\leq p^\alpha$, we count those that are not coprime with p^α, that is the multiples of p. They are $p, 2p, \ldots p^{\alpha-1}p$, so there are $p^{\alpha-1}$ of them. Then the number of those which are coprime with p^α is:

$$\phi(m) = p^\alpha - p^{\alpha-1} = p^{\alpha-1}(p-1) \text{ (3)}$$

This relation combined with the following theorem, will give us a method to compute $\phi(m)$.

Theorem 4.5: $(m, n) = 1 \Rightarrow \phi(mn) = \phi(m)\phi(n)$.

Proof: As $\phi(mn) = \# \left(U(\mathbb{Z}_{mn})\right)$ and

117

$$\phi(m)\phi(n) = \# \, (U(\mathbb{Z}_m)) \cdot \# \, (U(\mathbb{Z}_n)) = \# \, (U(\mathbb{Z}_m) \times (U(\mathbb{Z}_n))$$

will be enough proving that there is a bijection between $U(\mathbb{Z}_{mn})$ and $U(\mathbb{Z}_m) \times U(\mathbb{Z}_n)$.

Denoting by $-, \wedge, \sim$ the classes modulus nm, m, n respectively, define:

$$f: U(\mathbb{Z}_{mn}) \to U(\mathbb{Z}_m) \times U(\mathbb{Z}_n)$$

by $f(\bar{a}) = (\hat{a}, \tilde{a})$. We will show that f is a bijection. In first place, as:

$$(a, nm) = 1 \Leftrightarrow (a, m) = (a, n) = 1$$

we have that, really, f applies $U(\mathbb{Z}_{mn})$ in $U(\mathbb{Z}_m) \times U(\mathbb{Z}_n)$.

Moreover, $\bar{a} = \bar{b} \Leftrightarrow mn \mid a - b \Leftrightarrow m \mid a - b$ and $n \mid a - b \Leftrightarrow \hat{a} = \hat{b}$ and $\tilde{a} =$
$\tilde{b}$ (where in the second double implication we use the hypothesis $(m, n) = 1$). So:

$$\bar{a} = \bar{b} \Leftrightarrow \hat{a} = \hat{b} \text{ and } \tilde{a} = \tilde{b}$$

This proves that f is well defined, i.e., is a function (by $\Rightarrow$) and also proves that f is injective (by $\Leftarrow$).

It remains to verify that f is surjective, that is to say, to verify that given an arbitrary element $(\hat{b}, \tilde{c})$ in $U(\mathbb{Z}_m) \times U(\mathbb{Z}_n)$, there exists $\bar{a} \in U(\mathbb{Z}_{mn})$ such that $f(\bar{a}) = (\hat{b}, \tilde{c})$. As $(m, n) = 1$, there exist $s, t \in \mathbb{Z}$ such that $1 = sm + tn$, then $b - c = s'm + t'n$ where $s' = (b - c)s$ and $t' = (b - c)t$, hence $b - s'm = c + t'n$ and denoting by a to this common value, we obtain $a \equiv b(mod\ m)$ and $a \equiv c(mod\ n)$, i.e.: $f(\bar{a}) = (\hat{b}, \tilde{c})$. .∎

Example: Let us compute $\phi(540)$. As $540 = 2^2 \cdot 3^3 \cdot 5$ we have:

$$\phi(540) = \phi(2^2 \cdot 3^3 \cdot 5) = \phi(2^2 \cdot 3^3)\phi(5) = \phi(2^2)\phi(3^3)\phi(5)$$
$$= (2^2 - 2)(3^3 - 3^2)(5 - 1) = 144$$

and so, there are 144 invertible elements in $\mathbb{Z}_{540}$.

Corollary   **4.6**:   If $n = p_1{}^{\alpha_1} \ldots p_s{}^{\alpha_s}$ with the p_i distinct primes and the α_i natural numbers, then:

$$\phi(n) = n \left(1 - \frac{1}{p_1}\right) \ldots \left(1 - \frac{1}{p_s}\right) (4)$$

Proof: By (3) and the preceding theorem, we have:

$$\phi(n) = \phi(p_1{}^{\alpha_1})\ldots\phi(p_s{}^{\alpha_s}) = p_1{}^{\alpha_1}\left(1 - \frac{1}{p_1}\right)\ldots p_s{}^{\alpha_s}\left(1 - \frac{1}{p_s}\right) =$$
$$n\left(1 - \frac{1}{p_1}\right)\ldots\left(1 - \frac{1}{p_s}\right).\blacksquare$$

5 - FERMAT-EULER THEOREM

Theorem 5.1: (Euler) $a \in U(\mathbb{Z}_m) \Rightarrow \bar{a}^{\phi(m)} = \bar{1}$.

Proof: As $U(\mathbb{Z}_m)$ has $\phi(m)$ elements, we can write:

$$(U\mathbb{Z}_m) = \{\bar{1}, \bar{a}_2, \ldots, \overline{a_{\phi(m)}}\}, (1)$$

The elements $\overline{a_1}, \bar{a}\bar{a}_2, \ldots, \bar{a}\bar{a}_m$ in $U(\mathbb{Z}_m)$ are all distinct, for if $\bar{a}\bar{a}_i = \bar{a}\bar{a}_j$, multiplying by the inverse of $\bar{a}$, which exists by the hypothesis, we have $\bar{a}_i = \bar{a}_j$, then:

$$U(\mathbb{Z}_m) = \{\bar{a}\bar{1}, \bar{a}_2, \ldots, \overline{a_{\phi(m)}}\} (2)$$

Hence the product of the elements in (1), must coincide with the product of the elements in (2):

$$\overline{1a_2}\ldots\overline{a_{\phi(m)}} = \bar{a}\bar{a}\bar{a}_2\ldots\bar{a}\bar{a}_{\phi(m)} = \bar{a}_{\phi(m)}\bar{1}\bar{a}_2\ldots\bar{a}_{\phi(m)}$$

from where, being $a_2, \ldots, \bar{a}_{\phi(m)}$ invertible, results $\bar{a}_{\phi(m)} = 1.\blacksquare$

Corollary 5.2: (Fermat) If p is prime and a is an integer such that $p \nmid a$, then:

$$a^{p-1} \equiv 1(mod\,p)$$

Proof: If $p \nmid a$, we have $a \neq 0$ in Z_p and as $\phi(p) = p - 1$, this corollary is a particular case of the preceding theorem.$\blacksquare$

Corollary 5.2., called Fermat's Little Theorem, was obtained by Fermat along his research on perfect numbers. Actually, he obtains it first for $a = 2$, the case relevant for the study of perfect numbers, and later he generalized it. Which is the relationship between Fermat's Little Theorem and perfect numbers?. Recall that in order to determine the even perfect numbers, it is sufficient to find the Mersenne numbers, $M_p = 2_p - 1$ with p prime, such that they are prime. If q is a prime dividing M_p, we have:

$$q \mid 2_p - 1 \ (3)$$

Moreover, by Fermat's theorem, we have:

$$q \mid 2^{q-1} - 1 \ (4)$$

Putting $d = (p, q-1)$, by exercise 11 c, Chap. 4, we have: $(2^p - 1, 2^{p-1} - 1) = 2^d - 1$, then, by (3) and (4), we obtain: $q \mid 2^d - 1$, hence $d > 1$ so, being p prime, it must be $d = (p, q-1) = 1$ or p, and it follows that $(p, q-1) = p$, that is $p \mid q - 1$, or $q = hp + 1$ for some $h \in \mathbb{Z}$, but, assuming that $p \neq 2$, if h were odd, q would be even which contradicts (3), hence $h = 2k$ must be even. We have proved:

Corollary 5.3: If p, q are prime such that $q \mid M_p = 2^p - 1$, then $\exists k \in \mathbb{Z}$ such that:

$$q = 2kp + 1. \blacksquare$$

Example: Let us show that M_{29} is composite. By corollary 5.3, a prime factor q of M_{29} must be such that:

$$q = 2k29 + 1 = 58k + 1$$

Given values to k, we obtain the possible prime divisors of M_{29}: $k = 1$; $q = 59$, but, as we will see $59 \nmid M_{29}$.

$k = 2$; $q = 117$ which is not prime.

$k = 3$; $q = 175$ which is not prime.

$k = 4$; $q = 233$ and as we will see $233 \mid M_{29}$.

The checks $59 \nmid M_{29}$ and $233 \mid M_{29}$ are easily made by modular arithmetic. In fact,

$2^{29} = (2^6)^4 2^5 = 5^4 2^5 = (5^2 2)^2 2^3 \equiv (-9)^2 8 \equiv 22 \cdot 8 = 176 \equiv -1 (mod 59)$

then $59 \mid 2^{29}+1$ and so $59 \nmid 2^{29}-1 = M\_29$.

Similarly, we have,

$2^{29} = (2^8)^3 2^5 \equiv 23^3 2^5 = 23^2 \cdot 23 \cdot 32 \equiv 63 \cdot 37 = 2331 \equiv 1 (mod 233)$

then $233 \mid 2^{29}-1 = M_{29}$.

By Fermat's theorem, given a prime p and an integer a coprime with p, we have a^(p-1)≡1(modp), then, by well ordering, there exists a smallest natural r such that $a^r \equiv 1 (mod p)$; r is called in such a case, the order of a, modulus p and we have:

Proposition 5.4: If r is the order of a modulus p and if $a^s \equiv 1 (mod p)$, then $r \mid s$.

Proof: There are integers q, r' such that $s = rq + r'$ and $0 \leq r' < r$. Hence,

$$1 \equiv a^s = a^{rq+r'} = (a^r)^q a^{r'} \equiv a\char`\^\{r'\}(mod\, p)$$

and, by minimality of r, we have $r' = 0$. ∎

Corollary 5.5: If a prime p divides the n th Fermat's number $F_n = 2^{2^n} + 1$, then p is of the form:

$$p = 2^{n+1}q + 1$$

with q integer.

Proof: If p is   prime   and   divides F_n,   we   have $2^n \equiv -1(mod\, p)$, then $2^{n+1} \equiv 1(mod\, p)$ from where, if r is the order of 2 módulo p, it follows by prop. 5.4 that $r \mid 2^{n+1}$, so $r = 2^s$ for some integer s such that $0 \leq s \leq n + 1$. If it were $s < n + 1$, we would write $n = s + t$ with $t \in \mathbb{N} \cup \{0\}$ and we should have: $2^{2^n} = \left(2^{2^s}\right)^{2^t} \equiv 1(mod\, p)$, hence $1 \equiv -1(mod\, p)$, that is $p = 2$ which is contradictory. Then we must have $r = 2^{n+1}$ and, as by Fermat's theorem: $2^{p-1} \equiv 1(mod\, p)$, the above proposition implies: $2^{n+1} \mid p - 1$. ∎

Example: Let us show that $F_5 = 2^{2^5} + 1$ is composite. According to corollary 5.5, a prime p dividing F_5, must be of the form $p = 64q + 1$, and given values to q we have:

$q = 1 \Rightarrow p = 65$ not a prime.

$q = 2 \Rightarrow p = 129$ not a prime.

$q = 3 \Rightarrow p = 193$ which is prime but does not divide F_5.

$q = 4 \Rightarrow p = 257$ but 257 does not divide F_5.

$q = 5 \Rightarrow p = 321$ not a prime.

$q = 6 \Rightarrow p = 385$ not a prime.

$q = 7 \Rightarrow p = 449$ does not divide F_5 .

$q = 8 \Rightarrow p = 513$ not a prime.

$q = 9 \Rightarrow p = 577$ does not divide F_5.

$q = 10 \Rightarrow p = 641$, and $641 \mid F_5$, since from $641 = 2^4 + 5^4 = 5 \cdot 2^7 + 1$ it follows:

$$2^{32} = 2^4 2^{28} = (641 - 5^4)2^{28} \equiv -(5 \cdot 2^7)^4 \equiv -1(mod\, 641).$$

As an application of Euler-Fermat theorem, we will make a brief incursion into Cryptography.

By diplomatic, military, industrial and even personal needs, it is advisable to have an unbreakable cryptosystem, that is to say, to have a method that allows sending secret messages that only the receiver can decipher. We will deal only with one of these systems, due to Rivest, Shamir, and Adelman, referring to [3], for more information.

A message may be considered as a number M in the decimal system, by assigning, for example, to each letter of the alphabet, each punctuation mark and to each space, a pair of digits. In the system that concerns us, each receiver A has a public key consisting of a pair of natural numbers (n, r), where n is a product of two "large" primes (of about a hundred digits), $n = pq$ and $r > 1$ is coprime with $\phi(n) = (p - 1)(q - 1)$. We emphasize that n and r are publicly known, in particular by message senders and possible interceptors, but p and q are known only by A. With this knowledge, A computes $s \in \mathbb{N}$ such that:

$$sr \equiv 1 (mod\phi(n)) \text{ and } 1 \leq s < \phi(n)$$

such s exists since r is coprime with $\phi(n)$.

To send a message M to A, the sender encrypts it as M', using the public key, as follows:

$$M' \equiv M^r (mod n) \text{ with } 1 \leq M' < n \text{ and send } M'.$$

We can assume $M < n$, because otherwise the message can be divided into blocks, that fulfill that condition. When receiving M', A decodes it by calculating:

$$M'^s \equiv M^{rs} \equiv M (mod n) \; (*)$$

since, in fact, as $rs = \phi(n)t + 1$ for some $t \in \mathbb{N}$, by Euler's theorem it follows that, $M^{rs} = M^{\phi(n)t} M \equiv M (mod n)$. Since $M < n$, $(*)$ determines univocally M.

Note that knowing the factorization $n = pq$ of n, is equivalent to know $\phi(n)$. In fact, obviously if we know p and q, we know $\phi(n) = (p - 1)(q - 1)$; and if we know $\phi(n) = pq - (p + q) + 1 = n + 1 - (p + q)$, we know then $p + q$ and pq and so p and q.

In order to break the system, it is necessary to know $\phi(n)$ (to be able to find s and, hence M), which is equivalent to know the factorization of n. If, as we said, the primes p and q are taken in the range of a hundred digits, n results in a number of two hundred digits which, with the currently available methods, is impossible to factorize in a reasonable time.

The question that arises is: how to construct primes in the hundred-figure range?. Let us say that it is possible to do so, by taking random numbers in that range and applying to them certain primality criteria. Further details can be obtained from the above-mentioned bibliography.

7- WILSON'S THEOREM

This theorem was stated by Ibn al-Haitham (c.1000 A.D.) and by Wilson in the 18th century and bears his name, although it was previously known by Leibnitz.

Theorem 7.1: (Wilson) Let $p \in \mathbb{N}$ with $p > 1$. p is prime $\Leftrightarrow$ the following congruence is satisfied:

$$(p - 1)! \equiv -1 (mod \ p) \ (1)$$

Proof: ($\Leftarrow$): If $a \mid p$ then $|a| \leq p$. If $|a| = p$ we have $a = \pm p$, while, if $|a| < p$, we have $a \mid (p - 1)!$ from where by (1): $a \mid 1$, so $a = \pm 1$, and p es prime.

($\Rightarrow$): In any field, the only elements that are its own inverses are ± 1, in fact, $x^2 = 1 \Rightarrow 0 = (x - 1)(x + 1) \Rightarrow x = \pm 1$. In particular in $\mathbb{Z}_p$, $\bar{1}$ and $\overline{-1} = \overline{p - 1}$ are the only elements that are its own inverses, so multiplying all the elements of

$$\mathbb{Z}_p - \{\bar{0}\} = \{\bar{1}, \bar{2}, \bar{3}, \ldots, \overline{p - 2}, \overline{p - 1}\}$$

each element $a \neq \pm 1$ has an inverse b such that $b \neq a$ and $b \neq \pm 1$, then a and b are neutralized, hence:

$$\overline{(p - 1)!} = \overline{p - 1} = \overline{-1}$$

That is to say $(p - 1)! \equiv -1 (mod \ p)$.■

Let us exemplify this proof with $p = 11$. We have $\mathbb{Z}_{11} - \{\bar{0}\} = \{\bar{1}, \bar{2}, \ldots, \overline{10}\}$. The inverse of $\bar{2}$ is $\bar{6}$, that of $\bar{3}$ is $\bar{4}$, that of $\bar{5}$ is $\bar{9}$ and that of $\bar{7}$ is $\bar{8}$, then:

$$\overline{10!} = \bar{1}(\overline{26})(\overline{34})(\overline{59})(\overline{78})\overline{10} = \overline{10} = \overline{-1}.$$

Wilson's theorem is a primality criterion, but not of practical value, much faster primality test are known.

Corollary 7.2: Let p be an odd prime. -1 is a square in $\mathbb{Z}_p$ $\Leftrightarrow$ p is of the form $4n + 1$.

Proof: ($\Rightarrow$): Assume $-1 = x^2$ with $x \in \mathbb{Z}_p$. By Fermat theorem

$$1 = x^{p-1} = (x^2)^{\frac{p-1}{2}} = (-1)^{\frac{p-1}{2}}$$

and if p were of the form $4n + 3$, $\frac{p-1}{2}$ would be odd, and then we would have $1 = -1$ contradicting that p is odd, so p must be of the form $4n + 1$.

($\Leftarrow$): In the product $(p - 1)!$ we lump together the factors of the form a and $p - a = -a$, and as $p - ((p-1)/2) = (p+1)/2$, by Wilson theorem we obtain:

$$-1 = (-1)^{\frac{p-1}{2}} 1^2 2^2 \ldots (p-1)/2)^2$$

and, being p of the form $4n + 1$, we have $(-1)^{\frac{p-1}{2}} = 1$ and hence -1 is a square in $\mathbb{Z}_p$.∎

Imitating the proof of Euclid on the infinitude of primes, it is easy to prove that there are an infinite number of primes of the for $4n + 3$ (exercise 21 b), based on the above corollary we can also prove the infinity of primes of the form $4n+1$:

Corollary 7.3: There are infinitely many primes of the form $4n + 1$.
Proof: If $p_1, \ldots, p_r$ were all the primes of that form, consider:

$$a = 4(p_1 \ldots p_r)^2 + 1$$

if p is a prime divisor of a, we have in $\mathbb{Z}_p$:

$$(2p_1 \ldots p_r)^2 = -1$$

so, by the above corollary, p must be of the form $4n + 1$. As p divides a and must be one of $p_1, \ldots, p_r$, it follows that $p \mid 1$, a contradiction.∎

The infinitude of primes of the form 4n+1 or of the form $4n + 3$, are particular cases of a famous theorem of Dirichlet on the infinitude of primes in arithmetic progressions stated below:

Dirichlet's theorem: If a and b are coprime natural numbers, then there are infinitely many primes of the form $an + b$.

The proof of this theorem is beyond the scope of this book.

The following corollary of Wilson's theorem, characterizes the prime pairs $p, p + 2$, although, has no practical value in finding such a pair.

Corollary 7.4 (Clement): Let $p \in \mathbb{N}$ with $p > 1$. p and $p + 2$ are both primes $\Leftrightarrow$ the following congruence is satisfied:
$$4[(p - 1)! + 1] + p \equiv 0 (mod\ p(p + 2))\ (2)$$
Proof: ($\Rightarrow$) As p is prime, we have $(p - 1)! + 1 \equiv 0 (mod\ p)$, then:
$$4[(p - 1)! + 1] + p \equiv 0 (mod\ p)\ (3)$$
and as $p + 2$ is prime: $(p + 1)! + 1 \equiv 0 (mod\ p + 2)$ and since,
$$p(p + 1) \equiv 2 (mod\ p + 2)$$
we have:
$$0 \equiv (p + 1)! + 1 = p(p + 1)[(p - 1)! + 1] - p(p + 1) + 1 \equiv 2[(p - 1)! + 1] - 1 (mod\ p + 2)$$
hence:
$$4[(p - 1)! + 1] - 2 \equiv 4[(p - 1)! + 1] + p \equiv 0 (mod\ p + 2)\ (4)$$
As p and $p + 2$ are coprime, from (3) and (4) follows (2).
($\Leftarrow$) From (2) follows that $2 \mid p \Rightarrow 4 \mid p$, but if $4 \mid p$ putting $p = 4q$ we obtain from (2):
$$(4q - 1)! + 1 + q \equiv 0\ (mod\ q)$$
and   as $q \mid (4q - 1)!$, it results $q \mid 1$, then $p = 4$, but this value does not satisfy (2), hence $2 \nmid p$.
By (2) we have that $4[(p - 1)! + 1] \equiv 0 (mod\ p)$ and as $(4, p) = 1$ (since $2 \nmid p$), we obtain:
$$(p - 1)! + 1 \equiv 0 (mod\ p)$$
and p results prime by Wilson's theorem.
Also, from (2) follows: $4[(p - 1)! + 1] + p \equiv 0 (mod\ p + 2)$, that is $4(p - 1)! + 2 \equiv 0 (mod\ p + 2)$, then:

$$4(p + 1)! + 2p(p + 1) \equiv 0 (mod\ p + 2)$$

or $4(p + 1)! + 4 \equiv 0 (mod\ p + 2)$, and as $(4, p + 2) = 1$, we have:

$$(p + 1)! + 1 \equiv 0 (mod\ p + 2)$$

so then $p + 2$ is also prime.■

8 - FIRST DEGREE EQUATION IN $\mathbb{Z}_m$

In any field K the first degree equation:

$$ax = b$$

with given $a, b \in K$, is easily discussed:

If $a = 0$ and $b \neq 0$ there is no solution.

If $a = 0$ and $b = 0$, any $x \in K$ is a solution.

If $a \neq 0$; $x = ba^{-1}$ is the unique solution.

The study of the first-degree equation in the rings $\mathbb{Z}_m$, although a little more laborious, is also easy to do,

Theorem 8.1: The equation in $\mathbb{Z}_m$:

$$\bar{a}\bar{x} = \bar{b}$$

where $a, b \in \mathbb{Z}$ are given, has a solution if, and only if: $(a, m) \mid b$. In such a case the equation has exactly $d = (a, m)$ solutions, which are obtained from one of them $\overline{x_0}$, by:

$$\bar{x} = \bar{x}_0 + \frac{m}{d}\bar{r}, r = 0, 1, \ldots, d - 1$$

Proof: The equation $\bar{a}\bar{x} = \bar{b}$ is equivalent to the congruence $ax \equiv b(mod\ m)$ which in turn, is equivalent to the Diophantine equation $ax + my = b$, and we already proved (9, Chap. 4) that this last has a solution if, and only if: $(a, m) \mid b$ and that if x_0, y_0 is a solution, any solution x, y is obtained by:

$$x = x_0 + \left(\frac{m}{d}\right)t\ ;\ y = y_0 + \left(\frac{m}{d}\right)t$$

where $t \in \mathbb{Z}$. Obviating the information about y, which is not relevant here, we have:

$$x = x_0 + \left(\frac{m}{d}\right)t$$

with $t \in \mathbb{Z}$. Furthermore, there are integers q, r such that $t = dq + r$ and $0 \leq r < d$, then, as $\bar{m} = \bar{0}$:

$$\bar{x} = \bar{x}_0 + \left(\frac{m}{d}\right)\bar{r} \; ; \; r = 0, 1, \dots, d - 1$$

It remains to see that these d classes are distinct. In fact, if : $\bar{x}_0 + \left(\frac{m}{d}\right)\bar{r} = \bar{x}_0 + \left(\frac{m}{d}\right)\bar{s}$ with $0 \le s \le r < d$, results $m \mid \left(\frac{m}{d}\right)(r - s)$, then $d \mid r - s$, and hence $r = s$.∎

Example: Let us solve the equation $8x = 20$ in Z_{44}.

We have $d = (8, 44) = 4 = 44 - 5 \cdot 8$. As $4 \mid 20$ there are solutions. We have $20 = 5 \cdot 44 - 25 \cdot 8$, then $x_0 = -25$ is a solution and any solution is obtained by $x = x_0 + \left(\frac{m}{d}\right) = -25 + 11r$, so the four solutions in Z_{44} are:

$$x_0 = -25 = 19; \; x_1 = -14 = 30; \; x_2 = 41; \; x_3 = 8.$$

9 - POLYNOMIAL FUNCTIONS

Let A be a ring. A *polynomial function* In A, is a function $f\colon A \to A$ such that there exist $n \in \mathbb{N} \cup \{0\}$ and $a_n, \dots, a_1, a_0 \in A$ so that:

$$f(x) = a_n x^n + \dots + a_1 x + a_0 \; \forall x \in A$$

$a_n, \dots, a_1, a_0$ are said to be the *coefficients* of f, If, moreover, $a_n \neq 0$, f is said to have *degree* n, We do not assign a degree to the polynomial function with all of its coefficients equal to zero. One more definition: $\alpha \in A$ is said to be a root or zero of f iff $f(\alpha) = 0$.

Proposition 9.1: With the above notations and assuming $n \ge 1$ and $a_n \neq 0$, we have:

$\alpha \in A$ is a root of $f \Leftrightarrow$ There exist a polynomial function g of degree $n - 1$ such that:

$$f(x) = (x - \alpha)g(x) \; \forall x \in A$$

Proof: The implication $\Leftarrow$ is clear. Let us show $\Rightarrow$: if α is a root of f we have:

$$f(x) = f(x) - f(\alpha) = a_n(x^n - \alpha^n) + \dots + a_2(x^2 - \alpha^2) + a_1(x - \alpha)$$
$$=$$
$$= (x - \alpha)[a_n(x^{n-1} + \alpha x^{n-2} + \dots + \alpha^{n-1}) + \dots + a_2(x + \alpha) + a_1]$$

and denoting by $g(x)$ to the expression between square brackets, we obtain the proposition.∎

Proposition 9.2: Let A be a domain. A polynomial function f in A of degree $n \geq 1$, has at most n roots in A.

Proof: If f has a root α in A, then by the proposition above, there is a polynomial function g of degree $n - 1$ such that for any $x \in A$:

$$f(x) = (x - \alpha)g(x)$$

Hence $\beta \in A$ is a root of $f \Leftrightarrow 0 = f(\beta) = (\beta - \alpha)g(\beta) \Leftrightarrow \beta = \alpha$ or β is a root of g, where in the last implication we use the hypothesis of being A a domain.

By induction on n, for $n = 1$ it follows from the fact that $g(x)$ is a nonzero constant, so then g does not have roots and if $n > 1$ it follows from the inductive hypothesis. $\blacksquare$

The above result is not necessarily valid if A is not a domain, for example, in Z_4 The function f defined by $f(x) = 2x^3 + 2x$ has the four elements of Z_4 as roots.

Corollary 9.3: Let A be an infinite domain, f, g, h polynomial functions in A:

      a)   If $f(x) = a_n x^n + \ldots + a_1 x + a_0 = 0 \; \forall x \in A$, then $a_n = \ldots = a_1 = a_0 = 0$.

      b)   If $g(x) = b_r x^r + \ldots + b_1 x + b_0$, $h(x) = c_s x^s + \ldots + c_1 x + c_0$ with (say) $r \geq s$ and $g(x) = h(x) \forall x \in A$, then $b_r = \ldots = b_{s+1} = 0$ and $b_s = c_s, \ldots, b_0 = c_0$.

Proof: a) If there is an $a_i \neq 0$, we may assume that $a_n \neq 0$. If $n = 0$, f is a nonzero constant and has no roots, while if $n \geq 1$, by prop. 9.2, f has at most n roots in A, contradicting the hypothesis that all the elements of A are roots of f and that A has infinitely many elements.

b) follows readily from a. $\blacksquare$

It is important not to confuse the concepts of polynomial function and of that of polynomial. We do not define polynomials yet, for now, it is sufficient to say that a polynomial satisfies the conditions of corollary 9.3, almost by definition, even if the ring of coefficients is finite and even if it is not a domain. For infinite domains the concepts co-incide essentially, but not in general.

If A is finite the previous corollary is not valid, for example, we have $x^2 = x \; \forall x \in Z_2$.

Proposition 9.4: (Relations between the coefficients and the roots): If A is an infinite domain and the polynomial function in A, $f = x^n + \ldots + a_1 x + a_0$ with $= 1$, is a product of linear factors:
$$f(x) = (x - \alpha_1) \ldots (x - \alpha_n)$$
with $\alpha_i \in A$, we have the following relations between the coefficients and the roots:
$$a_{n-1} = -(\alpha_1 + \ldots + \alpha_n)$$
$$a_{n-2} = \sum_{i<j} \alpha_i \alpha_j$$
$$\ldots$$
$$a_1 = (-1)^n \alpha_1 \ldots \alpha_n$$

Proof: It follows by making the product and applying corollary 9.3, b.∎

Proposition 9.5: If A is a ring and in the set F of all functions from A to A, we define operations: $f + g$; fg by:
$$(f + g)(a) = f(a) + g(a), (fg)(a) = f(a)g(a)$$
then F is a ring, and the subset P of F formed by the polynomial functions is a subring of F.

Proof: Let f, g be the polynomial functions defined by:
$$f(x) = \sum_{i=0}^{n} a_i x^i \; ; \; g(x) = \sum_{j=0}^{m} b_j x^j$$
if we put $a_i = 0, b_j = 0$ for any $i > n$ and any $j > m$, we have:
$$(f + g)(x) = \sum_{i=0}^{max(n,m)} (a_i + b_i) x^i$$
$$; \; (fg)(x) = \sum_{k=0}^{m+n} \left(\sum_{i+j=k}^{max(n,m)} (a_i b_i) \right) x^k$$
and so the sum and product of polynomial functions are polynomial functions. We leave the rest of the proof as an exercise.∎

In any field K, the second degree equation $ax^2 + bx + c = 0$ with $a \neq 0$, is reduced, multiplying by the inverse of a, to one of type:

$$x^2 + px + q = 0$$

where $p, q \in K$ are given. If $1 + 1 = 2 \neq 0$ in K, we can "complete the square" to obtain:

$$\left(x + \frac{p}{2}\right)^2 = \frac{p^2}{4} - q$$

then if $\frac{p^2}{4} - q$ is not a square in K, there are no solution; if $\frac{p^2}{4} - q = 0$ there is a unique solution $x = -\frac{p}{2}$ and, if $\frac{p^2}{4} - q = r^2$ is a nonzero square in K there are exactly two solutions $-\frac{p}{2} + r$ and $-\frac{p}{2} - r$ (they are distinct, for if not we would have $r = -r$, or $2r = 0$, and as $2 \neq 0$, we would have $r = 0$).

It is then necessary to determine the squares in K. In $\mathbb{R}$ the squares coincide (as we will see) with the positive. In the field of complex numbers (Chap. 7) any element is a square, and so any second degree equation (and, in fact, of any degree ≥ 1) has roots. To solve, or at least discuss (establish whether there are or not solutions, and if there exist, how many are they) the second degree equation in the field Z_p, p prime, it is necessary a criteria to decide if an element of Z_p, is or is not a square in Z_p.

Note that in Z_2 a second degree equation, can easily been solved by direct replacement, for example $x^2 + x + 1 = 0$ does not have roots in Z_2, since neither $\bar{0}$ nor $\bar{1}$ satisfies it.

In what follows, we assume that p is an odd prime, and so $\bar{2} \neq \bar{0}$ in Z_p.

Proposition 10.1: (Euler's criteria) Let p be an odd prime:

a) $\bar{a}$ is a square in $Z_p - \{\bar{0}\} \Leftrightarrow \bar{a}^{\frac{p-1}{2}} = \bar{1}$.

b) $\bar{a}$ is not a square in $Z_p - \{\bar{0}\} \Leftrightarrow \bar{a}^{\frac{p-1}{2}} = -\bar{1}$.

Proof: (First proof) Any $\bar{a} \in Z_p - \{\bar{0}\}$ satisfies, by Fermat's theorem.

$$\bar{1} \equiv \bar{a}^{p-1} = \left(a^{\frac{p-1}{2}} - 1\right)\left(a^{\frac{p-1}{2}} + 1\right)$$

so any $\bar{a} \in Z_p - \{\bar{0}\}$ satisfies $\bar{a}^{\frac{p-1}{2}} = \bar{1}$ (1) or $\bar{a}^{\frac{p-1}{2}} = -\bar{1}$ (2). The squares satisfy the first, for if $\bar{a} = \bar{b}^2$ with $\bar{b} \in Z_p - \{\bar{0}\}$, then $\bar{a}^{\frac{p-1}{2}} = \bar{b}^{p-1} = 1$ by Fermat's theorem. Moreover, only the squares satisfies (1) for by prop.9.2, the equation (1) has at most $\frac{p-1}{2}$ roots in $Z_p - \{\bar{0}\}$ and the squares: $\bar{c}^2$, $\bar{c} = \bar{1}, \bar{2}, \ldots, \overline{\left(\frac{p-1}{2}\right)}$ are distinct, for if $\bar{c}^2 = \bar{d}^2$ with $1 \leq c, d \leq \frac{p-1}{2}$, then $\bar{c} = \bar{d}$ or $\bar{c} = -\bar{d} = \bar{p} - \bar{d}$, but in this last case, c would be out of range. This proves a), and as any element of $Z_p - \{\bar{0}\}$ must satisfy (1) or (2), then a) implies b).

(Second proof): We will show an argument, due to Dirichlet, which derives the theorems of Fermat, Wilson and Euler criteria from a common source.

Let $\bar{a} \neq \bar{0}$ be a square in, so there exists $b \in$ such that $b^2 = a$, then also $-b^2 = a$. We   have $\bar{b} \neq -\bar{b}$, for if $\bar{b} = -\bar{b}$ it would result, $\bar{b} + \bar{b} = 2\bar{b} = 2\bar{b} = \bar{0}$ and as $\bar{b} \neq \bar{0}$, it would be $2 = 0$.   Moreover $\bar{b}$ and $-\bar{b}$ are the only elements of whose square is a, since $\bar{c}^2 = \bar{a} \Rightarrow \bar{c}^2 = \bar{b}^2 \Rightarrow \bar{c} = \pm\bar{b}$. Then for each $\bar{x} \in \{\bar{0}, \bar{b}, -\bar{b}\}$ there is a single $\bar{x}' \in Z_p - \{\bar{0}, \bar{b}, -\bar{b}\}$, $\bar{x}' \neq \bar{x}$ such that $\bar{x}\bar{x}' = \bar{a}$. Taking $\bar{y} \in Z_p$ such that $\bar{y} \neq \bar{0}, \bar{b}, -\bar{b}, \bar{x}, \bar{x}'$ there exists, in the same manner, $\bar{y}' \neq \bar{0}, \bar{b}, -\bar{b}, \bar{x}, \bar{x}', \bar{y}' \neq \bar{y}$ such that $\bar{y}\bar{y}' = \bar{a}$ and so on. Multiplying member by member the $\frac{p-3}{2}$ equalities:

$$\bar{x}\overline{x'} = \bar{a}; \quad \overline{y}\overline{y'} = \bar{a}; \quad \ldots\ldots\ldots$$

with $\bar{b}\,\overline{-b} = -\bar{a}$, we obtain that if $\bar{a} \neq 0$ is a square, then:

$$(p-1)! \equiv -a^{\frac{p-1}{2}} \ (mod\, p) \ (1)$$

Similarly, if $\bar{a}$ is not a square, we have:

$$(p-1)! \equiv a^{\frac{p-1}{2}} \ (mod\, p) \ (2)$$

Taking $a = 1$ in (1), Wilson's theorem follows, and applying it in (1) and (2), Euler criterion follows. Finally, by squaring in (1) and (2), we obtain Fermat's theorem.∎

A numerical example can clarify the proof. Let $p = 17$ and $a = 4$. We have $\bar{a} = \bar{2}^2 = \overline{15}^2$ and:

$$\overline{2}\ \overline{15} = -\overline{4};\ \overline{14} = \overline{4};\ \overline{37} = \overline{4};\ \overline{5}\ \overline{11} = \overline{4};\ \overline{6}\ \overline{12} = \overline{4};\ \overline{8}\ \overline{9} = \overline{4};\ \overline{10}\ \overline{14}$$
$$= \overline{4};\ \overline{13}\ \overline{16} = \overline{4}$$

y multiplying them:

$$(p-1)! \equiv 2 \cdot 15 \cdot 1 \cdot 4 \cdot 3 \cdot 7 \cdot 5 \cdot 11 \cdot 6 \cdot 12 \cdot 8 \cdot 9 \cdot 10 \cdot 14 \cdot 13 \cdot 16$$
$$\equiv -4^8$$

Example: As an application of Euler criteria, we will show that there is an infinity of primes of the form $4n + 1$.

If $p_1, \dots, p_r$ were all primes of that form, consider:

$$a = 4(p_1 \dots p_r)^2 + 1$$

And let p be a prime divisor of a. Then $-1 \equiv 4(p_1 \dots p_r)^2 (mod\ p)$, so that -1 would be a square in $\mathbb{Z}_p$, but by Euler criteria, this is possible only in case p is of the form $4n + 1$.

For $\bar{a} \in Z_p - \{\bar{0}\}$, the *Legendre symbol* $\left(\frac{a}{p}\right)$, is defined by:

$$\left(\frac{a}{p}\right) = \begin{cases} 1\ if\ \bar{a}\ is\ a\ square\ in\ \mathbb{Z}_p \\ -1\ if\ \bar{a}\ is\ not\ a\ square\ in\ \mathbb{Z}_p \end{cases}$$

Corollary 10.2: If $a, b \in \mathbb{Z}_p - \{0\}$, where p is an odd prime, then:

1) $\left(\frac{a}{p}\right) \equiv a^{\frac{p-1}{2}} (mod\, p)$

2) $\left(\frac{ab}{p}\right) = \left(\frac{a}{p}\right)\left(\frac{b}{p}\right)$

Proof: 1) This is Euler's criterion in other words.

2) By 1), we have $\left(\frac{ab}{p}\right) \equiv (ab)^{\frac{p-1}{2}} = a^{\frac{p-1}{2}} b^{\frac{p-1}{2}} \equiv (a/p)(b/p)(mod\, p)$ and as the values that Legendre symbol can take are ± 1 and $1 \equiv -1 (mod\, p)$ is impossible since p is odd, 2) follows.∎

The following result is the famous *quadratic reciprocity law* conjectured independently by Euler, Legendre, and Gauss and proved for the first time by Gauss, who throughout his life published six proofs of it. We limit ourselves to the statement of this law, referring for its proof to any book on Number Theory, for example "[19], [27] or [37].

Quadratic reciprocity law: If p and q are distinct odd primes, we have:

$$\left(\frac{p}{q}\right) = (-1)^{\frac{p-1}{2}\frac{q-1}{2}}\left(\frac{p}{q}\right)$$

This relation combined with the following "complementary law":

$$\left(\frac{2}{p}\right) = (-1)^{\frac{p^2-1}{8}}$$

and with that deduced from Euler criterion:

$$\left(\frac{-1}{p}\right) = (-1)^{\frac{p-1}{2}}$$

allows to compute Legendre symbol. For example:

$$\left(\frac{30}{101}\right) = \left(\frac{2}{101}\right)\left(\frac{3}{101}\right)\left(\frac{5}{101}\right)$$

But:

$$\left(\frac{2}{101}\right) = (-1)^{\frac{101^2-1}{8}} = (-1)^{1275} = -1$$

$$\left(\frac{3}{101}\right) = (-1)^{50}\left(\frac{3}{101}\right) = \left(\frac{101}{3}\right) = \left(\frac{2}{3}\right)$$

$$= (-1)^{\frac{3^2-1}{8}} = -1$$

$$\left(\frac{5}{101}\right) = (-1)^{100}\left(\frac{101}{5}\right) = \left(\frac{101}{5}\right) = \left(\frac{1}{5}\right) = 1$$

then $\left(\frac{30}{101}\right) = 1$, so 30 is a square in Z_{101}.

Note that the reciprocity law allows us to compute if a given number is a square modulus a prime, but not to compute a number whose square is the given number.

The "complementary law": $\left(\frac{2}{p}\right) = (-1)^{\frac{101^2-1}{8}}$ can be obtained from Euler's criterion:

Proposition 10.3: If p is an odd prime, then: $\left(\frac{2}{p}\right) = (-1)^{\frac{101^2-1}{8}}$.

Proof: According to Euler criterion it is enough to prove that $2^{\frac{p-1}{2}} \equiv$ $(-1)^{\frac{p^2-1}{8}} \pmod{p}$. Note that if $p = 4n +$

1, then $\frac{p+1}{2}$ es impar and so: $(-1)^{\frac{p^2-1}{8}} = \left((-1)^{\frac{p-1}{2}}\right)^{\frac{p+1}{4}} =$ $(-1)^{\frac{p-1}{4}}$. Similarly, when $p = n + 3$ we have $(-1)^{\frac{p^2-1}{8}} = (-1)^{\frac{p+1}{4}}$. Therefore the statement is equivalent to:

$$2^{\frac{p-1}{2}} = \begin{cases} (-1)^{\frac{p-1}{4}} \ si \ p = 4n + 1 \\ (-1)^{\frac{p+1}{4}} \ si \ p = 4n + 3 \end{cases}$$

Let $p = 4n + 1$, we have:

$$2^{\frac{p-1}{2}}(1 \cdot 2 \cdots \cdot \frac{p-1}{2}) = \prod_{i=1}^{2n} 2i = \prod_{i=1}^{n} 2i \cdot \prod_{i=n+1}^{2n} 2i \ (1)$$

But (all congruences are *modulus p*),

$$\prod_{i=n+1}^{2n} 2i \equiv \prod_{i=n+1}^{2n}(2i - p) = (-1)^{\frac{p-1}{4}} \prod_{i=n+1}^{2n}(p - 2i) \ (2)$$

Putting $j = 2n + 1 - i$, we have $p - 2i = 2j - 1$, and when i varies between $n + 1$ and $2n$, j varies between 1 and n. Hence: $\prod_{i=n+1}^{2n}(p - 2i) = \prod_{j=1}^{n}(2j - 1)$, and replacing this in (2) and then in (1), we obtain:

$$2^{\frac{p-1}{2}}\left(1 \cdot 2 \cdots \cdot \frac{p-1}{2}\right) \equiv (-1)^{\frac{p-1}{4}} \prod_{i=1}^{n} 2i \prod_{j=1}^{n}(2j - 1)$$

and by cancelation, it follows that $2^{\frac{p-1}{2}} \equiv (-1)^{\frac{p-1}{4}}$
Similarly, if $p = 4n + 3$:

$$2^{\frac{p-1}{2}} \prod_{i=1}^{\frac{p-1}{2}} i = \prod_{i=1}^{n} 2i \prod_{i=n}^{2n+1} 2i \equiv$$

$$\equiv (-1)^{\frac{p+1}{4}} \prod_{i=1}^{n} 2i \prod_{j=1}^{2n+1}(2j - 1) = (-1)^{\frac{p+1}{4}} \prod_{i=1}^{\frac{p-1}{2}} i$$

and then, $2^{\frac{p-1}{2}} \equiv (-1)^{\frac{p+1}{4}}$. ∎

11 - CHINESE REMAINDER THEOREM

Sun-Tzu (1st century) in his "Arithmetic" poses the problem of finding a number such that when divided by $3, 5, 7$, the remainders are

2, 3, 2, respectively, and he solves it by a method which is, essentially, the same as in the proof of the following:

Theorem 11.1: (Chinese remainder theorem) If $m_1, \ldots, m_n$ are natural numbers pairwise coprime and if $a_1, \ldots, a_n$ are integers, then the system of congruences:

$$x \equiv a_i (mod\ m_i)\ i = 1, \ldots, n$$

has a solution, which is unique modulus $M = m_1 \cdot \ldots \cdot m_n$.

Proof: Since for each $j = 1, \ldots, n, \dfrac{M}{m_j}$ and m_j are coprime, there exist (theorem 8.1) b_j such that:

$$\frac{M}{m_j} b_j \equiv 1 (mod\ m_j)\ j = 1, \ldots, n$$

If we putt $x = \sum_{j=1}^{n} a_j b_j \dfrac{M}{m_j}$, as $\dfrac{M}{m_j} \equiv 0\ (mod\ m_i)\ if\ i \neq j$, we have $x \equiv a_i (mod\ m_i)$.

If y is a solution of the system, we will have $x \equiv y (mod\ m\_i) \forall i = 1, \ldots, n$, or $m\_i \mid x - y\ \forall i$, hence $M \mid x - y$ since the m_i are pairwise coprime (corollary 5.4, cap. 4). ∎

Example: The Chinese generals used the above theorem to count their soldiers. For example, a general knows that, roughly, he has more than 5.000 and lesser than 9.000 soldiers and wants to know exactly how many he has.

He orders them to form in rows of 11, and there are 5 soldiers left, then to form in rows of 13 and 3 have left. Finally, he orders them to form in rows of 17 and 2 have left. It is then about to solve the system:

$$x \equiv 5 (mod\ 11)$$
$$x \equiv 3 (mod\ 13)$$
$$x \equiv 2 (mod\ 17)$$

Proceeding as in the proof of the theorem, we find b_1, b_2, b_3 such that

$$221 b_1 \equiv 1 (mod\ 11), that\ is\ b_1 \equiv 1 (mod\ 11)\ since\ 221$$
$$\equiv 1 (mod\ 11)$$
$$187 b_2 \equiv 1 (mod\ 13), that\ is\ 5 b_2 \equiv 1 (mod\ 13)\ since\ 187$$
$$\equiv 5 (mod\ 13)$$

$$143b_3 \equiv 1(mod17), that\ is\ 7b_3 \equiv 1(mod17)\ since\ 143$$
$$\equiv 7(mod17)$$

we may take then $b_1 = 1, b_2 = 8, b_3 = 5$ and

$$x = 5 \cdot 1 \cdot 221 + 3 \cdot 8 \cdot 187 + 2 \cdot 5 \cdot 143 = 7023$$

As this solution is unique modulus $M = 11 \cdot 13 \cdot 17 = 2431$, it turns out that any other solution is lesser than 5000 and greater than 9000, hence the general has exactly 7023 soldiers.

Another way of solving a system of congruences which is useful even is the modulus are not coprime, is given by:

Proposition 11.2: The system of congruences:

$$x \equiv a_1(modm_1)$$
$$x \equiv a_2(modm_2)$$

has a solution if, and only if $(m_1, m_2) \mid a_1 - a_2$ and, in such a case a solution   is   unique modulus $[m_1, m_2]$ (the least common multiple of m_1 and m_2 (exercise 13, Chap. 4)).

Proof: Put $d = (m_1, m_2)$. If there is a solution x of the system, then $d \mid m_1 \mid x - a_1$ and $d \mid m_2 \mid x - a_2$, hence $d \mid a_1 - a_2 = (x - a_1) - (x - a_2)$.

Reciprocally, if $d \mid a_1 - a_2$ exists $c \in \mathbb{Z}$ such that $a_1 - a_2 = cd$ and as $d = sm_1 + tm_2$ for some integers s, t, we obtain:

$$a_1 - a_2 = cd = csm_1 + ctm_2$$

then $x = a_1 - csm_1 = a_2 + ctm_2$ satisfies the system.

If y is a solution of the system. we have $y \equiv x\ (mod\ m_1)$ and $y \equiv x\ (mod\ m_2)$, hence $y \equiv x\ (mod\ [m_1, m_2])$.∎

Example: Find the least natural number with remainders 5, 4, 3 when divided by 6, 5, 4 respectively.

If x is such a number, we have $x + 1 = 6u = 5v = 4w$, from where, $x + 1 = 60t$, so the least of them is 59. In this solution of the problem, we use the particularities of the numbers involved. Using the proposition above we can proceed as follows. It is about solving the system:

$$x \equiv 5 (mod 6)$$
$$x \equiv 4 (mod 5)$$
$$x \equiv 3 (mod 4)$$

First, we solve the system formed by the two first. We have $1 = (6,5) = 6 - 5$, then $5 - 4 = 6 - 5$, from where $-1 = 5 - 6 = 4 - 5$ is a solution. Any solution x must comply $x \equiv -1(mod 30)$, Adding the third, we obtain the system:

$$x \equiv -1 (mod 30)$$
$$x \equiv 3 (mod 4)$$

As $2 = (30,4) = 30 - 7 \cdot 4$, we have $-1 - 3 = (-2) \cdot 2 = (-2) \cdot 30 + 14 \cdot 4$ hence

$$59 = -1 + 2 \cdot 30 = 3 + 14 \cdot 4$$

is a solution of the system. As the solution is unique $modulus$ $[30, 2] = 60$, it follows that 59 is the least natural number solution of the system.

The Chinese remainder theorem is useful as well, in reducing the resolution of polynomial equations $modulus$ m, to the case of modulus a power of a prime:

Proposition 11.3: Let $m = p_1^{\alpha_1} \ldots p_r^{\alpha_r}$ where the p_i are distinct primes and the α_i natural numbers. Let f be a polynomial function in $\mathbb{Z}$. The equation $f(x) \equiv 0 (mod m)$ has a solution if, and only if, the equations $f(x) \equiv 0 (mod p_i^{\alpha_i})$ have solutions for every $i = 1, \ldots, r$.

Proof: It is clear that if $x \in \mathbb{Z}$ is a solution of $f(x) \equiv 0 (mod m)$, then it is a solution of $f(x) \equiv 0 (mod p_i^{\alpha_i})$ $\forall i$ too. For each i let x_i be a solution of $f(x_i) \equiv 0 (mod p_i^{\alpha_i})$. By the Chinese remainder theorem, exists a solution x of the system:

$$x \equiv x_i (mod p_i^{\alpha_i}) \; i = 1, \ldots, r$$

then $f(x) = a^n x^n + \ldots + a_1 x + a_0 \equiv a^n x_i^n + \ldots + a_1 x_i + a_0 = f(x_i)(mod p_i^{\alpha_i})$ $\forall i$, hence $f(x) \equiv 0 (mod p_i^{\alpha_i})$ $\forall i$ and, then $f(x) \equiv 0 (mod m)$.∎

Example: The equation $x^2 + 14x + 40 \equiv 0 (mod 55)$, completing the square takes the form:

$$(x + 7)^2 \equiv 9 (mod 55)$$

Putting $y = x + 7$, and since $55 = 11 \cdot 5$, according to the above proposition we consider the equations:

$$y^2 \equiv 9(mod11)$$
$$y^2 \equiv 9(mod5)$$

from where we obtain $y \equiv \pm3(mod11), y \equiv \pm3(mod5)$. We consider then the systems:

$$\begin{cases} y \equiv 3(mod11) \\ y \equiv -3(mod5) \end{cases} \begin{cases} y \equiv -3(mod11) \\ y \equiv 3(mod5) \end{cases} \begin{cases} y \equiv -3(mod11) \\ y \equiv -3(mod5) \end{cases}$$

whose respective solutions modulus 55 are: $y = 3, y = 47, y = 8, y = 52$. Finally, as $x = y - 7$, we obtain the solutions (mod55) of the equation: $x = 51, 40, 1, 45$.

EXERCISES

Ex. 1: $a \equiv b(modm) \Rightarrow (m, a) = (m, b)$. Is the converse valid?.

Ex. 2: Prove the following statement by Stifel (16th century): If x has remainders r and s when divided by a and $a + 1$ respectively, then x and $(a + 1)r + a^2s$ have the same remainder when divided by $a(a + 1)$.

Ex 3: If a is odd, the following congruences are satisfied:

$$a^4 \equiv 1(mod16), a^8 \equiv 1(mod32), a^{16} \equiv 1(mod64)$$

Ex. 4: Find the remainder in the division of 34^{1801} by 54.

Ex. 5: Find divisibility criteria by $7, 25$ and 101 in the decimal system.

Ex. 6: Find divisibility criteria by 37 and 13 in the system of base $b = 1000$.

Ex. 7: Justify the following divisibility criteria:

a) To check if a number is divisible by 7, multiply the first (on the left) digit by 3, add the result to the second and repeat the procedure.

b) To check if a number is divisible by 7, 11 or 13, group the digits in triples from right to left, add the components alternately and repeat the procedure. For example, if the number is 54.328.295, we have: $54 - 328 + 295 = 13$ then it is divisible by 13 but not by 7 nor by 11.

Ex. 8: A relation R in a set A is said circular iff:

$$aRb \text{ and } bRc \Rightarrow cRa$$

Prove that a relation in A is an equivalence relation if, and only if it is circular and reflexive.

Ex. 9: Which is the error in the following argument "proving" that symmetry and transitivity imply reflexivity: : From aRb it follows by symmetry that bRa, then by transitivity: aRa?

Taking $A = \{a, b, c\}$ and $R = \{(a, a); (a, c); (c, c); (c, a)\}$ verify that R is symmetric and transitive but not reflexive.

Ex 10: Let $f: A \to B$ a function. We define for $a, a' \in A$: $a \sim a'$ iff $f(a) = f(a')$. Prove that $\sim$ is a equivalence relation in A. Prove, furthermore, that any equivalence relation $\sim$ in a set A arises in that way, i.e., that there exist some set B and some function $f: A \to B$ such that $a \sim a' \Leftrightarrow f(a) = f(a')$. (hint. : take $B = (A/\sim$) and $f: A \to (A/\sim)$ the canonical function, that is: $f(a) = a$).

Ex. 11: In how many ways 6 people can sit around a circular table if what matters is the relative position and not the occupied position? (R.: 120).

Ex. 12: If A and B are rings, in the cartesian product $A \times B$ we define a sum and a product by:

$$(a, b) + (a', b') = (a + a', b + b')$$
$$(a, b) \cdot (a', b') = (aa', bb')$$

where we have denoted the sum (product) in A, in B and in $A \times B$ with the same symbol, since we recognize the operations by the elements to which are applied. Prove:

a) $A \times B$ is a ring.

b) $A \times B$ is not a domain (even in the case that A and B are domains).

c) $A' = \{(a, 0)/a \in A\}$ is a subring of $A \times B$ with a different identity (neutral of the product).

*****Ex. 13**: Let $a, b \in \mathbb{Z}$ prove: $11 | a^3 - b^3 \Leftrightarrow 11 | a - b$.

Ex. 14: Find $m, n \in \mathbb{N}$ that are not coprime and such that:

$$\phi(mn) = \phi(m)\phi(n)$$

or prove that that is not possible.

Ex. 15: Find all the $n \in \mathbb{N}$ such that $5n = 11\phi(n)$.

Ex. 16: The solutions of $\phi(x) = 2^n$ are $x = 2^r \, p_1 \ldots p_s$ where $p_1, .., p_s$ are distinct Fermat primes such that the sum of its exponents is n if x is odd and is $n - r + 1$ if x is even.

Ex. 17: If d is the greatest common divisor of two integers m and n, then:

$$\phi(mn) = ((d\phi(m)\phi(n))/(\phi(d))).$$

Ex. 18: if m|n, then $\phi(m)|\phi(n)$.

Ex. 19: Prove Fermat Little Theorem (a^{p}≡a(mod p) with p prime and a∈N) as Leibnitz (17th century) did: whichever by the integers $a_1, ..., a\_{r}$ we have:

$$a_1^p+...+a_r^p \equiv (a_1+...+a_r)^p (mod p)$$

and taking $r = a, a_1 =...= a_r = 1$ the result follows.

Ex. 20: Find the remainder when dividing 321^{1681} by 525 using Euler theorem.

Ex.   21:   Given $n = 19749361535894833$ and $\phi(n) = 19749361232517120$ and knowing that n is a product of two primes, find those primes.

Ex. 22: Probe Wilson theorem as Lagrange did: let p be a prime, Starting from the relation obtained in the last example of 8, Chap. 3:

$$(p-1)! = \sum_{k=0}^{p-2}(-1)^k \frac{p-1}{k}(p-1-k)^{p-1}$$

as by Fermat theorem we have $(p-1-k)^{p-1} \equiv 1(mod p)$ para $k = 0,...,p-2$, and considering the binomial expansion of $(1-1)^{p-1}$, Wilson's theorem follows.

Ex. 23: Solve the following congruences:

a) $30x \equiv 36(mod 42)$, b) $30x \equiv 36(mod 25)$

Ex. 24: The sum of three consecutive squares cannot be a multiple of 19.

Ex.   25:   Prove   that   equation $x^5 \equiv 300x(mod 101)$ has one and only one solution $modulus$ 101.

Ex. 26: The equation $x^2 \equiv 819(mod 935)$ has no solution.

Ex. 27: Solve the problem of Sun-Tzu: find a number that leaves reminders $2, 3, 2$ when divided by $3, 5, 7$ respectively.

Ex. 28: A problem of Bhàscara (12th century): Find a number with remainders $1, 2, 3$, when divided by $2, 3, 5$ and such that the respective quotients when divided by $2, 3, 5$ yield residues $1, 2, 3$.

Ex. 29: Find the solutions of the equation: $x^3 + 2x - 3 \equiv 0(mod 15)$. (R.: 1; 3; 6; 8; 11; 13$(mod 15)$)

Ex. 30: a) $7|a^2 + b^2 \Leftrightarrow 7 | a$ and $7 | b$.

b) The equation $x^2 + y^2 = 7z^2$ has no integral solutions except the trivial one $x = y = z = 0$.

CHAPTER 6
LEAST UPPER BOUND

In this chapter, the axiomatic of the real numbers is completed with the statement of the least upper bound axiom, and its application in order to obtain the important properties of archimedeanity, the existence of an integral part, density of the rationals and existence of roots of positive real numbers. In addition, the exponential function is defined, and its characteristic property is proved. Finally, some fundamental properties of real numbers, related to sequences are considered.

1 - RATIONAL NUMBERS

By definition, a *rational number* is a quotient of two integers. We will denote by $\mathbb{Q}$ (quotient) to the set of rational numbers:

$$\mathbb{Q} = \{x \in \mathbb{R} / \text{ exist } a, b \in \mathbb{Z}, b \neq 0, \text{ such that } x = a/b\}$$

It is clear that if $\frac{a}{b}$ is a rational number $(a, b \in \mathbb{Z})$, we can assume that $b > 0$ and that a and b are coprime, for if $d = (a, b)$, we have $a = da', b = db', 1 = (a', b')$ and $\frac{a}{b} = \frac{a'}{b'}$.

Proposition 1.1: $\mathbb{Q}$ is a subfield of $\mathbb{R}$.

Proof: Since by prop. 1.1, Chap. 4, the sum and product of integers are integers; it follows that the sum and product of rationals are rationals. Also, $0 = \frac{0}{1}$ and $1 = \frac{1}{1}$ are rationals, as they are the opposite $\frac{-a}{b}$ of $\frac{a}{b}$ and the multiplicative inverse $\frac{b}{a}$ of $\frac{a}{b}$ if $a \neq 0$.∎

Furthermore, the restriction of the relation $<$ to the elements of $\mathbb{Q}$, satisfies obviously the properties of trichotomy, transitivity, and

141

compatibility with the sum and product, so $\mathbb{Q}$ satisfies all properties enounced for $\mathbb{R}$. This does not mean that $\mathbb{Q} = \mathbb{R}$ since we still have to state an axiom, which will allow us, among other things, to check that $\mathbb{Q} \neq \mathbb{R}$.

Let A be a subset of $\mathbb{R}$. A real number c, is said to be an *upper bound* of A iff $c \geq a \ \forall a \in A$. Of course, if c is an upper bound of A, any $d > c$ is an upper bound of A as well.

A is said to be *bounded from above* iff it has an upper bound.

Examples: 1) Any $c \in \mathbb{R}$ is an upper bound of the empty set $\emptyset$, for, if not, there must exist $a \in \emptyset$ such that $a > c$, but $\emptyset$ does not have elements.

2) 1 is an upper bound of $\{\frac{1}{n} / n \in \mathbb{N}\}$, since, as $n \geq 1 \ \forall n \in \mathbb{N}$, then $1 \geq \frac{1}{n} \ \forall n \in \mathbb{N}$.

3) $\mathbb{R}$ does not have an upper bound, for if $c \in \mathbb{R}$, then $c + 1 \in \mathbb{R}$ and $c + 1 > c$.

Even though we will soon prove that $\mathbb{N}$ is not bounded from above, it is advisable to prove independently the following:

Proposition 1.2: No rational can be an upper bound of $\mathbb{N}$.

Proof: If $\frac{a}{b} \in \mathbb{Q}$, with a, b integers and $b > 0$, then there exist $q, r \in \mathbb{Z}$ such that $a = bq + r$ and $0 \leq r < b$. Then $b(q + 1) = bq + b > bq + r = a$, that is $\frac{a}{b} < q + 1$.∎

Corollary   1.3:   Given $r, s \in \mathbb{Q}$ with $r > 0$, there exists $n \in \mathbb{N}$ such that $nr > s$.

Proof: $\frac{s}{r}$ can not be an upper bound of $\mathbb{N}$.∎

Corollary 1.4: Let A be a non-empty subset of $\mathbb{Q}$, such that it has an upper   bound $c \in \mathbb{Q}$. Given $\varepsilon \in \mathbb{Q}$ with $\varepsilon > 0$,   there   exists $a \in A$ such that $a + \varepsilon$ is an upper bound of A.

Proof: If $a + \varepsilon$ is not an upper bound of A, $\forall a \in A$, taking $a_1 \in A$, as $a_1 + \varepsilon$ is not an upper bound of A, exists $a_2 \in A$ such that $a_1 + \varepsilon < a_2$. In the same way, exists $a_3 \in A$ such that $a_2 + \varepsilon < a_3$, then $a_1 + 2\varepsilon < a_3$. Reiterating the process, it follows that $a_1 + n\varepsilon$ is not an upper bound of A for any $n \in \mathbb{N}$, from where: $a_1 + n\varepsilon \leq c \, \forall n \in \mathbb{N}$, but by 1.3 exists $n \in \mathbb{N}$ such that $n\varepsilon > c - a_1$.■

We now will prove "multiplicative versions" of 1, 3 and 1, 4.

Proposition 1.5: Given $t, u \in \mathbb{Q}$ with $t > 1$, there exists $n \in \mathbb{N}$ such that $t^n > u$.

Proof: Put $t = 1 + c$ with $c > 0$, by exercise 5, Chap. 3, we have $t^n = (1 + c)^n \geq 1 + nc \, \forall \, n \in \mathbb{N}$, then by corollary 1.3, it is enough to take n such that $nc > u - 1$, to obtain $t^n > u$.■

Proposition 1.6: Let A be a subset of $\mathbb{Q}$ that contains at least a positive number, and that it has an upper bound $q \in \mathbb{Q}$. Given $t \in \mathbb{Q}, t > 1$, there exists $a \in A$ such that ta is an upper bound of A.

Proof: If ta is not an upper bound of A $\forall \, a \in A$, taking any $a \in A, a > 0$, there exists $a_1 \in A$ such that $ta < a_1$, then it exists $a_2 \in A$ such that $ta_1 < a_2$. Continuing in that way we conclude that whichever be $n \in \mathbb{N}$, there exists $a_n \in A$ such that $t^n a < a_n$, hence $t^n a < q \, \forall \, n$. But by prop. 1.5 there exist $n \in \mathbb{N}$ such that $t^n > \frac{q}{a}$. This contradiction proves the result.■

2- LEAST UPPER BOUND AXIOM

Let A be a subset of $\mathbb{R}$. A real number s, is said to be the *least upper bound* or *supremum* of A iff the following two conditionds are fulfilled:

    1) s is an upper bound of A.

    2) If t is an upper bound of A, then $s \leq t$,

in other words, s is the least of the upper bounds of A.

Proposition 2.1: If s and s' are least upper bounds of A, then $s = s'$.

Proof: exercise.∎

The supremum of A, if there is one, is then unique and denoted by $supA$.

It is clear that if A is not bounded from above, it does not have a supremum. The empty set $\emptyset$ also does not have a supremum, since any real number is an upper bound of $\emptyset$. We postulate that these are the only exceptions:

III - Axiom of the supremum or least upper bound property: Any non-empty and bounded from above subset of $\mathbb{R}$, has a least upper bound.

Example: The set $\left\{ \left(1 + \frac{1}{n} \right)^n / n \in \mathbb{N} \right\}$ is, by Chap.3, ex.22.e, bounded from above and so has a least upper bound denoted universally by e which is one of the most important numbers of Mathematics. According to that exercise, we have $2 < e \leq 3$.

With the help of the axiom of the supremum, 1.4; 1.2; and 1.3 can be generalized:

Proposition 2.2: Let A be a non-empty and bounded from above subset of $\mathbb{R}$. Given $\varepsilon > 0$, there exists $a \in A$ such that $a + \varepsilon$ is an upper bound of A.

Proof: Let s be the least upper bound of A, as $s - \varepsilon$ cannot be an upper bound of A, there exists $a \in A$ such that $s - \varepsilon < a$, then $a + \varepsilon$ is an upper bound of A.∎

Theorem 2.3: $\mathbb{N}$ is not bounded from above.

Proof: For if it were bounded from above, it would have a supremum s. As s is an upper bound of $\mathbb{N}$, we would have $s \geq n \, \forall n \in \mathbb{N}$, then $s \geq n + 1 \, \forall n \in \mathbb{N}$, that is $s - 1 \geq n \, \forall n \in \mathbb{N}$ and $s - 1$ should result in an upper bound of N, from where, since s is the least upper bound, should follow $s \leq s - 1$ which is an absurd.∎

The following corollary was used in geometric form by Archimedes (given two segments (or magnitudes), a multiple of one of them will exceed the other) and bears his name, although it was stated explicitly and used before in Euclid's Elements.

Corollary 2.4: (Archimedeanity) Given $a, b \in \mathbb{R}$ with $a > 0$, $\exists n \in \mathbb{N}$ such that $na > b$.

Proof: By the above theorem, b/a cannot be an upper bound of $\mathbb{N}$, then there is $n \in \mathbb{N}$ such that $n > \frac{b}{a}$. ■

Example: Let us show that the least upper bound of $A = \{1 - (1/n) \, / \, n \in N\}$ is 1. Clearly, 1 is an upper bound of A. Let t be an upper bound of A, that is $t \geq 1 - (1/n)$ $\forall n \in \mathbb{N}$. If it were $t < 1$, we should have $1 - t > 0$ and by the corollary, it should exists $n \in \mathbb{N}$ such that $n(1 - t) > 1$, or $1 - (1/n) > t$, contradicting that t is an upper bound of A. We must have then $t \geq 1$ and 1 is the supremum of A.

Theorem 2.5: (Existence and uniqueness of the integral part) Given $x \in \mathbb{R}$, there is one, and only one, integer m such that:
$$m \leq x < m + 1$$
(such m is called the *integral part* of x and is usually denoted by $[x]$).

Proof: Consider first the case in which $x \geq 1$. As $\mathbb{N}$ is not bounded from above, the set:
$$\{n \in \mathbb{N}/n > x\}$$
is not empty, then, by well-ordering, it has a smallest element $r \in \mathbb{N}$ which must be > 1, since $x \geq 1$, Hence $m = r - 1 \in \mathbb{N}$ and we have $m \leq x < m + 1 = r$.

In cases: $0 \leq x < 1$ or $-1 \leq x < 0$ there is nothing to prove. Assume then $x < -1$, that is $-x > 1$ and, by the case just proved: $\exists m \in \mathbb{N}$ such that $m \leq -x < m + 1$, hence $-m \geq x > -m - 1$. If $x = -m, -m$ is the integral part of x, while if $x < -m$: $[x] = -m - 1$.

Let us show the uniqueness. Take $m, m' \in \mathbb{Z}$ such that $m \leq x < m + 1$ and $m' \leq x < m' + 1$. If it were $m < m'$ we would have by Prop.

1.1, e, Chap. 4: $m + 1 \leq m'$, and then $x < m + 1 \leq m' < x$, a contradiction. By symmetry, we have then $m = m'$. ∎

Note that, restricting to rational numbers, the existence of the integral part of $\frac{a}{b}$ $(a, b \in \mathbb{Z}, b > 0)$ is equivalent to the theorem on the division algorithm (there are $q, r \in \mathbb{Z}$ such that $a = bq + r$ and $0 \leq r < b$). In fact, if m=[a/b], we have $m \leq a/b < m + 1$, then $a = bm + r$ with 0≤r=a-bm<b. Conversely, from $a = bq + r$ with $0 \leq r < b$, it follows $a/b = q + \frac{r}{b}$, then $q \leq \frac{a}{b} < q + 1$ since $\frac{r}{b}$<1.

Although, as we will see soon $\mathbb{Q} \neq \mathbb{R}$, real numbers can be approached by rationals with any degree of approximation:

Theorem 2.6: (*Density* of $\mathbb{Q}$ in $\mathbb{R}$)   Given $x, y \in \mathbb{R}$ with $x < y$, there exists $r \in \mathbb{Q}$ such that $x < r < y$.

Proof: By Archimedeanity $\exists n \in \mathbb{N}$ such that $n(y - x) > 1$, then:
$$ny > 1 + nx \geq 1 + [nx] > nx$$
hence $y > (1 + [nx])/n > x$ and $(1 + [nx])/n \in \mathbb{Q}$. ∎

3- ROOTS

Lemma 3.1:   If a, b are   positive   real   numbers   and $n \in \mathbb{N}$, we have: $a < b \Leftrightarrow a^n < b^n$. ∎

This lemma is exercise 5.a, Chap. 3 and from it follows immediately, the uniqueness in the following,

Theorem 3.2: (*Existence of* n-*th roots of positive real numbers*) Given $a \in \mathbb{R}, a > 0$ and $n \in \mathbb{N}$, there exists one, and only one, $s \in \mathbb{R}, s > 0$ such that $s^n = a$.

Proof: Assume first that $a > 1$ and let:
$$A = \{b \in \mathbb{R}/b > 0 \text{ and } b^n < a\}$$
$A \neq \emptyset$ since, for example, $1 \in A$ and A is bounded from above since, for example, a is an upper bound of A, for if $b \in A$ we have $b^n < a$, but as $a > 1$, also $a^n > a$, then $b^n < a^n$ and by the above lemma $b < a$. By the axiom of the supremum, there exists $s = supA$. Clearly $s >$

146

0 . In order to prove that $s^n = a$, we will show that both $s^n < a$ and $s^n > a$ lead to contradiction.

Assume $s^n < a$, take ε such that $0 < \varepsilon < 1$. We have,

$$(s + \varepsilon)^n - s^n \leq \varepsilon \sum_{i=1}^{n} \binom{n}{i} s^{n-i} \varepsilon^i \leq \varepsilon \sum_{i=1}^{n} \binom{n}{i} s^{n-i}$$

Hence, if we put $M = \sum_{i=1}^{n} \binom{n}{i} s^{n-i}$, it is clear that $M > 0$ and as M does not depend on ε, we can choose ε such that (in addition to the condition $0 < \varepsilon < 1$): $\varepsilon \leq \frac{a-s^n}{M}$),

then$(s + \varepsilon)^n - s^n < a - s^n$ or, $(s + \varepsilon)^n < a$, but this means that $s + \varepsilon \in A$ which contradicts the fact that s is the least upper bound of A.

In case $s^n > a$ we proceed similarly: take ε such that $0 < \varepsilon < 1$ and we have,

$$s^n - (s - \varepsilon)^n \leq \varepsilon \sum_{i=1}^{n} \binom{n}{i} s^{n-i} (-\varepsilon)^i \leq \varepsilon \sum_{i=1}^{n} \binom{n}{i} s^{n-i}$$

putting $M = \sum_{i=1}^{n} \binom{n}{i} s^{n-i}$ and taking $\varepsilon \leq (s^n - a)/M$ we have,

$$s^n - (s - \varepsilon)^n \leq s^n - a$$

or $(s - \varepsilon)^n \geq a$. It follows then by the lemma, that $s - \varepsilon$ is an upper bound of A, which contradicts the fact that s is the supremum of A.

Then we must have $s^n = a$, so that the existence in case $a > 1$ is proven. If $a = 1$ the result is obvious. It remains to consider the case $a < 1$ (being allways $a > 0$). If $a < 1$, then $a^{-1} > 1$ and by the case just proved, there is an $s' > 0$ such that $s'^n = a^{-1}$, hence $(s'^{-1})^n = a$. ∎

Notation: If $a > 0$ and $n \in \mathbb{N}$, $\sqrt[n]{a}$ denotes the unique positive real number s such that $s^n = a$. This notation is not universal and often the symbol $\sqrt[n]{a}$ is used if a is not positive and even if it is not real. To avoid confusions (for example, the great mathematician Euler wrote: $\sqrt{-2}\,\sqrt{-3} = \sqrt{6}$ instead of $-\sqrt{6}$. Moreover, this kind of mistakes were systematic in Euler's works and they passed on to his disciples) we will use $\sqrt[n]{a}$ only with the given meaning, unless explicitly specified otherwise.

Proposition 3.3: Let a, b be positive real numbers, $n \in \mathbb{Z}, m, s \in \mathbb{N}$, then:

1) $\sqrt[m]{a^n} = \left(\sqrt[m]{a}\right)^n$.

2) $\sqrt[m]{\sqrt[s]{a}} = \sqrt[ms]{a}$.

3) $\sqrt[m]{ab} = \sqrt[m]{a}\sqrt[m]{b}$. ∎

Proof: 1) Put $x = \sqrt[m]{a^n} = \sqrt[m]{a}$, then $x^m = a^n$ and $y^m = a$, so $x^m = y^m$ hence by lemma 3.1, $x = y^n$.

The proofs of 2) and 3) are left as exercises.∎

Corollary 3.4: Let $a, n \in \mathbb{N}$; $\sqrt[n]{a}$ is rational $\Leftrightarrow$ a is a nth power of a natural.

Proof: Let $\sqrt[n]{a} = \frac{b}{c}$ with $b, c \in \mathbb{N}$, b and c pairwise prime, then $ac^n = b^n$ and the result follows from the Fundamental Theorem of Arithmetic.∎

From this corollary follows that $\sqrt{2} \notin \mathbb{Q}$, then $\mathbb{Q} \neq \mathbb{R}$ as we had anticipated. Real numbers which are not rational are called *irrationals*. Let us show that the irrationals are dense in $\mathbb{R}$:

Proposition   **3.5**:   *If $x, y \in \mathbb{R}$ with $x <$* y, there is some irrational z such that:

$$x < z < y$$

Proof: By archimedeanity, there is some $n \in \mathbb{N}$ such that $n\sqrt{2}(y - x) > 1$, then:

$$n\sqrt{2}y > n\sqrt{2}x + 1 \geq [n\sqrt{2}x] + 1 > n\sqrt{2}x$$

hence,

$$y > \frac{[n\sqrt{2}x] + 1}{n\sqrt{2}} > x$$

and $z = \frac{[n\sqrt{2}x]+1}{n\sqrt{2}}$ is irrational (if it were rational it would result, by a simple calculation, $\sqrt{2}$ rational).∎

Example: We will show a family of subfields of $\mathbb{R}$ that play a leading role in Algebraic Number Theory and (as we shall see) in the classical problems of geometric constructions with ruler and compass.

Let K be a subfield of R and $\alpha \in K$ such that $\alpha > 0$. Assume that $\sqrt{\alpha} \notin K$. We define:

$$K(\sqrt{\alpha}) = \{a + b\sqrt{\alpha} \ / a, b \in K\}$$

then $K(\sqrt{\alpha})$ becomes a subfield of $\mathbb{R}$.

We will prove only that each $a + b\sqrt{\alpha} \neq 0 (a, b \in K)$ has a multiplicative inverse in $K(\sqrt{\alpha})$, leaving the other verifications as an exercise. Note that $a + b\sqrt{\alpha} = 0 \Leftrightarrow a = b = 0$, since if $b \neq 0$, it would result $\sqrt{\alpha} = -a/b \in K$. Then let $a + b\sqrt{\alpha} \neq 0$, that is $a \neq 0$ or $b \neq 0$, we   have $a^2 - b^2\alpha \neq 0$ (if   not $a = b = 0$ or $\sqrt{\alpha} \in K$) and then the inverse of $a + b\sqrt{\alpha}$ is:

$$\frac{a - b\sqrt{\alpha}}{a^2 - b^2\alpha} \in K(\sqrt{\alpha}).$$

4- POWERS WITH RATIONAL EXPONENTS:

Let $a \in \mathbb{R}$ with $a > 0$, and let $n, m \in \mathbb{Z}$ with $m > 0$. We define $a^{\frac{n}{m}}$ by:

$$a^{\frac{n}{m}} = \sqrt[m]{a^n}$$

Note that this definition does not depend on the representant of the fraction $\frac{n}{m}$ since if $\frac{n}{m} = \frac{r}{s}$ with $r, s \in \mathbb{Z}, s > 0$, putting $x = \sqrt[m]{a^n}, y = \sqrt[s]{a^r}$, we have $x^m = a^n, y^s = a^r$, then $x^{ms} = a^{ns} = a^{rm} = y^{sm}$ and by lemma 3.1., results $x = y$.

Proposition 4.1: If a, b are positive real numbers and q, q' rational numbers, then:

1) $a^q a^{q'} = a^{q+q'}$.
2) $(a^q)^{q'} = a^{qq'}$.
3) $(ab)^q = a^q b^q$.
4) If $a > 1$, then $q < q' \Leftrightarrow a^q < a^{q'}$.

If $a < 1$, then $q < q' \Leftrightarrow a^q > a^{q'}$.

Proof: It is about, in each case, of reducing the property to the corresponding one with whole exponents (prop. 2.2, Chap. 4). For example,   for   1)   put $q = \frac{n}{m}, q' = \frac{r}{s}$ $(n, r \in \mathbb{Z}, m, s \in \mathbb{N})$ and let $x = a^q, y = a^{q'}$, so then $x^m = a^n$ and $y^s = a^r$, hence

$$(xy)^{ms} = x^{ms} y^{ms} \, a^{ns} = a^{mr} = a^{ns+mr}$$

then $a^q a^{q'} = xy = \sqrt[ms]{a^{ns+mr}} = a^{q+q'}$.

2) and 3) follow easily from prop. 3.3.

With the same notations, let us show that if $a > 1$ and $q < q'$ then $a^q < a^{q'}$. We have $\frac{n}{m} < \frac{r}{s}$, that is $ns < rm$ and by the corresponding property for integral exponents: $a^{ns} < a^{mr}$, but $a^{ns} = x^{ms}$, $a^{rm} = y^{ms}$ from where $x^{ms} < y^{ms}$ and, by lemma 3.1., $x < y$. The proof of the remainder of the proposition is left as an exercise.∎

5- MULTIPLICATIVE VERSIONS

In this section, we consider what we can call multiplicative versions of archimedeanity, integral part and density of $\mathbb{Q}$ in $\mathbb{R}$.

Theorem 5.1: (Multiplicative archimedeanity) Given the real numbers $a > 1$ and b, there exists $n \in \mathbb{N}$ such that $a^n > b$.

Proof: Put $a = 1 + c$ with $c > 0$. By exercise 5.b., Chap. 3 $a^n = (1 + c)^n \geq 1 + nc$, for any $\mathbb{N}$, then to have $a^n > b$, it is enough to take n such that $nc > b - 1$ and such n exists by ordinary archimedeanity.∎

Theorem 5.2 (multiplicative version of the existence of an integral part): Let $a > 1$. Given $y > 0$, there exists $n \in \mathbb{Z}$ such that:

$$a^n \leq y < a^{n+1}.$$

Proof: Assume first that $y \geq 1$, by the previous theorem, there are natural numbers m such that $a^m > y$. Let $n + 1$ be the least of such m ($n \in \mathbb{N} \cup \{0\}$). Then, if $n \in \mathbb{N}$ we have $a^n \leq y$ by minimality of m and, if $n = 0$, we have $a^0 = 1 \leq y$ by hypothesis. If $y < 1$, as $y^{-1} > 1$, by the case just proved we have $a^n \leq y^{-1} < a^{n+1}$ for some $n \in \mathbb{Z}$, and so $a^{-n} \geq y > a^{-n-1}$ and from this the theorem follows readily.∎

Theorem 5.3 (multiplicative density): Let $a > 1, 0 < x < y$. There is some $q \in \mathbb{Q}$, such that:

$$x < a^q < y$$

Proof: As $y > x$ and $x > 0$, we have $yx^{-1} > 1$ and by multiplicative archimedeanity $\exists n \in \mathbb{N}$ such that $(yx^{-1})^n > a$, or $y^n > ax^n$, but by the above theorem, as $x > 0$, there exists $r \in \mathbb{Z}$ such that
$$a^r \leq x^n < a^{r+1}$$
then $y^n > ax^n \geq a^{r+1} > x^n$ and by lemma 3.1., results:
$$y > a^{\frac{r+1}{n}} > x. \blacksquare$$

6 - EXPONENTIAL FUNCTION

Let $a, x \in \mathbb{R}$ with $a > 0$. If $a \geq 1$ we define:
$$a^x = sup\{a^q/q \in \mathbb{Q}, q \leq x\}$$
This supremum exists, since the set
$$\{a^q/q \in \mathbb{Q}, q \leq x\}$$
Is obviously non-empty, and it is bounded from above, since any $a^{q'}$ with rational q' such that $q' \leq x$ is an upper bound of it by prop. 4.1.4.

In case $a < 1$ (being allways $a > 0$), we define:
$$a^x = [(a^{-1})^x]^{-1}$$
If x is rational, it is clear that these definitions are consistent with the notation of powers of rational exponent by prop. 4.1.

Lemma 6.1: Let A, B be subsets of $\mathbb{R}_{>0}$ (the positive real numbers). Assume that there exist $s = supA$ and $t = supB$ and set:
$$C = \{ab/a \in A, b \in B\}$$
then $supC$ exists and $supC = st$.

Proof: If $ab \in C$ with $a \in A$ and $b \in B$, we have $a \leq s, b \leq t$ and as they are all positive, it follows that $ab \leq st$, so st is an upper bound of C.

Let r be an upper bound of C and assume, by contradiction, that $st > r$. We shall prove that there exist $a \in A, b \in B$ such that $r < ab$, reaching a contradiction. Put $\varepsilon = st - r > 0$, and take $\delta > 0$ such that the following conditions are verified:

$$\delta < s, \delta < t, \delta < \frac{\varepsilon}{t+s}$$

By prop.2.2, as $s = supA$, there is an $a \in A$ such that $s-\delta<a$ and as $t = supB$, there is a $b \in B$ such that $t - \delta < b$, then:

$$(s - \delta)(t - \delta) = st - \delta(t + s - \delta) < ab$$

and hence:

$$r = st - \varepsilon < st - \delta(t + s) < st - \delta(t + s - \delta) < ab. \blacksquare$$

Theorem 6.2: If $a, x, y \in \mathbb{R}$ with $a > 0$, then:

$$a^x a^y = a^{x+y}$$

Proof: Assume first that $a > 1$. We have:

$$a^x = sup\{a^q/q \in \mathbb{Q}, q \le x\}$$
$$a^y = sup\{a^{q'}/q' \in \mathbb{Q}, q' \le y\}$$
$$a^{x+y} = sup\{a^{q''}/q'' \in \mathbb{Q}, q'' \le x + y\}$$

By the lemma above, we have

$$a^x a^y = sup\{a^{q+q'}, \in \mathbb{Q}, q \le x, q' \le y\}$$

and it will be sufficient to verify that the sets:

$$A = \{a^{q+q'}, q, q' \in \mathbb{Q}, q \le x, q' \le y\} \text{ and } B = \{a^{q'}/q' \in \mathbb{Q}, q' \le x + y\}$$

have the same supremum. As $A \subset B$, it is enough to verify that $s = supA = a^x a^y$ is an upper bound of B. Assuming that is not the case, i.e., that exist $q' \in Q, q' \le x + y$ such that $s < a^{q'}$; we will reach a contradiction. Put $2\varepsilon = x + y - q' > 0$ (if it were $q'' = x + y$, we replace it by a q''' such that $s < a^{q'''} < a^{q'}$ which is possible by multiplicative density and so $\varepsilon > 0$), by ordinary density, there are $q, q' \in \mathbb{Q}$ tales que

$$x - \varepsilon \le q \le x, y - \varepsilon \le q' \le y$$

then $q'' = x + y - 2\varepsilon \le q + q' \le x + y$. Hence by prop.4.1, 4):

$$a^{q+q'} \ge a^{q''} > s$$

Which contradicts the selection of s as supremum of A.

The proof for the case $a \le 1$ is left as an exercise.$\blacksquare$

Theorem 6.3: Let $a > 0$ with $a \ne 1$. The function $f : \mathbb{R} \to \mathbb{R}_{>0}$, defined by

$$f(x) = a^x$$

is a bijection (f is called *exponential function* of base a). Moreover, it is *strictly increasing* $(x < y \Rightarrow f(x) < f(y))$ if $a > 1$ and strictly decreasing if $a < 1$.

Proof: We will do the case $a > 1$, leaving the other $(a < 1)$ as an exercise.

Let us first show that f is strictly increasing. Let $x < y$. By definition we have:

$$a^x = sup\{a^q/q \in \mathbb{Q}, q \leq x\}$$
$$a^y = sup\{a^{q'}/q' \in \mathbb{Q}, q' \leq x\}$$

As $\mathbb{Q}$ is dense in $\mathbb{R}$, there exists $q'' \in \mathbb{Q}$ such that $x < q'' < y$, then by prop. 4.1,4 $a^{q''}$ is an upper bound of $\{a^q/q \in \mathbb{Q}, q \leq x\}$ and then $a^x \leq a^{q''}$. Moreover, $a^{q''} < a^y$, since as a^y is an upper bound of $\{a^{q'}/q' \in Q, q' \leq y\}$, taking $q' \in \mathbb{Q}$ such that $q'' < q' \leq y$, by prop. 4.1,4 we obtain $a^{q''} < a^{q'} \leq a^y$. So we have proved that f is strictly increasing, and from that follows that f is injective, for if $f(x) = f(y)$, that is $a^x = a^y$, we have $x = y$ (if it were $x < y$ it would be $a^x < a^y$).

Let us see that f is surjective. Let $z \in \mathbb{R}_{>0}$, we have:

$$z = sup\{a^q/q \in \mathbb{Q}, a^q \leq z\}$$

since $A = \{a^q \in Q / a^q \leq z\} \neq \emptyset$ by 5.2, z is an upper bound of A and if t is an upper bound of A, if it were $t < z$, by multiplicative density it would exists $q \in \mathbb{Q}$ such that $t < a^q < z$ in contradiction with the election of t as an upper bound of A, then $z = supA$.

Putting $B = \{q \in \mathbb{Q} / a^q \leq z, \}$we have that B is not empty and is bounded from above, for by multiplicative density $\exists n \in \mathbb{N}$ such that $a^n > z \geq a^q \ \forall q \in B$, and so $n > q$. Then there exists $x = supB$ and let:

$$C = \{a^{q'}/q' \in \mathbb{Q}, q' \leq x\}$$

in such a way that (by definition) $a^x = supC$.

Finally, we will check that $A = C$, from where $a^x = z$ will follow. We have $A \subset C$, for $a^q \in A \Rightarrow a^q \leq z \Rightarrow q \in B \Rightarrow q \leq x \Rightarrow a^q \in C$.

Let us show that $C \subset A$. Let $a^{q'} \in C$ with $q' \leq x, q' \in Q$ and assume $a^{q'} \notin A$ i. e., $a^{q'} > z$. By multiplicative density, there exists $q'' \in Q$ such that $a^{q'} > a^{q''} > z$, then $q' > q''$ and since from $a^{q''} > z$, it follows that $q'' > q \ \forall q \in B$, and then $q'' > x$ and $q' > x$, which is a contradiction. ∎

Since the exponential function of base a is a bijection from $\mathbb{R}$ onto $\mathbb{R}_{>0}$, it admits an inverse function called a logarithmic function which is denoted by $\log_a$. We have $\log_a$ $: \mathbb{R}_{>0} \to \mathbb{R}$ and:

Corollary 6.4: If x, y are positive real numbers, then:

$$\log_a xy = \log_a x + \log_a y$$

Proof: It follows readily from definition and 6.2.■

7 - SEQUENCES

In order to prove some other fundamental properties of real numbers, we need the notions of sequences and its limits.

A sequence in a field K is a function $f: \mathbb{N} \to K$. If $f(n) = a_n$ we denote the sequence f by (a_n) or $(a_1, a_2, \ldots, a_n, \ldots)$. If there are more than one index within the parenthesis, we denote the sequence by $(a_n)_n$. Given two sequences: $(a_n), (b_n)$ we define its sum and its product by:

$$(a_n) + (b_n) = (a_n + b_n) \;;\; (a_n)(b_n) = (a_n b_n)$$

and we have:

Proposition 7.1: The set S of all sequences in a field K, with the operations just defined, is a ring with $(0, 0, \ldots, 0, \ldots)$ as the neutral element of the sum and $(1, 1, \ldots, 1, \ldots)$ as the neutral element of the product.

Proof: Exercise. ■

We have already defined, implicitly or explicitly, what is an *ordered field*: a field in which there is an order relation that satisfies the transitive, compatibility with the operations and trichotomy properties stated in Ch. 2. As has been done there, in any ordered field, the order determines an absolute value defined by:

$$|a| = \begin{cases} a \text{ if } a \geq 0 \\ -a \text{ if } a < 0 \end{cases}$$

Also, any ordered field contains $\mathbb{N}$ and $\mathbb{Q}$ for that is essentially what we have done in early chapters.

Any subfield of $\mathbb{R}$ is an ordered field, so are $\mathbb{Q}$ and the fields $F(\sqrt{a})$ of the example at the end of section 3.

A sequence (a_n) in an ordered field K, is said to be convergent iff there exists $a \in K$ such that for any $\varepsilon > 0$ ($\varepsilon \in K$) there exists $n_0 \in \mathbb{N}$ such that: $n \geq n_0 \Rightarrow |a_n - a| < \varepsilon$. In this case (a_n) is said to *converge* to a, and we write $(a_n) \to a$ when $n \to \infty$, or simply $(a_n) \to a$ and also $\lim_{n \to \infty} a_n = a$, and a is called the limit of the sequence.

Proposition 7.2: Every convergent sequence in an ordered field K is bounded (from above and from below).

Proof: Let (a_n) be a convergent sequence in K, and let $(a_n) \to a \in K$ when $n \to \infty$. Then $\exists\, n_0$ such that $n \geq n_0 \Rightarrow |a_n - a| < 1$, from where $|a_n| < 1 + |a|$, and simply take $M = max\{1 + |a|, |a_1|, \ldots, |a_{n_0}|\}$ to obtain $|a_n| \leq M \ \forall n \in \mathbb{N}.\blacksquare$

If not explicitly specified, we consider in this and the following section, sequences in an ordered field K. A sequence can not have more than one limit.

Proposition 7.3: If $(a_n) \to a$ and $(a_n) \to b$ then $a = b$.

Proof: Assuming that $a > b$, then there exist n_0 and n_1 such that:

$$|a_n - a| < \frac{a - b}{2} \ / \ \forall n \geq n_0 \ and \ |a_n - b| < \frac{a - b}{2} \ \forall n \geq n_1,$$

and therefore, $a - b \leq |a - a_n| + |a_n - b| < a - b$ which is contradictory.$\blacksquare$

Proposition 7.4: The set S' of convergent sequences in an ordered field K, is a subring of the ring S of sequences in K and $\lim_{n \to \infty}$ is a morphism of rings, which means that if $(a_n) \to a$ and $(b_n) \to b$ then:

$$(a_n) + (b_n) \to a + b \ ; \ and \ (a_n)(b_n) \to ab.$$

Proof: Let us make the limit of the product, leaving the remainder as an exercise. Let: $(a_n) \to a$ and $(b_n) \to b$. By prop. 7.2 $\exists M$ such that $|a_n| < M \; \forall n \in \mathbb{N}$. Given $\varepsilon > 0$, if $b = 0$, simply take n_0 so that $n \geq n_0 \Rightarrow |b_n| < \frac{\varepsilon}{M}$, to obtain $|a_n b_n| < \varepsilon$. If $b \neq 0$ take n_1 such that $|b_n - b| < \frac{\varepsilon}{2M}$, $\forall n \geq n_1$ and take n_2 such that $|a_n - a| < \varepsilon/(2|b|)$, $\forall n \geq n_2$ and taking $n_3 = max\{n_1, n_2\}$ we have: $n \geq n_3$ implies:

$$|a_n b_n - ab| \leq |a_n||b_n - b| + |b||a_n - a| < \frac{M\varepsilon}{2M} + \frac{|b|\varepsilon}{2|b|} = \varepsilon. \; \blacksquare$$

Note that, for the purposes of its convergence, it does not matter whether we replace a finite number of terms, by any values in a convergent sequence, or if we take certain subsequences, etc., for example, in $\mathbb{R}$ the sequences $\left(\frac{1}{n}\right), \left(\frac{1}{2^n}\right)$ and $(0, 0, \ldots, 0, \frac{1}{m}, \frac{1}{m+1}, \ldots)$ have the same limit.

In principle we could say that two convergent sequences are equivalent, if they have the same limit, and that is an equivalence relation in the set S' of convergent sequences, but note, that to say that two convergent sequences have the same limit, is equivalent to say that their difference tends to 0, but the latter defines an equivalence relation in the set S of all sequences, whether or not convergent. For this reason, we adopt the second definition, i. e., $(a_n), (b_n)$ are sequences in an ordered field K we say that they are equivalent, and write $(a_n) \sim (b_n)$, iff $(a_n - b_n) \to 0$. It is immediate to check that $\sim$ is an equivalence relation in S. We have:

Proposition 7.5: Let $(a_n) \to a$ with $a \neq 0$. There is some sequence (b_n) such that $(b_n) \sim (a_n)$; $b_n \neq 0 \; \forall n$ and that $\left(\frac{1}{b_n}\right) \to (1/a)$.

Proof: $\exists n_0$ such that $|a - a_n| < \frac{|a|}{2} \; \forall n \geq n^0$, then $|a| - |a_n| < \frac{|a|}{2}$ and therefore $|a_n| > \frac{|a|}{2}$ so that $a_n \neq 0 \; \forall n \geq n_0$. Define $b_n = a_n$ if $n \geq n_0$ and $b_n = 1$ if $n < n_0$. It is clear then that $(b_n) \sim (a_n)$ and that $b_n \neq 0 \; \forall n$. To check that $\frac{1}{b_n} \to \frac{1}{a}$, let $n \geq n_0$ then (since in that case $b_n = a_n$ and $\frac{1}{|a_n a_{\{n\}}|} < \frac{2}{|a|}$:

$$\left|\frac{1}{b_n} - \frac{1}{a}\right| = \left|\frac{1}{a_n} - \frac{1}{a}\right| = \frac{1}{|a_n a|}|a_n - a| < \frac{2}{|a|^2}|a_n - a|$$

Then, given $\varepsilon > 0$, it is enough to take n_1 such that $|a_n - a| \leq \frac{\varepsilon|a|^2}{2}$ $\forall n \geq n_1$ to obtain

$$\left|\frac{1}{b_n} - \frac{1}{a}\right| < \varepsilon \ \forall n \geq max\{n_0, n_1\}. \blacksquare$$

If in the quotient set $\frac{S'}{\sim}$ we define the sum and the product of classes of sequences by:

$$\overline{(a_n)} + \overline{(b_n)} = \overline{(a_n + b_n)} \ ; \ \overline{(a_n)}\overline{(b_n)} = \overline{(a_n b_n)}$$

it is easily verified that these operations are well defined, and we have:

Proposition 7.6: With the just defined operations $\frac{S'}{\sim}$ is a field.

Proof: We will see that each nonzero element has a multiplicative inverse. Let $(a_n) \neq 0$, that is, (a_n) does not converge to 0, but since (a_n) is convergent, there exists $a \in K$ such that $(a_n) \to a$, with $a \neq 0$, so by prop. 7.6, there exists a sequence (b_n) such that $(\frac{1}{b_n}) \to \frac{1}{a}$, then $(a_n)(\frac{1}{b_n})) \to 1$, so that $(\frac{1}{b_n})$ is the multiplicative inverse of (a_n). $\blacksquare$

8 - CAUCHY SEQUENCES

A sequence (a_n) in an ordered field K is said to be a Cauchy sequence, or that it satisfies the condition of Cauchy, iff for any $\varepsilon > 0$ ($\varepsilon \in K$), exists n_0 such that $n, m \geq n_0 \Rightarrow |a_n - a_m| < \varepsilon$.

We can say, informally speaking, that a sequence is convergent, if as n grows, the values of the sequence are closer and closer to a number which is its limit. In the same way, we can say that a sequence is Cauchy if, as n grows, its values are closer to each other. As we shall see, for sequences in $\mathbb{R}$, being Cauchy and being convergent is equivalent and hence the utility of Cauchy's condition, because it allows convergence to be established without the need to know the limit. However, this equivalence is not valid in general.

Lemma 8.1: Every Cauchy sequence is bounded.

Proof: There exists n_0 such that: $n, m \geq n_0 \Rightarrow |a_n - a_m| < 1$, in particular, $|a_n - a_{n_0}| < 1$, so then $|a_n| < 1 + |a_{n_0}|$.∎

Proposition 8.2: Every convergent sequence is Cauchy's.

Proof: Let $(a_n) \to a$ and let $\varepsilon > 0, \exists\, n_0$ such that $|a_n - a| < \frac{\varepsilon}{2}, \forall\, n \geq n_0$. Then if $n, m \geq n_0$ we have:

$$|a_n - a_m| \leq |a_n - a| + |a - a_n| < \varepsilon. \blacksquare$$

Proposition 8.3: The set S'' formed by the Cauchy sequences is a subring of S.

Proof: It is enough to verify that the sum and product of two Cauchy sequences are Cauchy, and that is left as an exercise.∎

Proposition 8.4: If (a_n) is Cauchy and $(a_n) \nsim 0$ (is not equivalent to zero or, what is the same, does not tend to zero) then there exists $(b_n) \sim (a_n)$ such that $\left(\frac{1}{b_n}\right)$ is Cauchy.

Proof: As (a_n) does not tend to zero, there exists $a > 0$ such that $\forall\, n_0$ exists $r > n_0$ such that $|a_r| \geq a$, but since (a_n) is   Cauchy,   exists n_1 such that $n, m \geq n_1 \Rightarrow |a_n - a_m| < \frac{a}{2}$, then taking on the above $n_0 = n_1$,

$$|a_r| - |a_m| \leq |a_r - a_m| < \frac{a}{2} \,\forall\, m \geq n_1$$

Therefore $a \leq |a_r| < |a_m| + \frac{a}{2}$ so   that $|a_m| > \frac{a}{2} \,\forall\, m \geq n_1$. If we define $b_n = a_n$ if $n \geq n_1$ and $b_n = 1$ if $n < n_1$, it is clear that $(b_n) \sim (a_n)$, and that $b_n \neq 0 \,\forall\, n$ and for $n, m \geq, n_1$, we have:

$$\left|\frac{1}{b_n} - \frac{1}{b_m}\right| = \left|\frac{1}{a_n} - \frac{1}{a_m}\right| = \frac{1}{|a_n a_m|}|a_n - a_m| < \frac{4}{a^2}|a_n - a_m|$$

Then given $\varepsilon > 0$, there exists n_2 such that $n \geq n_2 \Rightarrow |a_n - a_m| < \frac{\varepsilon|a^2|}{4}$ and finally:

158

$$\left| \frac{1}{b_n} - \frac{1}{b_m} \right| < \varepsilon. \blacksquare$$

The above proposition states that the class of a Cauchy sequence not equivalent to zero is invertible, then:

Proposition 8.5: The quotient set $\dfrac{S''}{\sim}$ is a field. $\blacksquare$

9 - SEQUENCES IN $\mathbb{R}$

We now are in a position to prove some fundamental properties of real numbers:

Theorem 9.1: (monotone sequences) In $\mathbb{R}$ any increasing sequence (a_n) (increasing means $a_n \leq a_{n+1}$ $\forall n \in \mathbb{N}$) which is bounded from below, that is the set $A = \{a_n/n \in \mathbb{N}\}$ of the values of the sequence is bounded from below, converges to $a = supA$.

Proof: Given $\varepsilon > 0$, by prop. 2.2, Chap. 6, $\exists\, n_0$ such that $a_{n_0} + \varepsilon > a$, then if $n \geq n_0$ we have $a - a_n < a - a_{n_0} < \varepsilon. \blacksquare$

Of course, the dual of the above theorem is valid, that is in $\mathbb{R}$ any decreasing sequence bounded from below is convergent. Prove it.

Example 1: The sequence $\sqrt[n]{n}$ is decreasing if $n \geq 3$, for $\sqrt[n]{n} > \sqrt[n+1]{n+1} \Leftrightarrow n^{n+1} > (n+1)^n \Leftrightarrow n > \left(1 + \frac{1}{n}\right)^n$ and the later is valid for $n \geq 3$. Furthermore, 1 is clearly a lower bound of the set of the values of the sequence, then it must be convergent to a number $a \geq 1$. If it were $a > 1$, it would exists $m \in \mathbb{N}, m > 3$ such that $a > 1 + \frac{1}{m}$ (by archimedeanity), then taking $n = m^2$, we have:

$$a^n = a^{m^2} > \left(1 + \frac{1}{m}\right)^{m^2} > 2^m > m^2 = n$$

That is, there exists $n \in \mathbb{N}$ such that $a^n > n$, so that it would result in $a > \sqrt[n]{n}$ which contradicts the fact that a is the least lower bound of the terms of the sequence.

Example 2: In section 2 we have defined the number e as the supreme of the set:

$$A = \left\{\left(1+\frac{1}{n}\right)^n /n \in \mathbb{N}\right\}$$

As we shall see, the sequence $\left(1+\frac{1}{n}\right)^n$ is increasing, so then converges to e. Moreover, if we consider this sequence in $\mathbb{Q}$, it turns out to be non-convergent as it will be seen that e is not rational.

Theorem 9.2: (nested intervals) Any sequence of nested closed intervals in $\mathbb{R}$:

$$I_1 \supset I_2 \supset ... \supset I_n \supset ...$$

with $I_n = [a_n, b_n]$ has a non-empty intersection. Moreover, if $(b_n - a_n) \to 0$, then the intersection has only one element.

Proof: The sequence (a_n) is increasing and the sequence (b_n) is decreasing, since $I_{n+1} \subset I_n$ implies $a_n \le a_{n+1} \le b_{n+1} \le b_n$. (a_n) is bounded from above (b_1 is an upper bound) and (b_n) is bounded from below, so by prop. 7.2, $(a_n) \to a = sup\{a_n\}$ and $(b_n) \to b = inf\{b_n\}$. As $a = sup\{a_n\}$, given $\varepsilon > 0$ there exists n_0 such that $a - a_{n_0} < \varepsilon$, and as $b = inf\{b_n\}$, there exists n_1 such that $b_{n_1} - b < \varepsilon$, then for any $n \ge n_2 = max\{n_0, n_1\}$ we have:

$$a - b + b_n - a_n < 2\varepsilon$$

We must have $a \le b$, for if $a > b$, taking above $\varepsilon = \frac{a-b}{2}$ we obtain $b_n a_n < 0$, a contradiction. Hence for any $n \in \mathbb{N}$:

$$a_n \le a \le b \le b_n$$

so that, both a and b belong to the intersection of all intervals, and moreover any c such that $a_n \le c \le b_n \, \forall n$, must comply: $a \le c \le b$, and then if $(b_n - a_n) \to 0$, we have that $a = b$ is the only element of the intersection.■

We have seen that in any ordered field, any convergent sequence is a Cauchy sequence. If the converse is valid, the field is said to be *complete* or *Cauchy complete*. An ordered field K is said to be *Archimedean* iff given $a, b \in K$ with $a > 0$, there exists $n \in \mathbb{N}$ such that $na \ge b$.

Theorem 9.3: Every Archimedean, complete, ordered field K satisfies the least upper bound property.

Proof: Let A be a non-empty, bounded from above, subset of K. Denote by B the set of all upper bounds of A. Take $a \in A, b \in B$ and let $m = b - a$, and $c = \frac{b+a}{2}$. If c is an upper bound of A, we set $a_1 = a, b_1 = c$, whereas if c is not an upper bound of A, $\exists a' \in A, a' \geq c$ and we set $a_1 = a', b_1 = b$. In any case, we have:

$$a_1 \in A, b_1 \in B; \ b_1 - a_1 \leq \frac{m}{2}; \ b - b_1 \leq \frac{m}{2}$$

Let us recommence the process with a_1, b_1 and reiterate, we obtain two sequences $(a_n), (b_n)$ such that:

$$a_n \in A, b_n \in B; \ b_n - a_n \leq \frac{m}{2^n}; \ b_{n-1} - b_n \leq \frac{m}{2^n} \ (*)$$

Given $\varepsilon > 0$, by archimedeanity, there exists $n_0 \in \mathbb{N}$ such that $n_0 \varepsilon \geq m$, then for $n \geq n_0$ we have:

$$\frac{m}{2^n} \leq \frac{m}{2^{n_0}} \leq \frac{m}{n_0} \leq \varepsilon$$

Which proves that $(b_n - a_n) \to 0$ and from where it follows that (b_n) is Cauchy.

By the hypothesis of completeness, it results (b_n) convergent and call s to its limit. We assert that $s = \sup A$. Indeed, s is an upper bound of A, since if $a \in A$ we have $a \leq b_n \ \forall n$, then $a \leq s$. In addition, if t is an upper bound of A we have $t \geq a_n \ \forall n$, then $b_n - t \leq b_n - a_n$, from where it follows that if there exists n such that $b_n < t$, is clear that $s \leq t$, whereas if $b_n \geq t \ \forall n$, results $(b_n) \to t$ and therefore $t = s$. ∎

Theorem 9.4: If an ordered field K satisfies the incresing sequences property, then it satisfies the least upper bound property.

Proof: Let A be a non-empty, bounded from above, subset of K. We repeat the first part of the above theorem until $(*)$. By construction (a_n) is an increasing sequence bounded from above, then by hypothesis, there exists $a \in K$ such that $(a_n) \to a$. We will prove that $a = \sup A$. a must be an upper bound of A, for if not, there is an n_0 such that $a < a_{n_0}$, and as the sequence is increasing we have: $a < a_{n_0} \leq a_n \ \forall n \geq n_0$. Put $\varepsilon = a_{n_0} - a$, then there exists n_1 such that $n \geq n_1 \Rightarrow a_n - a < \varepsilon = a_{n_0} - a$, then $a_n < a_{n_0}$ for any $n \geq n_0$, a contradiction.

Hence a is an upper bound of A. Let b be any upper bound of A. If it were $b < a$, it would exists n_2 such that $n \geq n_2 \Rightarrow a - a_n < a - b$, then $a_n > b$ which is absurd.∎

Theorem 9.5: The field $\mathbb{R}$ of real numbers is complete.

Proof: Let (a_n) be a Cauchy sequence in $\mathbb{R}$ and consider the sets:
$$A = \{x/x \leq a_n \ \forall n \geq n_1\}$$
$$B = \{y/y \geq a_n \ \forall n \geq n_2\}$$
More precisely A consists of the real numbers x such that there exists some n_1 (which depends on x) such that $x \leq a_n \ \forall n \geq n_1$, and analogously B. Since every Cauchy sequence is bounded, A is bounded from above and B is bounded from below. Then there exist:
$$a = supA \ ; \ b = infB$$
And since $x \leq y$ whichever be $x \in A, y \in B$, we have $a \leq b$. We assert that $a = b$, since if it were $a < b$ let $\varepsilon = \frac{b-a}{3}$, as (a_n) is Cauchy ∃ n_0 such that:
$$|a_n - a_m| < \varepsilon \ \forall n, m \geq n_0 \ (1)$$
But since $a = supA$, ∃ $x \in A$ such that $x + \varepsilon \notin A$ (prop.2.2), but this means that for any n_1 exists $n \geq n_1$ such that $a_n < x + \varepsilon$, taking in particular $n_1 = n_0$, we have:
$$\exists n \geq n_0 \text{ such that } a_n < x + \varepsilon \ (2)$$
Proceeding in the same way with B, we obtain:
$$\exists m \geq n_0 \text{ such that } y - \varepsilon < a_m \ (3)$$
but, as $y - x \geq b - a$, from (2) and (3) follows:
$$a_m - a_n > y - x - 2\varepsilon \geq b - a - 2\varepsilon = \varepsilon$$
Which contradicts (1) and therefore $a = b$.

Let us finally see that $(a_n) \to a$. Now let ε be any real and positive number (not necessarily the previous one). Let $x \in A$ be such that $a - x < \varepsilon$, and let $y \in B$ such that $y - a < \varepsilon$. Since $x \in A$ exists n_1 such that $x \leq a_n \ \forall n \geq n_1$, and likewise there exists n_2 such that $a_n \leq y \ \forall n \geq n_2$. Taking $n_3 = max\{n_1, n_2\}$ yields $\forall n \geq n_3$:
$$a - \varepsilon < x \leq a_n \leq y < a + \varepsilon.∎$$

Ex. 1: Prove: a) $1 < \sqrt{2} < \sqrt{3} < 2$.

b) $\sqrt{8} + \frac{1}{\sqrt{8}} \in \mathbb{Q}(\sqrt{2})$ where $\mathbb{Q}(\sqrt{2}) = \{a + b\sqrt{2}/a, b \in \mathbb{Q}\}$.

Ex. 2: If $a, b > 0$:

a) $a < b \Leftrightarrow \sqrt{a} < \sqrt{b}$

b) $\sqrt{ab} = \sqrt{a}\sqrt{b}$

c) $\sqrt{a^2 b} = a\sqrt{b}$

Ex. 3: If $a > b^2$, prove:

$$\sqrt{a + 2b\sqrt{a - b^2}} + \sqrt{a - 2b\sqrt{a - b^2}} = \begin{cases} 2b \ si \ a < 2b^2 \\ 2\sqrt{a - b^2} \ si \ a \geq 2b^2 \end{cases}$$

Ex. 4: The following numbers are irrational:

a) $\sqrt{2} + \sqrt{3}$, b) $\sqrt[6]{2} + \sqrt{3}$, c) $(1 + \sqrt{2})^3$

Ex. 5: If x is irrational and a, b, c, d are rational. then:

a) $a + bx = c + dx \Leftrightarrow a = c$ and $b = d$

b) $\frac{a+bx}{c+dx}$ is rational $\Leftrightarrow$ $ad = bc$

Ex. 6: A segment AB is said to be divided in golden section by C iff $AB \cdot AC = BC^2$. Prove that $\frac{AC}{BC}$ is irrational.

Ex. 7: a, b, c rationals and $a\sqrt{2} + b\sqrt{3} + c\sqrt{5} = 0 \Rightarrow a = b = c = 0$.

Ex. 8: a) Let $a_0, a_1, \ldots, a_n\}$ integers and $\frac{r}{s}$ with $r, s \in \mathbb{Z}$ coprime a rational root of:

$$a_n x^n + \ldots + a_1 x + a_0 = 0$$

then $s \mid a_n$ and $r \mid a_0$.

b) Find the rational roots of: $9x^3 - 6x^2 + 15x - 10$.

Ex. 9: Find all the integers a, b such that:

$$\sqrt[3]{7 + 5\sqrt{2}} = a + b\sqrt{2}$$

*Ex. 10: Given $a > 0$, we define recursively:

$$x_1 = \sqrt{a}, \ x_{n+1} = \sqrt{a + x_n}$$

Prove that $A = \{x_n / n \in \mathbb{N}\}$ has a least upper bound and find it.

Ex. 11: Define lower bound and greatest lower bound and prove that any non-empty set of real numbers bounded from below has a greatest lower bound.

Ex. 12: If there exist a greatest lower bound, it is unique.

*Ex. 13: Given $a, b \in \mathbb{R}$ such that $0 < a < b$, we define recursively:

$$a_1 = a, b_1 = b, a_{n+1} = \sqrt{a_n b_n}, b_{n+1} = \frac{a_n + b_n}{2}$$

Prove:

　　1) $a_n < a_{n+1}$ and $b_{n+1} < b_n$ $\forall n \in \mathbb{N}$.

　　2) $A = \{a_n / n \in \mathbb{N}\}$ is bounded from above and $B = \{b_n / n \in \mathbb{N}\}$ is bounded from bellow.

　　3) $\sup A = \inf B$ (this common value is called *arithmetic-geometric mean* of a and b).

Ex. 14: Find necessary and sufficient conditions on a rational x in order that $3x^2 - 7x$ be an integer.

Ex. 15: Imitating the proof of 1.4, show its multiplicative version:

Let A be a set of positive rational numbers such that it has an upper bound in $\mathbb{Q}$. Given $t \in \mathbb{Q}$ with $t > 1$, there exists $a \in A$ such that ta is an upper bound of A.

Ex. 16: Idem with the multiplicative version of 2.2:

If A is a non-empty and bounded from above set. of positive real numbers and $t > 1$, then exists $a \in A$ such that ta is an upper bound of A.

*Ex. 17: There are one, and only one, function $f : \mathbb{N} \to \mathbb{N}$ such that, for $m, n \in \mathbb{N}$, the following conditions are verified:

　　1) $f(mn) = f(m)f(n)$.

　　2) $m \neq n$ and $m^n = n^{\{m\}} \Rightarrow f(m) = n$ or $f(n) = m$.

　　3) $m, n \geq 3$ and $m^n < n^m \Rightarrow f(n) < f(m)$.

*Ex. 18: Let $a_1, \ldots, a_n$ be positive real numbers, prove the following inequality between its generalized geometric and arithmetic means:

$$\sqrt[n]{a_1 \ldots a_n} \leq \frac{a_1 + \ldots + a_n}{n}$$

*Ex. 19: Find all x such that: $\log_x 2 + 3 \log_{2x} 2 - 10 \log_{4x} 2 = 0$.

Ex. 20: $\log_2 3$ is irrational.

CHAPTER 7
COMPLEX NUMBERS

Complex numbers were created by the Italian algebraists in the 16th century, in order to solve polynomial equations. In this chapter complex numbers are defined. We study then binomic equations, solution by radicals of the second, third and fourth-degree equations, and we discuss the impossibility of continuing in this way for equations of higher degree.

1 - INTRODUCTION:

In chapter 5, we discussed the first and second-degree equations with coefficients in any field. The resolution formula for the second-degree equation was known by the Chaldeans, the Hindus, and Arabs since ancient times.

In the 16th century, the Italian algebraists achieved the solution by radicals (only using the four elementary operations and the extraction of roots) of the equations of the third and fourth degrees and introduced the numbers they called imaginaries. For instance, Cardano in his "Ars Magna" [7] noted that the problem of dividing 10 into two parts whose product is 40, when solved in the usual manner, the solutions $5 + \sqrt{-5}$ and $5 - \sqrt{-5}$ are reached. Although he called these numbers "sophistic", he points out that whatever they are, operating according to ordinary algebraic rules, their sum and product behaves as required by the "evidently impossible" problem. The need for complex numbers, actually appears when solving the third degree equation, where there are cases in which the equation has real roots, but in order to find them, according to the Tartaglia-Cardano method (this is not a lack of the method, for it can be proved that, to solve by radicals some kind of third-degree equations with real roots, it is essential to go outside the reals), it is necessary to solve an auxiliary second degree equation with no real roots. Although these equations were rejected by Cardano, they were considered by Bombelli a few decades later.

In these problems, it suffices the existence of a square root of -1 to obtain the solutions, since they are solvable by radicals and if the usual formal properties are valid, we will have, for example, $\sqrt{-5} = \sqrt{5}\sqrt{-1}$.

We, therefore, see the convenience of having (if possible) a ring A (since the usual formal properties must be valid) which must contains $\mathbb{R}$ as a subring, and in which there is an element $i \in A$ such that $i^2 = -1$.

It is usual in Mathematics that, when the existence of an object is suspected (or desired), it is provisionally assumed, to deduce from this assumption a contradiction or an idea to prove its existence.

Assume then, that there exists such a ring A and consider the following subset of A:

$$C = \{a + bi/a, b \in \mathbb{R}, i^2 = -1\}$$

Note that, if $a, b, c, d \in \mathbb{R}$ we have:

1) $a + bi = c + di \Leftrightarrow a = c$ and $b = d$.

In fact, $a + bi = c + di \Rightarrow (a - c)^2 = -(b - d)^2$ and this equality in real numbers is possible (Chap. 2 ex. D, 6) only if $a = c$ and $b = d$.

Moreover, C is a subring of A, since the sum is closed in C:

2) $(a + bi) + (c + di) = (a + c) + (b + d)i$ as it is the product:

3) $(a + bi)(c + di) = (ac - bd) + (ad + bc)i$

$0 + 0i$ is the neutral element of the sum and $1 + 0i$ that of the product.

Since C is a ring which contains $\mathbb{R}$ as a subring and contains i, it suffices to prove the existence of an object like $\mathbb{C}$. Now, 1), 2) and 3) suggest how to construct it, since 1) tells us that $a + bi$ behaves as an ordered pair (a, b) of real numbers and 2) and 3) tell us how to define operations in the set of such ordered pairs. This will be formally developed in the next section.

2 - DEFINITION OF $\mathbb{C}$

In the set (suggested by 1)) $\mathbb{R} \times \mathbb{R}$ of ordered pairs of real numbers, we define a sum (suggested by 2)):

$$(a, b) + (c, d) = (a + c, b + d)$$

and a product (suggested by 3)):

$$(a,b) \cdot (c,d) = (ac - bd, ad + bc)$$

With these definitions, we have:

Proposition 2.1: $C = (\mathbb{R} \times \mathbb{R}, +, \cdot)$ is a field.

Proof: To verify the associativity of the product we proceed as follows:

$$[(a,b) \cdot (c,d)] \cdot (e,f) = (ac - bd, ad + bc) \cdot (e,f) =$$
$$((ac - bd)e - (ad + bc)f, (ac - bd)f + (ad + bc)e) =$$
$$= (a(ce - df) - b(de + cf), a(cf + de) + b(ce - df))$$
$$= (a,b) \cdot [(ce - df), de + cf)] =$$
$$= (a,b) \cdot [(c,d) \cdot (e,f)].$$

The neutral element of the sum is $(0,0)$, and that of the product is $(1,0)$.

Let us see that each $(a, b) \neq (0,0)$ has a multiplicative inverse. In fact, by exercise D, Chap. 2: $(a,b) \neq (0,0) \Leftrightarrow a^2 + b^2 \neq 0$ and we have:

$$(a,b) \cdot \left(\frac{a}{a^2 + b^2}, \frac{-b}{a^2 + b^2}\right) = (1,0)$$

(which can be obtained posing $(a, b) \cdot (x, y) = (1, 0)$, i.e., the system of equations $ax - by = 1, ay + bx = 0$).

We leave finding the rest of the proof as an exercise.∎

Note that it would be more easy to define a product in $\mathbb{R} \times \mathbb{R}$ coordinate to coordinate, that is to define a product $\cdot$ by: $(a, b) \times (c, d) = (ac, bd)$ and, as can be easily verified $(\mathbb{R} \times \mathbb{R}, +, \times)$ would result also a ring, however this does not respond to what was posed in the introduction, since with this structure of ring, the neutral element of the product is $(1, 1)$, its additive inverse is (-1, -1) and there are no element (a, b) such that $(a, b)^2 = (a^2, b^2) = (-1, -1)$. Furthermore, with those operations, $\mathbb{R} \times \mathbb{R}$ is not a field, nor even a domain, since $(a, 0) \times (0, b) = (0, 0)$ even if $a \neq 0$ and $b \neq 0$.

Instead C respond to what was posed in the introduction, since:

$$(0, 1)^2 = (0, 1) \cdot (0, 1) = (-1, 0) = -(1, 0) \text{ i.e.,}$$

putting $i = (0, 1)$ we have $i^2 = -1$.

Moreover, although it is true that $\mathbb{R}$ is not a subring of $\mathbb{C}$, simply because $\mathbb{R}$ is not a subset of $\mathbb{R} \times \mathbb{R}$, this is not an algebraic difficulty but of set theory, which can be solved noting that if we put $\mathbb{R}' = \left\{\frac{a,0}{a} \in \mathbb{R}\right\}$, $\mathbb{R}'$ results in a subring of $\mathbb{C}$ and the function $f\colon \mathbb{R} \to \mathbb{R}'$ defined by $f(a) = (a, 0)$ is a bijection that preserves the operations:

$$f(a + b) = f(a) + f(b), f(ab) = f(a)f(b)$$

and through f all properties of $\mathbb{R}$ (those of an ordered field that satisfies the least upper bound property) are transferred to $\mathbb{R}'$ and this can be considered as the real number field. If one wants to be more meticulous, the set $\mathbb{R} \cup (\mathbb{C} - \mathbb{R}')$ can be taken as the set to define the complex, and define operations there, replacing each $a \in \mathbb{R}$ by $(a, 0)$.

We have proved:

Theorem 2.2: There exists a field $\mathbb{C}$ (called the *field of complex numbers*) containing $\mathbb{R}$ as a subfield and containing an element i such that $i^2 = -1$ and that:

$$\mathbb{C} = \{a + bi \mathbin{/} a, b \in \mathbb{R}\}$$
$$(a + bi) + (c + di) = (a + c) + (b + d)i$$
$$(a + bi)(c + di) = (ac - bd) + (ad + bc)i. \ \blacksquare$$

Note that $\mathbb{C}$ is not an ordered field, since, as we saw, in any ordered field the squares are > 0 and $1 > 0$, so that $-1 < 0$, which are incompatible with the relation $i^2 = -1$.

By the above discussion, it is clear that complex numbers are neither sophistical nor imaginary, they are elements of $\mathbb{R} \times \mathbb{R}$, that is, points in the Cartesian plane. This interpretation of complex numbers as points of the plane was first given by Wessel, Argand, and Gauss, at the beginning of the 19th century and their formalization as ordered pairs, was presented shortly afterward by W.R. Hamilton, giving in that way, an end to the mystery of the imaginaries.

3 - MODULUS AND CONJUGATION

The distance from a point in the plane to the origin, suggests defining the *modulus* of a complex number $z = a + bi$, by:

$$|z| = \sqrt{a^2 + b^2}$$

If $b = 0$, that is if $z = a$ is real, this definition coincides with that of a real number given in chapter 2, since $\sqrt{a^2}$ denotes the unique positive real number c such that $c^2 = a^2$, i.e.. a If $a \geq 0$ or $-a$ if $a < 0$.

The reflection on the real axis, suggests the definition of the conjugate z of $z = a + bi$, by:

$$\bar{z} = a - bi$$

We have $z\bar{z} = |z|^2$, which shows that the inverse of $z \neq 0$ is $\dfrac{\bar{z}}{|z^2|}$.

The abscissa of the point z is called the *real part* of z and it is denoted *Re z*, while its ordinate, the *imaginary part* of z, is denoted *Im z*. Both *Re z* and *Im z* are real numbers.

Proposition 3.1: If z and w are complex numbers, we have:

1) $\bar{\bar{z}} = z$
2) $\overline{z + w} = \bar{z} + \bar{w}$
3) $\overline{zw} = \bar{z}\bar{w}$
4) $z\bar{z} = |z|^2$
5) $|zw| = |z||w|$
6) $|z^n| = |z|^n \; \forall n \in \mathbb{N}.$
7) $Re(zw) \leq |z||w|$
8) $|z + w| \leq |z| + |w|$

Proof: We will prove 5), 7) and 8) leaving the others as exercises.

5) $|zw|^2 = zwzw = zwzw = |z|^2|w|^2.$

7) For any $u \in \mathbb{C}$ we clearly have: $Re\, u \leq |u|$. Putting $u = z\bar{w}$ it follows that, $Re(z\bar{w}) \leq |z\bar{w}| = |z||w|$.

8) $|z + w|^2 = (z + w)(\overline{z + w}) = (z + w)(\bar{z} + \bar{w}) = z\bar{z} + z\bar{w} + \bar{z}w + w\bar{w} =$

$$= |z|^2 + z\bar{w} + \overline{z\bar{w}} + |w|^2 = |z|^2 + 2Re(z\bar{w}) + |w|^2 \leq$$
$$\leq |z|^2 + 2|z||w| + |w|^2 = (|z| + |w|)^2. \blacksquare$$

4 - SECOND DEGREE EQUATION IN $\mathbb{C}$

As we already said, the solution of the second-degree equation is known from ancient times. For example in one of the oldest text of Babylon a problem [6] asks for the side of a square such that its area minus the side is 870, i.e., to solve the equation: $x^2 - x = 870$ and the solution is presented thus: Take half of 1, which is 0.50, and multiply 0.50 by 0.50, which is 0.25, add this to 870 to get 870.25. This is the square of 29.50. Now add 0.50 to 29.50, and the result is 30 the side of the square. The babylonean solution is exactly the application of the quadratic formula familiar from school, but expressed in words. To solve such equations they constructed extensive tables of squares.

According to the discussion of 9, Chap. 5, to solve a second-degree equation with coefficients in a field, it is necessary to know the squares in that field. In the case at hand, that of the complex, we have:

Proposition 4.1: In $\mathbb{C}$ any element is a square.

Proof: Given $z \in \mathbb{C}$, we must prove that there is $w \in \mathbb{C}$ such that $w^2 = z$. Putting $z = a + bi, w = x + yi$ with a, b, x, y real numbers, we must prove the existence of $x, y \in \mathbb{R}$ such that:

$$(x + yi)^2 = a + bi \text{ (1)}$$

comparing the real and imaginary parts it must be,

$$x^2 - y^2 = a \text{ (2)}$$
$$2xy = b \text{ (3)}$$

Taking modulus in (1), we have:

$$x^2 + y^2 = \sqrt{(a^2 + b^2)} \text{ (4)}$$

Adding and subtracting (2) and (4), results:

$$x^2 = \frac{\sqrt{a^2+b^2}+a}{2} \quad y^2 = \frac{\sqrt{a^2+b^2}-a}{2} \text{ (5)}$$

Since for any $a, b \in \mathbb{R}$, we have $|a| \leq \sqrt{(a^2 + b^2)}$, it follows that $\sqrt{a^2 + b^2} + a \geq 0$ and that $\sqrt{a^2 + b^2} - a \geq 0$, hence (3.1 cap.6) there exist x, y that satisfy (5) and choosing the signs to satisfy (3), the values of w are obtained.∎

Examples: 1) Solve the equation: $z^2 + (1 - i)z - i = 0$.
If z is a solution, by "completing the square" we have:

$$(z + ((1 - i)/2))^2 = (i/2)$$

Putting $w = z + ((1 - i)/2)$ and, if $w = x + yi$ with $x, y \in \mathbb{R}$, we have $w^2 = (x + yi)^2 = \left(\frac{i}{2}\right)$, from where $2xy = \frac{1}{2}$ and

$$x^2 - y^2 = 0$$
$$x^2 + y^2 = 1/2$$

then $x = \pm\frac{1}{2}, y = \pm\frac{1}{2}$, but as $2xy = \frac{1}{2}$, x, y must have the same sign, hence there are two values of w: $w_1 = \frac{1}{2} + \frac{1}{2}i$, $w_2 = -\frac{1}{2} - \frac{1}{2}i$, and so two values of z:

$$z_1 = w_1 - \frac{1-i}{2} = i \quad z_2 = w_2 - \frac{1-i}{2} = -1$$

2) Find the roots of the equation: $z^2 - z + 1 = 0$.
Completing the square, we obtain:

$$(z - \frac{1}{2})^2 = -(3/4) = (\frac{\sqrt{3}}{2}i)^2$$

then the roots are: $z_1 = \frac{1}{2} + \frac{\sqrt{3}}{2}i, z_2 = \frac{1}{2} - \frac{\sqrt{3}}{2}i$.

We have constructed $\mathbb{C}$ with the aim of having roots of the equation $x^2 + 1 = 0$ and we have just seen, that not only that equation, but any second degree equation, has roots in $\mathbb{C}$. More striking is the fact that any polynomial equation of degree ≥ 1 with complex coefficients has roots in $\mathbb{C}$, as stated by the so-called Fundamental Theorem of Algebra. For its proof we refer, for example, to [15], [26] (books on Algebra), or [1], [8] (books on Complex Variable).

5 - ARGUMENT AND TRIGONOMETRIC FORM

As we have seen, a complex number is a point of the Cartesian plane. Such a point can be view as a vector from the origin of coordinates to the point. Seeing it like this, the sum of complex numbers, has a simple geometric interpretation given by the "parallelogram rule." The product also has a simple geometric interpretation if we express the numbers in function of its polar coordinates instead of its cartesian coordinates. That is to say, in function of its distance to the origin (modulus) and of the angle that forms with the semi-axis of the abscissas. This expression depends on trigonometric functions, so we will begin by reviewing them.

For a solid basis of the trigonometric functions, it is advisable to wait for the development of mathematical tools such as series or integrals. This is why we will accept the notions of trigonometry learned at school; including the measure of angles, the number π as the ratio of the length of a circumference to its diameter and the existence of two functions: sine (sin) and cosine (cos) with domain the set of real numbers and codomain the closed interval $[-1.1]$:

$$\sin \ : \mathbb{R} \to [-1, 1], \cos \ : \mathbb{R} \to [-1, 1]$$

that verify the following list of properties (where α and β are real numbers):

1) $\sin$ and $\cos$ are surjective or onto $[-1, 1]$.
2) $\sin^2 \alpha + \sin^2 \alpha = 1$ ($\cos^2 \alpha$ is an abbreviation of $(\cos \alpha)^2$.
3) $\sin \alpha = \sin \beta \Leftrightarrow \begin{cases} \beta = \alpha + 2k\pi \text{ for some } k \in \mathbb{Z}, \text{or} \\ \beta = \pi - \alpha + 2k\pi \text{ for some } k \in \mathbb{Z} \end{cases}$
4) $\cos \alpha = \cos \beta \Leftrightarrow \begin{cases} \beta = \alpha + 2k\pi \text{ for some } k \in \mathbb{Z}, \text{or} \\ \beta = -\alpha + 2k\pi \text{ for some } k \in \mathbb{Z} \end{cases}$
5) $\sin(\alpha + \beta) = \sin \alpha \cos \beta + \sin \beta \cos \alpha$
6) $\cos(\alpha + \beta)(\alpha + \beta) = \cos \alpha \cos \beta - \sin \alpha \sin \beta$
7) $\sin(-\alpha) = -\sin \alpha$
8) $\cos(-\alpha) = \cos \alpha$
9) The usual tables of sines and cosines are valid, in particular:

α	$\sin \alpha$	$\cos \alpha$
0	0	1
$\pi/6$	$1/2$	$\sqrt{3}/2$
$\pi/4$	$\sqrt{2}/2$	$\sqrt{2}/2$
$\pi/3$	$\sqrt{3}/2$	$1/2$
$\pi/2$	1	0

Proposition 5.1: The elements of the unit circumference (center 0 and radius 1) are of the form: $\cos \alpha + i \sin \alpha$, for some $\alpha \in \mathbb{R}$.

Proof: If $z = a + bi$ has modulus 1, we have $a^2 + b^2 = 1$, then $-1 \leq a, b \leq 1$. As $\sin$ and $\cos$ are surjective, there exist $\beta, \gamma \in \mathbb{R}$ such that $\cos \beta = a$ and $\sin \gamma = b$, but as $a^2 + b^2 = 1$, it follows $\cos^2 \beta + \sin^2 \gamma = 1 = \cos^2 \beta + \sin^2 \beta$, hence $\sin^2 \gamma = \sin^2 \beta$, i.e.; $\sin \gamma = \sin \beta$ or $\sin \gamma = -\sin \beta$. If $\sin \gamma = \sin \beta$ we take $\alpha = \beta$, while if $\sin \gamma = -\sin \beta$ we take $\alpha = -\beta$, obtaining in both cases:

$$\cos \alpha + i \sin \alpha = z. \blacksquare$$

The properties 3 and 4 of the following proposition justify, at least in part, the adoption of the following notation: (being α a real number)

$$e^{i\alpha} = \cos\alpha + i\sin\alpha \quad (*)$$

Proposition 5.2: If $\alpha, \beta \in \mathbb{R}$, we have:

1) $\left|e^{i\alpha}\right| = 1$.
2) $e^{i\alpha} = e^{i\beta} \Leftrightarrow \exists k \in \mathbb{Z}$ such that $\beta = \alpha + 2k\pi$.
3) $e^{i\alpha}e^{i\beta} = e^{i(\alpha+\beta)}$.
4) $\left(e^{i\alpha}\right)^n = e^{in\alpha}$ $(n \in \mathbb{N})$ (De Moivre's formula).

Proof: 1) $\left|e^{i\alpha}\right|^2 = \cos^2\alpha + \sin^2\alpha = 1$.

2) $e^{i\alpha} = e^{i\beta} \Leftrightarrow \cos\alpha = \cos\beta$ and $\sin\alpha = \sin\beta$, but,

$$\cos\alpha = \cos\beta \Leftrightarrow \begin{cases} \exists k \in \mathbb{Z} \text{ such that } \beta = \alpha + 2k\pi \text{ or} \\ \exists h \in \mathbb{Z} \text{ such that } \beta = -\alpha + 2h\pi \end{cases}$$

$$\sin\alpha = \sin\beta \Leftrightarrow \begin{cases} \exists k \in \mathbb{Z} \text{ such that } \beta = \alpha + 2k\pi \text{ or} \\ \exists l \in \mathbb{Z} \text{ such that } \beta = \pi - \alpha + 2l\pi \end{cases}$$

If it were $\beta \neq \alpha + 2k\pi$ to any $k \in \mathbb{Z}$, we would have: $\beta = -\alpha + 2h\pi$ and $\beta = \pi - \alpha + 2l\pi$, for some integers h, l, then $2h = 1 + 2l$ which is absurd.

3) $\quad e^{i\alpha}e^{i\beta} = \cos\alpha\cos\beta - \sin\alpha\sin\beta + i(\sin\alpha\cos\beta + \sin\beta\cos\alpha) =$

$$= \cos(\alpha + \beta) + i\sin(\alpha + \beta) = e^{i(\alpha+\beta)}.$$

4) It follows readily from 3 by induction on n.■

Although in this chapter we will use it as mere notation, it is convenient to say that $(*)$ establishes a deep relationship between the exponential function and the trigonometric functions, discovered by Euler in the 18th century. Starting from the well-known developments in power series for x real:

$$e^x = 1 + x + \frac{x^2}{2!} + \ldots + \frac{x^n}{n!} + \ldots$$

$$\cos x = 1 - \frac{x^2}{2!} + \frac{x^4}{4!} - \ldots + (-1)^n\frac{x^{2n}}{(2n)!} + \ldots$$

$$\sin x = x - \frac{x^3}{3!} + \frac{x^5}{5!} - \ldots + (-1)^n\frac{x^{2n+1}}{(2n+1)!} + \ldots$$

Euler (without rigorous justification) had the idea of extrapolating these relations to complex values of the variable and, putting $x = i\alpha$ with α real, got (*). Putting $\alpha = \pi$ in (*), he obtained the following relationship between the five most important numbers in mathematics:

$$e^{i\pi} + 1 = 0$$

The following proposition expresses that a nonzero complex number is determined by its polar coordinates, that is, by the angle formed by the semi-axes of the abscissas and the vector (from the origin) determined by the number, and its modulus.

Proposition 5.3: (Trigonometric or polar form) If $z \in \mathbb{C} - \{0\}$, there exist $\alpha, r \in \mathbb{R}$ with $r > 0$ such that

$$z = re^{i\alpha}$$

Moreover, there is uniqueness in that expression, in the following sense $(r, s, \alpha, \beta \in \mathbb{R}, r, s > 0)$:

$$re^{i\alpha} = se^{i\beta} \Rightarrow r = s \text{ and } \exists k \in \mathbb{Z} \text{ such that } \beta = \alpha + 2k\pi$$

Proof: If $z \neq 0, \dfrac{z}{|z|}$ belongs to the unit circumference, so by .4.1. there exists $\alpha \in \mathbb{R}$ such that $\dfrac{z}{|z|} = \cos\alpha + i\sin\alpha = e^{i\alpha}$ and putting $r = |z|$, we obtain $z = re^{i\alpha}$.

For the uniqueness, taking modulus in $re^{i\alpha} = se^{i\beta}$, it follows that $r = s$, then $e^{i\alpha} = e^{i\beta}$, and by prop.5.2, $\beta = \alpha + 2k\pi$ for some $k \in \mathbb{Z}$.∎

If $z = |z|e^{i\alpha} \neq 0$, α is said to be an *argument* of $\mathbb{Z}$ which, according to the above proposition, is determined excepting integral multiples of 2π.

If $z = |z|e^{i\alpha}$ and $w = |w|e^{i\beta}$ are two nonzero complex numbers, then $zw = |zw|e^{i(\alpha+\beta)}$, and this is the promised geometric interpretation of the product: the angles are added and the modulus are multiplied.

6 - PLANE GEOMETRY AND COMPLEX NUMBERS

Since a point (x, y) in the cartesion plane may be identified with the complex number $z = x + yi$, plane geometry can be developed in terms of complex numbers. Thus the (straight) line joining two (distinct) complex numbers u, v can be described by the set:

$$L(u, v) = \{z \in \mathbb{C}/\exists \lambda \in \mathbb{R} \text{ such that } z = u + \lambda(v - u)\}.$$

The parallel to $L(u, v)$ (when writing $L(u, v)$ we will assume tacitly that $u \neq v$) passing through z is $L(z, z + v - u)$, (this can be viewed, for example, proving that the lines $L(u, v)$ and $L(z, z + v - u)$ coincide if they have a point in common), then the lines $L(u, v)$ and $L(z, w)$ $(u \neq v, z \neq w)$ are parallel $\Leftrightarrow L(z, w) = L(z, z + v - u) \Leftrightarrow w - z = a(v - u)$ for some $a \in \mathbb{R}$. In particular, two lines $L(u, v)$ and $L(u, w)$ with a point u in common are the same, that is the three points u, v, w are collinear, if and only if, $w - u = a(v - u)$ for some $a \in \mathbb{R}$.

For two nonzero complex numbers $u = |u|e^{i\alpha}, v = |v|e^{i\beta}$ we may consider $\beta - \alpha$ (or $\alpha - \beta$) as "the" angle between them. In particular, u, v (or the vectors which they define) are *orthogonal* or *perpendicular* when $\beta - \alpha = \pm\frac{\pi}{2}$, that is when:

$$\frac{v}{u} = |\frac{v}{u}|e^{\pm i\left(\frac{\pi}{2}\right)} = \pm|\frac{v}{u}|i$$

i.e., when v/u is purely imaginary, that is it belongs to $\mathbb{R}i = \{ai/ a \in \mathbb{R}\}$. Two lines $L(u, v)$ *and* $L(z, w)$ are *orthogonal* when the vectors $v - u$ and $w - z$ are ortogonal.

Summarizing, we have:

Proposition 6.1: If u, v, w, z are points in the complex plane such that $u \neq v, z \neq w$, then:

1) $L(u, v)$ is parallel to $L(z, w)$ $\Leftrightarrow \frac{w-z}{v-u} \in \mathbb{R}$.

2) $L(u, v)$ is ortogonal to $L(z, w) \Leftrightarrow \frac{w-z}{v-u} \in \mathbb{R}i$.

3). u, v, w are collinear $\Leftrightarrow \frac{w-u}{v-u} \in \mathbb{R}$. ∎

Let us analyze further the condition $\frac{v}{u} \in \mathbb{R}i$, for two nonzero vectors to be orthogonal. Since for any $z \in \mathbb{C}$, we have $z \in \mathbb{R}i \Leftrightarrow z = -\bar{z}$, then $\frac{v}{u} \in \mathbb{R}i \Leftrightarrow u\bar{v} + \bar{u}v = 0$. In terms of coordinates we have: $u\bar{v} + \bar{u}v = 2(u_1v_1 + u_2v_2)$, where the index 1 indicate the real part, and the index 2 the imaginary part of the corresponding complex number. To be in accordance with the habitual use of the ordinary inner product, we define the *inner product* $< u, v >$ of the vectors, or complex numbers, u, v as:

$$< u,v >= u_1 v_1 + u_2 v_2 = \frac{u\bar{v} + \bar{u}v}{2}$$

This inner product verifies:

Proposition 6.2: 1) For nonzero u, v we havè: u is ortogonal to $v \Leftrightarrow$ $< u, v >= 0$.

2) $< u, v >=< v, u >$.

3) $< u, u >= |u|^2$.

4) $< u, v + w >=< u, v > +< u, w >$ and $< u + z, v >=< u, v > +< z, v >$ for any $u, v, w, z \in \mathbb{C}$.

5) $< au, v >= a < u, v >=< u, av >$ for any $a \in \mathbb{R}, u, v \in \mathbb{C}$.

Proof: 1) follows from the considerations below. The others are straightforward.∎

The line $L(u, v)$ is parallel to $L(0, v - u)$ and this is orthogonal to $L(0, i(v - u))$, which in turn is parallel to $L(z, z + i(v - u))$, then $L(z, z + i(v - u))$ is the line ortogonal to $L(u, v)$ which passes through z. We have: $x \in L(z, z + i(v - u)) \Leftrightarrow x - z = ai(v - u)$ with $a \in \mathbb{R} \Leftrightarrow x - z$ is ortogonal to $v - u \Leftrightarrow < x - z, v - u >= 0$. We have shown:

Lemma 6.3: The perpendicular to $L(u, v)$ which passes trough z, has the two characterizations:

$$L(z, z + i(v - u)) = \{x \in \mathbb{C}/< x - z, v - u >= 0\}. \blacksquare$$

Proposition 6.4: The lines determined by the altitudes of a triangle intersect at a point h, the *orthocenter* of the triangle.

Proof: A triangle is, of course, a set of three noncollinear points u, v, w, with obvious definitions of its vertices, sides and angles. The lines determined by the altitudes of the triangle are:

$$L_1 = L(u, u + i(v - w)); \quad L_2 = L(v, v + i(w - u)); \quad L_3$$
$$= L(w, w + i(v - u))$$

If $L_1 \cap L_2$ were empty, then L_1, L_2 would be parallel, so by prop. 6.1, we would have $\frac{v - w}{w - u} \in \mathbb{R}$ which contradicts the fact that u, v, w are noncollinear. By a similar argument, L_1 and L_2 do not coincide, so there is a unique point h in $L_1 \cap L_2$. It remains to prove that $h \in L_3$, or, by the lemma, that $< h - w, v - u >= 0$. As $h \in L_1 \cap L_2$, we have:

$$0 \;=\; < h - u, v - w > \;=\; < h, v > \; - < u, v > \; - < h, w > \; + < u, w >$$
$$0 \;=\; < h - v, w - u > \;=\; < h, w > \; - < v, w > \; - < h, u > \; + < v, u >$$

and adding them:

$$0 =< h, v > \; + < u, w > \; - < v, w > \; - < h, u >=< h - w, v - u >. \;\blacksquare$$

Other notable points of a triangle are the centroid and the circumcenter (among others). The *centroid* is the intersection of the medians, that is the intersection of the segments joining each vertex with the midpoint of the opposite side. The *circumcenter* is the center of the circumference which passes through the three vertices of the triangle. These points exist as are stated by the following propositions that are left as exercises.

Proposition 6.5: In the triangle determined by three noncollinear points u, v, w, the centroid is the point $\frac{1}{3}(u + v + w)$. $\blacksquare$

Proposition 6.6: The circumcenter of a triangle, is the orthocenter of the triangle determined by the middle points of the sides of the first triangle.$\blacksquare$

Thus, the existence of the circumcenter is reduced to the existence of the orthocenter. Conversely, the existence of the orthocenter can be derived from the existence of the circumcenter as is shown below.

Proposition 6.7: (Euler line) Denote by f the circumcenter and by g the centroid of the triangle determined by u, v, w, then the orthocenter h is the point:

$$h = j + 3(g - f)$$

It follows that in any triangle, the circumcenter, the centroid, and the orthocenter lie on the same line, *Euler line*.

Proof: By definition of circumcenter we have $|u - f| = |v - f| = |w - f|$. To verify that $j + 3(g - f)$ is the orthocenter, we must verify, for example, that:

$$< j + 3(g - f) - u, w - v >= 0$$

but, by prop. 6.5, $g = (1/3)(u + v + w)$ then:

$$< j + 3(g - f) - u, w - v >=< v + w - 2f, w - v >=$$
$$< (v - f) + (w - f), (w - f) - (v - f) >$$
$$= -|v - f|^2 + |w - f|^2 +< v - f, w - f > -$$
$$< w - f, v - f >= 0. \blacksquare$$

Euler line reduces to a point if and only if the triangle is equilateral (exercise). Although the centers of a triangle were studied since Euclid, it is intriguing that just in the 18th century Euler discovered that the circumcenter, the centroid, and the orthocenter are collinear.

Until now our use of complex numbers in geometry involve only properties of them as vectors. More interesting is that there are results, where geometric facts are reflected as algebraic properties of complex numbers, that does not have a counterpart as vectors. The following is an example which we leave as a (starred) exercise:

Consider a triangle determined by the points u, v, w. The triangle is equilateral if, and only if the following equality is verified:

$$u^2 + v^2 + w^2 - uv - uw - vw = 0.$$

More information about the use of complex numbers in geometry can be found in [2] or [27].

7 - ROOTS OF COMPLEX NUMBERS

Theorem 7.1: Given $w \in \mathbb{C}, w \neq 0$ and $n \in \mathbb{N}$; the equation $z^n = w$ admits exactly n solutions:

$$z_k = \sqrt[n]{|w|} e^{i\left(\frac{\alpha + 2k\pi}{n}\right)}, k = 0, \ldots, n.$$

being α an argument of w.

Proof: If z is a solution of the equation, it must be $z \neq 0$ and putting $z = |z|e^{i\theta}$ with $\in \mathbb{R}$, we have by De Moivre's formula (prop.5.2),

$$|z|^n e^{in\theta} = |w|e^{i\alpha}$$

then by the above proposition, $|z|^n = |w|$ and $n\theta = \alpha + 2h\pi$ for some $h \in \mathbb{Z}$, and replacing in the last one:

$$h = nq + k, 0 \le k \le n - 1$$

we obtain,

$$\theta = \frac{\alpha}{n} + \frac{2k\pi}{n} + 2q\pi$$

hence,

$$z = \sqrt[n]{we}^{i\left(\frac{\alpha}{n} + \frac{2k\pi}{n} + 2q\pi\right)} = \sqrt[n]{we}^{i\left(\frac{\alpha + 2k\pi}{n}\right)} \text{ with } k = 0, \dots, n-1$$

Moreover, if $z_l = z_k$ with $0 \le k \le l \le n - 1$, then:

$$e^{i\left(\frac{\alpha + 2l\pi}{n}\right)} = e^{i\left(\frac{\alpha + 2k\pi}{n}\right)}$$

from where by prop.5.2. $\frac{2l\pi}{n} = \frac{2k\pi}{n} + 2j\pi$ for some $j \in \mathbb{Z}$, hence $n \mid l - k$, and so $l = k$.

Finally, each z_k is a root of the equation, since:

$$z_k{}^n = |w| + 2k\pi e^{i(\alpha + 2k\pi)} = |w|e^{i\alpha}. \blacksquare$$

Example: We will find the roots of the equation $z^3 = i$.

As $i = e^{i\frac{\pi}{2}}$, putting $z = |z|e^{i\theta}$, it follows $|z| = 1$ and $\theta = \frac{\pi}{6} + \frac{2k\pi}{3}$ with $k = 0, 1, 2$, hence:

$$z_0 = e^{i\frac{\pi}{6}} = \cos\frac{\pi}{6} + i\sin\frac{\pi}{6} = \frac{\sqrt{3}}{2} + \frac{1}{2}i$$

$$z_1 = e^{i\frac{5\pi}{6}} = \cos\frac{5\pi}{6} + i\sin\frac{5\pi}{6} = -\frac{\sqrt{3}}{2} + \frac{1}{2}i$$

$$z_3 = e^{i\frac{3\pi}{2}} = \cos\frac{3\pi}{2} + i\sin\frac{3\pi}{2} = -i$$

A nth root of unity that is not a mth root of unity is called a *primitive n - root of unity*. In other words, a primitive n - root of unity is a complex number ε such that $\varepsilon^n = 1$, but $\varepsilon^m \ne 1 \; \forall \; m \in \mathbb{N}, m < n$. According to th. 6.1, the n - roots of unity are: $z_k = e^{i\frac{2k\pi}{n}}, k = 0, 1, \dots, n - 1$ and we have:

Proposition 7.2: $z_k = e^{i\frac{2k\pi}{n}}, k = 0, \dots, n - 1$ is a primitive n - root of unity $\Leftrightarrow gcd(k, n) = 1$.

Proof: Put $d = gcd(k,n), k = dk', n = dn'$, then $z_k = e^{i\frac{2k'\pi}{n'}}, k' = 0, \ldots, n' - 1$, so that z_k is a n' – root of unity. Hence it is primitive $\Leftrightarrow n' = n \Leftrightarrow d = 1$. ∎

Examples: 1 is the primitive 1 – root of unity; -1 the primitive 2 – root of unity; $\frac{-1+\sqrt{3}i}{2}, \frac{-1-\sqrt{3}i}{2}$ are the primitive 3 – roots of unity; $\pm i$ the primitive 4 – roots of unity.

8 – THIRD-DEGREE EQUATIONS

As we have said, the Babylonians constructed tables of squares to solve quadratic equations. They also build up tables of cubes and of $x^2 + x^3$. These last they used to solve third degree equations of the form $ax^3 + bx^2 = c$ (*), since multiplying by $\frac{a^2}{b^3}$) it follows $\left(\frac{a}{b}x\right)^3 + \left(\frac{a}{b}x\right)^2 = \frac{a^2c}{b^3}$ and the solution is reduced to look in the table. There is no evidence that the Babylonians had knowledge that any third degree equation can be reduced to one of the type (*) but it is possible to do such reduction (ex. 2).

The solution of the third-degree equation by radicals was achieved by the Italian algebraists Del Ferro, Tartaglia and Cardano. A concise version of the discovery is like follows (for more details see [6] or [39]. At the beginning of the 16th century Del Ferro, professor of Mathematics at Bologna University, managed to solve equations of the type:

$$x^3 + px = q$$

but did not publish his method. Tartaglia, being aware of that discovery, achieved a solution to those type of equations and also of others like:

$$x^3 + px^2 = q$$

Tartaglia, revealed the method to Cardano with the promise of not publishing it until Tartaglia did. Cardano, with such information, was able to solve all cases of the equation (thirteen cases are presented in the Ars Magna) and, moreover, his disciple Ferrari was successful in solving the fourth dfourth-degree. In view of Tartaglia's delay in publishing the solution, Cardano travels to Bologna where a relative of Del Ferro, gives him access to papers of Del Ferro where the solution appears. Cardano then feels relieved of his promise and publishes his Ars Magna (1545) where he exposes the complete solution of the third and fourth-degrees with the due credit to Del Ferro, Tartaglia, and Ferrari. However, Tartaglia feels betrayed, and a sour dispute arises with Cardano.

In order to solve the third-degree equation, note in first place, that the equation:

$$x^n + a_{n-1}x^{n-1} + \ldots + a_1 x + a_0 = 0$$

can be transformed into one of the same degree, but with the coefficient of the $(n-1)$th power null (the "reduced" equation), by the substitution:

$$x = z - \frac{a_{n-1}}{n} \ (1)$$

being the "completion of the square" a particular case with $n = 2$.

In the case of the third-degree equation, the reduced equation is one of the type:

$$z^3 + pz + q = 0 \ (2)$$

The procedure of Del Ferro-Tartaglia-Cardano to solve this equation consists, essentially, to make the following replacing in (2): $z = u + v$, so that:

$$u^3 + v^3 + (3uv + p)(u + v) + q = 0 \ (3)$$

Since a variable z has been replaced by two u, v, these have some degree of freedom and can be imposed to them the simplifying condition:

$$3uv + p = 0 \ (4)$$

Replacing in (3) we obtain the system:

$$u^3 + v^3 = -q$$

$$u^3 v^3 = -\frac{p^3}{27}$$

then u^3 and v^3 are the roots of the second degree equation (the resolvent):

$$x^2 + qx - \frac{p^3}{27} = 0 \ (5)$$

Solving this we obtain u^3 and v^3, from where, by the methods of the previous section, we obtain at most three values of u and three of v and, as a consequence, at most nine values of z, which, using (4), are reduced to a maximum of three.

Some observations are in order. For the Italian algebraist of the 16th century, an equation like (2) had no meaning since they consider only positive numbers, so they considered separately equations like

$z^3 + pz = q$ or $z^3 = pz + q$. For the same reason, the substitution $z = u + v$ in some cases was $z = u - v$ with $u > v$. They argued geometrically, the substitution $z = u + v$ had the purpose of decompose a cube z^3 (a geometrical cube of side z) in an union of cubes u^3, v^3, and three parallelepipeds of sides $u, v, z = u + v$.

Advances achieved in algebraic notation, made that geometric methods (Greek tradition) were gradually replaced by algebraic ones. A few decades after Cardano, Viète solved equation (2) making the following substitution in (2) $z = w - \frac{p}{3w}$, obtaining:

$$(w^3)^2 + qw^3 - \frac{p^3}{27} = 0$$

that is the same resolvent as before. It is enough to find a single root of this, to obtain at most three values of w and an equal number of values of z.

Example: To find the roots of

$$z^3 + 3\left(\frac{\sqrt{3}}{2} + \frac{1}{2}i\right)z + (1 - i) = 0$$

we form the resolvent $x^2 + qx - \frac{p^3}{27} = 0$, in our case:

$$x^2 + (1 - i)x - i = 0$$

whose roots are i and -1 (section 4). According to the first procedure, we pose the equations: $u^3 = i, v^3 = -1$, from where

$$u_1 = -i \quad v_1 = -1$$

$$u_2 = \frac{\sqrt{3}}{2} + \frac{1}{2}i \quad v_2 = \frac{1}{2} + \frac{\sqrt{3}}{2}i$$

$$u_3 = -\frac{\sqrt{3}}{2} + \frac{1}{2}i \quad v_3 = \frac{1}{2} - \frac{\sqrt{3}}{2}i$$

As the condition: $3uv = -p$ must be satisfied, the roots are:

$$z_1 = u_2 + v_1 \quad z_2 = u_1 + v_3 \quad z_3 = u_3 + v_2$$

Following the method of Viète, we choose one (for example -1) of the roots of the resolvent and solve $w^3 = -1$, obtaining the roots v_1, v_2, v_3, and the values of z will be: $z_j = v_j - \frac{p}{3v_j}, j = 1, 2, 3$.

9 - EQUATIONS BEYOND THE THIRD-DEGREE

Without further details, we shall describe a method (taking from [5]), to solving by radicals the fourth-degree equation (in exercise 7 there is another). Given the equation in the form:

$$x^4 = ax^2 + bx + c$$

we determine z such that the second member of:

$$(x^2 + z)^2 = (a + 2z)x^2 + bx + (c + z^2)$$

be a perfect square, that is, such that:

$$b^2 = 4(a + 2z)(c + z^2)$$

which gives a third degree equation in z. Solving this and finding for each value of z, solutions u, v of $a + 2z = u^2$ and $v^2 = c + z^2$, we obtain

$$(x^2 + z)^2 = (ux + v)^2$$

Finally, the values of x are obtained from $x^2 + z = \pm(ux + v)$.

In view of the success obtained with the third and fourth-degree equations, the algebraists of the 16th, 17th and 18th centuries, tried to solve by radicals the equations of higher degree, but their efforts were unsuccessful. Then, at the beginning of the 19th century, Abel proves the impossibility of solving by radicals the general equation of degree $n \geq 5$.

At this point, it is advisable to make some precisions. Solving an equation by radicals, is to find its solutions carrying out on its coefficients only rational operations, that is addition, subtraction, multiplication, and division, and additionally extracting roots of any order. On the other hand by "general equation" we understand the equation considered with "indeterminate" coefficients", That is, that the coefficients may be substituted independently by any numbers, i.e., the solution does not depend on particularities of the coefficients neither on the relation between them. Thus the solutions of the equation $x^2 + ax + b = 0$, can be obtained from:

$$x = -\frac{a}{2} \pm \sqrt{\frac{a^2}{4} - b} \ (6)$$

where $\sqrt{c}$ (in this section we depart from our convention established in section 4 of Chap. 6) designates one of the real or complex numbers whose square is c (the other is then $-\sqrt{c}$). The resolution has been made by radicals, since on the coefficients only rational operations and an extraction of a square root have been made. In addition, the general equation was solved, since for any particular equation with numerical coefficients, it is enough substituting in (6) to obtain the roots.

With the third degree equation, something similar happens. The roots x_1, x_2 of (5) can be expressed by:

$$x^1 = -\frac{q}{2} + \sqrt{\frac{q^2}{4} + \frac{p^3}{27}}, x^2 = -\frac{q}{2} - \sqrt{\frac{q^2}{4} + \frac{p^3}{27}}$$

and as $u^3 = x_1, v^3 = x_2$ and $z = u + v$, the roots of (4) follows from:

$$z = \sqrt[3]{-\frac{q}{2} + \sqrt{\frac{q^2}{4} + \frac{p^3}{27}}} + \sqrt[3]{-\frac{q}{2} - \sqrt{\frac{q^2}{4} + \frac{p^3}{27}}} \quad (7)$$

where $\sqrt[3]{d}$ must be replaced by each of the complexes whose cube is d. Here too, on the coefficients were carried out only rational operations and extraction of square and cubic roots, as well the coefficients were treated as "indeterminates": for each particular numerical equation, its solutions are obtained substituting in (7) p and q by the numerical coefficients.

Abel's theorem on the impossibility of solving the equation of degree ≥ 5, refers to solution by radicals, using only rational operations and admitting as auxiliaries only the equations of the type $x^m = a$, and refers to the "general" equation, treating the coefficients as indeterminates.

The failed attempts to solve by radicals the equations of higher degree, led to consider the possibility that such a solution was impossible. Ruffini studies the work of his predecessors, especially Lagrange, and was convinced of the impossibility of solving by radicals the equations of degree ≥ 5. He gives a proof with some gaps. Then Abel, also with some gaps (Wantzel, reviewing the proofs, describes as vague that of Abel and of very vague that of Ruffini) proves the theorem that bears his name. The same result is obtained by Galois, but the theory developed by him, allows us to analyze other equations (not only the general), and to study the resolution of an equation admitting as auxiliary any equation not necessarily of the type $x^m = a$.

EXERCISES

Ex. 1: Compute:

a)i^n $(n \in \mathbb{N})$, b)$\sum_{j=1}^{n} i^j$ c)$(1 + i)^{125}$

Ex. 2: Make a graph of:

184

a) $\{z \in \mathbb{C}/Im_z \geq 0 \ y \ Re_z < 0\}$

b) $\{z \in \mathbb{C}/Re_z + 3Im_z = 5\}$

c) $\{z \in \mathbb{C}/Re_z \geq 1 \ and \ |z| \geq 2\}$

Ex. 3: Let $z, w \in \mathbb{C}$. Prove:

a) $|z|^{-1}|z - w||w|^{-1} = |z^{-1} - w^{-1}|$ with $z \neq 0$ and $w \neq 0$.

b) $|z - w|^2 + |z + w|^2 = 2(|z|^2 + |w|^2)$

Ex. 4: Solve the equations:

a) $z^2 = 1 + i$, b) $z^2 = -2i$, c) $z^2 + (1 - 4i)z + (-5 + i) = 0$

Ex. 5: a) Construct a table of $n^2 + n^3$ from $n = 1$ to $n = 10$ and use it to find a root of $x^3 + 2x^2 - 3136 = 0$.

b) A problem of a Babylonian text from about 1800 $B. C$, asks for the solution of the system:

$$xyz + xy = \frac{7}{6}$$

$$y = \frac{2x}{3}$$

$$z = 12x$$

solve it using the table constructed in a).

Ex. 6: Prove that any third-degree equation can be reduced to one of the type $ax^3 + bx^2 = c$ and one of the second degree.

Ex. 7: The following is a method of Euler (taken from [35]) to solve the fourth-degree equation. Starting from the reduced form of the equation:

$$x^4 + ax^2 + bx + c = 0. \ (1)$$

By analogy with the reduced third-degree equation in which the roots take the form $\sqrt[3]{p} + \sqrt[3]{q}$ with p, q roots of a second degree equation, Euler puts $x = \sqrt{p} + \sqrt{q} + \sqrt{r}$ to construct a third degree equation having p, q, r as roots. We have:

$$x^2 - (p + q + r) = 2(\sqrt{pq} + \sqrt{pr} + \sqrt{qr})$$

and squaring again:

$$x^4 - 2(p + q + r)x^2 + (p + q + r)^2 = 4(pq + pr + qr) + 8\sqrt{pqr}x \ (2)$$

p, q, r are roots of the third-degree equation:

$$y^3 + \alpha y^2 + \beta y + \gamma = 0$$

where:

$$\alpha = p + q + r; \quad \beta = pq + pr + qr; \quad \gamma = pqr$$

and substituting in (2):

$$x^4 - 2\alpha x^2 - 8\sqrt{\gamma x} + \alpha^2 - 4\beta$$

Compare this expression with (1) to find the coefficients α, β y γ.

Ex. 8: Compute the modulus and one argument of:

a) $\left(-\frac{1}{2} + \frac{\sqrt{3}}{2}i\right)^{83}$ b) $\left(\frac{1}{\sqrt{2}} - \frac{\sqrt{3}}{\sqrt{2}}i\right)^{100}$

c) $(-\sqrt{2} - \sqrt{2}i)^{12}$ d) $(-i)^{316}$

Ex. 9: Solve the equations:

a) $z^3 = -1$, b) $z^6 = 1$, c) $z^4 = -1 + \sqrt{3}i$,

d) $(z + 1)^3 = z^3$, e)$(z^2 - 3z + 1)^3 = 1$

Ex. 10: Determine the complex z such that:

a) $z = z^{-1}$, b) $z^2 \in \mathbb{R}$, c) $z^2 = z^3$,

d) $z^3 = z^3$, e) $z^n = z^m$ $(n, m \in \mathbb{N})$

Ex. 11: Graph all $z \in \mathbb{C}$ such that:

a) $z^3 = z$ and $Rez \le \frac{1}{2}$

b) $|z| = |\sqrt{2} - \sqrt{2}i|$ and $arg(z^2) = arg\left(\frac{-1+\sqrt{3}i}{1+\sqrt{3}i}\right)$

c) $|z| \le 1$ and $Arg(z) \le \frac{\pi}{4}$

Where $arg(u) = arg(v)$ weans that some argument of u equals some argument of v and where $Arg(z)$ A designates the principal argument of z, i.e., the unique which verifies: $0 \le Arg(z) < 2\pi$.

Ex. 12: In the triangle determined by three noncollinear points u, v, w, the centroid is the point $\frac{1}{3}(u + v + w)$.

Ex. 13: The circumcenter of a triangle, is the orthocenter of the triangle determined by the middle points of the sides of the first triangle.

Ex. 14: Euler line reduce to a point if, and only if the triangle is equilateral.

***Ex. 15:** Let u, v, w be distinct complex numbers. They determine an equilateral triangle if, and only if:

$$u^2 + v^2 + w^2 - uv - uw - vw = 0.$$

CHAPTER 8
POLYNOMIALS

1 - HOMOMORPHISMS

If A and B are rings, a ring *homomorphism* from A to B is a function $f: A \to B$ such that:
$$f(a + b) = f(a) + f(b)$$
$$f(ab) = f(a)f(b)$$
Homomorphism means equal form, it is a function that preserves the form, i.e., the operations, and we will also call it simply a *morphism*.

A morphism of rings $f: A \to B$, is called:

— *monomorphism* iff f is injective.
— *epimorphism* iff f is surjective.
— *isomorphism* iff f is bijective.

The existence of an isomorphism between two rings tells us that they have the same structure or that they differ only in notations. For example, when we define the complex numbers, we used ordered pairs of real numbers, and we identified a real number a with the pair $(a, 0)$. In fact, the function $f, f: R \to C$ defined by $f(a) = (a, 0)$ is a ring monomorphism, and so it is an isomorphism from $\mathbb{R}$ onto its image which is a subring of $\mathbb{C}$.

Examples: 1) For any natural number n, the function $f: \mathbb{Z} \to \mathbb{Z}_n$ defined by $a \to a$, where a is the class of $a \bmod n$, is an epimorphism of rings.

2) The only morphism f from $\mathbb{Q}(\sqrt{2})$ to $\mathbb{Q}$, is the null morphism $f(x) = 0 \; \forall \, x \in \mathbb{Q}(\sqrt{2})$. In fact, as $f(1) = f(1 \cdot 1) = f(1)f(1)$, it follows that $f(1) = 0$ or $f(1) = 1$. In the first case f is the null morphism. In the second, it follows that $f(2) = 2$ and if we

put $\alpha = \sqrt{2}$, Then $2 = f(\alpha^2) = (f(\alpha))^2$, hence $f(\alpha) = \pm\sqrt{2}$ which is not rational.

3) The conjugation $z \to \bar{z}$ is an isomorphism of $\mathbb{C}$ onto $\mathbb{C}$.

4) In this example, we set aside our convention that "ring" means "commutative" ring. The function $f: \mathbb{C} \to M_2(\mathbb{R})$ defined by:

$$f(a + bi) = \begin{pmatrix} a & b \\ -b & a \end{pmatrix}$$

is a morphism of rings (verify) and is injective, then it is an isomorphism onto its image. This allows us to define, alternatively, $\mathbb{C}$ as the subring of $M_2(\mathbb{R})$ formed by the matrices of the type:

$$\begin{pmatrix} a & b \\ -b & a \end{pmatrix}$$

The proof of the next proposition is straightforward and is left as an exercise.

Proposition 1.1: Let $f: A \to B$ be a morphism of rings, then:

a) $f(0) = 0$; $f(-a) = -f(a)$; $f(a - b) = f(a) - f(b)$; $f(na) = nf(a) \forall n \in \mathbb{Z}$. Moreover, $f(a^m) = f(a)^m \ \forall \ m \in \mathbb{N}$.

b) If B is a domain, then $f(1) = 0$ or $f(1) = 1$ (there are reasons in the Theory of Categories to demand in the definition of morphism the condition of preserving identities, that is $f(1) = 1$, but we will not make such assumption). If $f(1) = 0$, then $f(a) = 0 \forall a \in A$, and f is called the null or zero morphism ($f = 0$).

c) If A, B are fields, $f \neq 0$, and $a \in A, a \neq 0$, then: $f(a^{-1}) = f(a)^{-1}$; $f\left(\frac{b}{a}\right) = \frac{f(b)}{f(a)}, f(a^n) = f(a)^n \ \forall \ n \in \mathbb{Z}.\blacksquare$

Often occurs in Algebra the following situation: we have two algebraic structures (rings, fields, etc.) A, B and a monomorphism from A to B and want to have an "extension" C of A isomorphic to B. For example, that situation was presented to us when defining complex numbers and, will be presented again, when building the field of quotients of a domain, or when defining rings of polynomials. We can solve it in two ways, one admitting that an extension of A is, by definition, a pair (B, f) where is the same class of structure that A and f a monomorphism from A to B, or, as we will see in the next lemma, building C isomorphic to B and containing A.

Lemma 1.2: Let $f: A \to B$ a monomorphism of rings. Then there exists a ring C that contains A as subring and such that is isomorphic to B.

Proof: Let D be a set such that exist a bijection $g: D \to B - f(A)$ and such D and A are disjoint (such D exists by Set Theory) and let $C = A \cup D$. We have another bijection: $f: A \to f(A)$. We paste the two, defining $h: C \to B$ by: restricted to A is f and restricted to D is g. Clearly, h is a bijection. We transfer to C the ring structure of B, defining for $x, y \in C$:

$$x + y = h^{-1}(h(x) + h(y))$$
$$xy = h^{-1}(h(x)h(y))$$

and with these operations, C is a ring that contains A as subring and h is an isomorphism of rings. ∎

2 FIELD OF QUOTIENTS OF A DOMAIN

Let A be a subring of a field F, so A is a domain, and let:

$$K = \{ab^{-1}/a, b \in A, b \neq 0\}$$

K is a subfield of F. In fact, $0, 1 \in K$ and, being $a, b, c, d \in A$ with $b, d \neq 0$, we have $bd \neq 0$ and:

$$ab^{-1} + cd^{-1} = (ad + bc)(bd)^{-1} \text{ (1)}$$
$$(ab^{-1})(cd^{-1}) = (ac)(bd)^{-1} \text{ (2)}$$

Note also that:

$$ab^{-1} = cd^{-1} \Leftrightarrow ad = bc \text{ (3)}$$

K is called the field of quotients of A contained in F.

Start now with a domain A, and we will construct a field that contains A as a subring. The relation (3) suggests to define in the set $A \times (A - \{0\})$ a relation $\sim$ by:

$$(a, b) \sim (c, d) \text{ iff } ad = bc$$

$\sim$ is an equivalence relation in $A \times (A - \{0\})$, and if $\overline{(a, b)}$ denotes the class of (a, b) under $\sim$, we have:

$$\overline{(a, b)} = \overline{(c, d)} \Leftrightarrow ad = bc$$

The relations (1) and (2), suggest to define addition and multiplication in the quotient set $(A \times (A - \{0\}))/\sim$ by:

$$\overline{(a, b)} + \overline{(c, d)} = \overline{(ad + bc, bd)}$$
$$\overline{(a, b)} \cdot \overline{(c, d)} = \overline{(ac, bd)}$$

Let us verify that this sum is well defined. Let $\overline{(a, b)} = \overline{(a', b')}$ and $\overline{(c, d)} = \overline{(c', d')}$, then $ab' = ba'$ and $cd' = dc'$, hence $adb'd' = a'd'bd$ and $bcb'd' = b'c'bd$ and, adding: $(ad + bc)b'd' = (a'd' + b'c')bd$.

Similarly, the product is well defined.

It is easy to verify that with these operations, $(A \times (A - \{0\}))/{\sim}$ is a field, with $\overline{(0, 1)}$ as the neutral element of the sum, $\overline{(1, 1)}$ as the neutral element of the product and, if $\overline{(a, b)} \neq \overline{(0, 1)}$, that is if $a \neq 0$, the multiplicative inverse of $\overline{(a, b)}$ is $\overline{(b, a)}$.

Moreover, the function $f: A \to (A \times (A - \{0\}))/{\sim}$ defined by $f(a) = \overline{(a, 1)}$ is a monomorphism of rings and so by lemma 1.1, exists a field which contains A as a subring. We have proved:

Proposition 2.1: Any domain is a subring of some field.∎

3 - POLYNOMIALS

Let A be a subring of a ring B. An element X in B is said to be *transcendental* or an *indeterminate* over A iff it fulfills the following condition:

$$n \in \mathbb{N} \cup \{0\}; \; a_0, \ldots, a_n \in A; \; \sum_{i=0}^{n} a_i X^i = 0 \Rightarrow a_i = 0 \; \forall i = 0, \ldots, n.$$

For now, we assume the existence of such B and X and make a shallow study of their behavior, to obtain ideas in order to prove their existence. Consider the following subset of B:

$$A[X] = \{f \in B / f = \sum_{i=0}^{n} a_i X^i \; ; \; n \in \mathbb{N} \cup \{0\}; \; a_i \in A\}$$

$A[X]$ is a subring of B. Check, for example, that the sum and product of elements of $A[X]$ are again elements of $A[X]$:

$$\sum_i a_i X^i + \sum_i b_i X^i = \sum_i (a_i + b_i) X^i \; (1)$$
$$\left(\sum_i a_i X^i\right) \cdot \left(\sum_j b_j X^j\right) = \sum_k \sum_{i+j=k} a_i b_j X^k \; (2)$$

where the sums with sub-indexes i, j, we may assume that varies in the same range, for if not, it suffices to add terms with null coefficients in one of them.

With the same convention, note that, as X is trascendental over A, we have:

$$\textstyle\sum_i a_i X^i = \sum_i b_i X^i \Rightarrow a_i = b_i \forall i \ (3)$$

We now remove the hypothesis of the existence of B and X and we start from a ring A (commutative with identity) and will prove:

Proposition 3.1: If A is a ring, then there exist a ring B, which contains A as a subring, and an element $X \in B$ transcendental over A.

Proof: Condition (3) tells us that $\sum_{i=0}^{n} a_i X^i$ behaves as a $(n+1)-$ uple: $(a_0, a_1, \ldots, a_n)$, but since n may take any value in $\mathbb{N} \cup \{0\}$, rather it must be though as an indefinite uple, that is a sequence, where the terms are zero from one onwards, in other words, as a function $f : \mathbb{N} \cup \{0\} \to A$ such that $f(i) = 0$ from a certain index onwards.

Consider, as is suggested by (3), the cartesian product of $\mathbb{N} \cup \{0\}$ copies of A, i.e.: $A^{\{\mathbb{N} \cup \{0\}\}} = \{f : \mathbb{N} \cup \{0\} \to A\}$

In this set we define a sum and a product. If $f, g \in A^{\{\mathbb{N} \cup \{0\}\}}$, define $f + g : \mathbb{N} \cup \{0\} \to A$ and $f \cdot g : \mathbb{N} \cup \{0\} \to A$ by:

$$(f + g)(k) = f(k) + g(k)$$

$$(f \cdot g)(k) = \sum_{i+j=k} (f(i)g(k)$$

these definitions are not fanciful, but are suggested by (1) and (2).

It is straightforward to verify that $A^{\{\mathbb{N} \cup \{0\}\}}$ with these operations is a ring (commutative with identity). The neutral element of the sum is the zero function, which applies each $k \in \mathbb{N} \cup \{0\}$ in 0, the neutral element of the addition in A, The neutral element of the product is the function which applies 0 in 1, neutral element of the product in A, and applies each $k \in \mathbb{N}$ in 0. As an example we verify the associativity of the product:

Let $f, g, h : \mathbb{N} \cup \{0\} \to A$, we have:

$$[(f \cdot g) \cdot h](k) = \sum_{i+j=k} (f \cdot g)(i)h(j) = \sum_{i+j=k} \sum_{r+s=i} f(r)g(s)h(j)$$

$$= \sum_{r+s+j=k} f(r)g(s)h(j)$$

$$[f \cdot (g \cdot h)](k) = \sum_{r+t=k} f(r)(g.h)(t) = \sum_{r+t=k}\sum_{s+j=t} f(r)g(s)h(j)$$

$$= \sum_{s+j=t} f(r)g(s)h(j)$$

The ring $A^{\{\mathbb{N}\cup\{0\}\}}$ we have just defined, is called the ring of *formal series* on A, and we will denote it briefly by C.

The   function $\alpha: A \to C$ which applies each $a \in A$ in the function $\alpha(a)$ defined by:

$$\alpha(a)(k) = \begin{cases} a \; si \; k = 0 \\ 0 \; si \; k \neq 0 \end{cases}$$

that is: $a \to (a, 0, 0, \ldots, 0, \ldots)$ is a monomorphism of rings and, hence, an isomorphism of A onto its image in C. According to 6.1, there exist a ring B, which contains A as subring, and an isomorphism $\alpha': B \to C$ whose restriction to A is α.

Let $g \in C$ the function defined by:

$$g(k) = \begin{cases} 1 \; si \; k = 1 \\ 0 \; si \; k \neq 1 \end{cases}$$

that is $g = (0, 1, 0, 0, \ldots, 0, \ldots)$. Proceeding by induction on h it follows that:

$$g^h(k) = \begin{cases} 1 \; si \; k = h \\ 0 \; si \; k \neq h \end{cases}$$

Let $X = \alpha'^{-1}(g)$ and we will see that X is trascendental over A. In fact, if $\sum_{h=0}^{n} a_h X^h = 0$ with $a_h \in A$, then we have:

$$0 = \alpha'\left(\sum_{h=0}^{n} a_h X^h\right) = \sum_{h=0}^{n} \alpha(a_h)\, g^h \Rightarrow \left(\sum_{h=0}^{n} \alpha(a_h)\, g^h\right)(k) =$$

$$= \sum_{h}^{n}(\alpha\,(a_h)g^h)(k) = 0 \; \forall k \in \mathbb{N} \cup \{0\}$$

i.e., for any $k \in \mathbb{N} \cup \{0\}$ we have:

$$0 = \sum_{h=0}^{n}\sum_{i+j=k}^{n} \alpha(a_h)(i)g^h(j) \quad = \sum_{h=0}^{n} a_h g^h(k) = a_k. \blacksquare$$

Theorem 3.2: Given a ring A, there exists a ring $A[X]$ containing A as a subring and containing an element X trascendental over A, in such a way that:

$$A[X] = \{f \in B/f = \sum_{i=0}^{n} a_i X^i \; ; \; n \in \mathbb{N} \cup \{0\}; \; a_i \in A\}$$

$$\sum_i a_i X^i + \sum_i b_i X^i = \sum_i (a_i b_i) X^i \; (1)$$

$$\left(\sum_i a_i X^i\right)\left(\sum_i b_i X^i\right) = \sum_k \sum_{i+j=k} a_i b_j X^k \quad (2)$$

and $A[X]$ is called the *polynomial ring* in the *indeterminate* X with coefficients in A.∎

Let us see an example which shows the convenience of disposing of a polynomial ring.

Example: In $\mathbb{Z}[X]$, from the obvious relation:

$$(X + 1)^{n+m} = (X + 1)^n (X + 1)^m$$

with $n, m \in \mathbb{N}$, follows:

$$\sum_{k=0}^{n+m} \binom{n+m}{k} X^k = \left(\sum_{i=0}^{n} \binom{n}{i} X^i\right)\left(\sum_{j=0}^{m} \binom{m}{j} X^j\right) =$$
$$\sum_{k=0}^{n+m} \left(\sum_{i+j=k} \binom{n}{i}\binom{m}{j}\right) X^k$$

from where, as X is trascendental over $\mathbb{Z}$, we obtain:

$$\binom{n+m}{k} = \sum_{i+j=k} \binom{n}{i}\binom{m}{j}$$

for any k such that $0 \le k \le n + m$. In particular, we obtain the useful identity:

$$\binom{2n}{n} = \sum_{i=0}^{n} \binom{n}{i}^2.$$

A characteristic property of the polynomial ring $A[X]$ is that the "indeterminate" X, may be "specialized" or "determined." More precisely:

Proposition 3.3 : Let A be a subring of a ring A' and let $x \in A'$. The function $\phi_x \colon A[X] \to A'$, of *specialization* of X in x, defined by:

$$\phi_x\left(\sum_i a_i X^i\right) = \sum_i a_i x^i$$

is a morphism of rings.

Proof: The transcendence of X over A allows us to prove that ϕ_x is well defined, being the rest a straightforward verification.∎

If, moreover, X' is trascendental over A, we have the morphism $\psi \colon A[X] \to A[X']$ of specialization of X in X',

$$\psi\left(\sum_i a_i X^i\right) = \sum_i a_i X'^i$$

ψ is an isomorphism, since the transcendence of X' over A allows us to define, in an obvious way, the inverse of ψ. For this reason, we refer to $A[X]$ as "the" polynomial ring in an indeterminate with coefficients in A.

If $\phi_x \colon A[X] \to A$ is the morphism of specialization in $x \in A$, each $f(X) \in A[X]$ determines a function $f \colon A \to A$ defined by $f(x) = \phi_x(f)$. A straightforward verification, shows that the function so defined: $f(X) \to f$, is an epimorphism of rings, from the polynomial ring $A[X]$, onto the ring of polynomial functions defined in chapter 5, sec.9. Moreover, if A is an infinite domain, this epimorphism is an isomorphism, since it remains to verify the injectivity, so let:

$f(X) = \sum_i a_i X^i\} \to f$ with $f(x) = \sum_i a_i x^i$

$g(X) = \sum_i b_i X^i \to g$ with $g(x) = \sum_i b_i x^i$

if $f = g$, then $\sum_i a_i x^i = \sum_i b_i x^i \; \forall x \in A$, hence by prop. 9.3, b, Chap. 5, we have $a_i = b_i \; \forall i$, so then $f(X) = g(X)$. If A is not an infinite domain the injectivity is not necessarily valid, for example, if $A = \mathbb{Z}_2$, we have $X^2 \neq X$, but $x^2 = x \; \forall x \in \mathbb{Z}_2$.

From now on, we will use f, g to denote polynomials, instead of $f(X), g(X)$.

If $f \in A[X]$ is nonzero, we can write:

$$f = \sum_{i=0}^{n} a_i X^i$$

with $a_n \neq 0$. In such a case, n is called the degree of f, denoted $d(f)$, a_n its principal coefficient and, if this equals 1, f is said to be monic. Note that the degree is not defined for the zero polynomial. $x \in A$ is said to be a root of f iff $f(x) = 0$

Proposition 3.4: Let A be a ring and f=$\sum_{i=0}^{n}$a_{i}X^{i}$\in A[X]$ a polynomial of degree n $(a_n \neq 0)$. $x \in A$ is a root of $f \Leftrightarrow \exists\, g \in A[X]$ of degree $n - 1$ such that: $f = (X - x)g$.

Proof: Is the same as the proof of prop. 9.1 of Chap. 5. That is, if x is a root of f, i. e., $f(x) = 0$, then:

$$f = f - f(\alpha) = a_n(X^n - \alpha^n) + \ldots + a_2(X^2 - \alpha^2) + a_1(X - \alpha) =$$
$$= (X - \alpha)[a_n(X^{n-1} + \alpha X^{n-2} + \ldots + \alpha^{n-1}) + \ldots + a_2(X + \alpha) + a_1]$$

and denoting by $g(x)$ the expression between square brackets, we obtain the implication for the right.

For the converse, applying ϕ_x to $f = (X - x)g$, we have $f(x) = \phi_x(f) = \phi_x(X - x)\phi_x(g) = (x - x)g(x) = 0$. ∎

Proposition 3.5: If $f, g \in A[X]$ are non zero, then:

a) $f + g = 0$ or $d(f + g) \leq max\{d(f), d(g)\}$
b) $fg = 0$ or $d(fg) \leq d(f) + d(g)$
c) If A is a domain, then $A[X]$ also is a domain and, in such a case, $d(fg) = d(f) + d(g)$.

Proof: a) and b) are left as exercises. For c), if $d(f) = n$ and $d(g) = m$, we have:

$$f = \sum_{i=0}^{n} a_i X^i, g = \sum_{j=0}^{m} b_j X^j$$

then

$$fg = \sum_{k=0}^{n+m} (a_i b_j) X^k$$

but as $\sum_{i+j=n+m} a_i b_j = a_n b_m \neq 0$ since $a_n \neq 0, b_m \neq 0$ and A is a domain, result in $fg \neq 0$ and $d(fg) = n + m$. ∎

If K is a field, the polynomial ring $K[X]$ has some properties similar to that of the ring Z of integers, in particular:

Theorem 3.6: (Division Algorithm) Given $f, g \in K[X]$ where K is a field and $g \neq 0$, there exist polynomials $q, r \in K[X]$ such that:

$$f = gq + r, \text{ with } r = 0 \text{ or } d(r) < d(g)$$

Moreover, such q (quotient) and r (residue) are unique.

Proof: If $f = 0$ or if $d(f) < d(g)$ it is enough to take $q = 0$ and $r = f$. Let then $d(f) \geq d(g)$, putting $f = a_n X^n + a_{n-1} X^{n-1} + \ldots + a_1 X + a_0$, $g = b_m X^m + b_{m-1} X^{m-1} + \ldots + b_1 X + b_0$, with $a_n \neq 0, b_m \neq 0$ and $n \geq m$, the polynomial:

$$f - a_n b_m^{-1} X^{n-m} g$$

is zero or its degree is $< gr(f) = n$. Proceeding by induction on n, if $f - a_n b_m^{-1} X^{n-m} g = 0$ by we have said at the beginning of the proof, and if $d(f - a_n b_m^{-1} X^{n-m} g) < n$ by the induction hypothesis, it follow that there exist $q', r \in K[X]$ such that:

$$f - a_n b_m^{-1} X^{n-m} g = q'g + r, \text{ with } r = 0 \text{ or } d(r) < d(g)$$

Then it is sufficient taking $q = q' + a_n b_m x^{n-m}$ to obtain the existence.

For the uniqueness, from:

$$f = gq + r = gq' + r' \text{ with } r = 0, r' = 0, \text{ or } d(r), d(r') < d(g)$$

follows that $g(q - q') = r' - r$, so then if $q - q' \neq 0$ and $r' - r \neq 0$ we   obtain $d(g) + d(q - q') = d(r' - r) \leq max\{d(r'), d(r)\} < d(g)$, a contradiction, hence $q = q'$ and $r = r'$. ∎

Corollary   3.7:   Let K be   a   field, $a \in K$ and $\in$ $K[X]$. The residue when dividing f by $X - a$ is $f(a)$.

Proof: By the division algorithm, we have: $f = (X - a)q + r$, with $r = 0$ or $d(r) < d(X - a) = 1$ so $r \in K$, and as the specialization is a morphism that leaves fixed the elements of K, we obtain: $f(a) = \phi_a(r) = r$. ∎

The similarity between the division algorithm of integers and polynomials over a field, suggest the study of other domains with a similar algorithm. Actually, the impulse to such study came, from two sources: Fermat Last Theorem and reciprocity laws of higher order.

Gauss at the beginning of 19th century, proves the Euclideanity of the ring, now called, of Gaussian integers, as an auxiliary result to his proof of the biquadratic reciprocity law.

In 1847 at the Academy of Sciences of Paris, several sessions were dedicated to Fermat Last Theorem. In one of them, Lame, who already had proved the theorem for exponent 7, announced he was about to get a general proof using an extension of the property of unique factorization to rings of complex numbers (cyclotomic integers). Cauchy announced that he, using similar methods, was at the edge of obtaining the same result too. Liouville pointed out that unique factorization should be proved. Weeks later, Liouville reads a letter by Kummer showing that unique factorization in such rings, is valid in some cases but it is not valid in others. Kummer had reached these conclusions throughout his researches on reciprocity laws. Lame noted that to prove unique factorization, it suffices to have a division algorithm. This led to the study of primes, unique factorization, division algorithm, in rings other than the integers or polynomials.

4 - EUCLIDEAN DOMAINS

A domain A is called *Euclidean* iff exists a function, called the *Euclidean function* or *norm*, $N: A - \{0\} \to \mathbb{N} \cup \{0\}$ such that:

1) $N(a) \leq N(ab)$ whichever   be $a, b \in A - \{0\}$, and $N(a) < N(ab)$ if b is not invertible in A.

2) Given $a, b \in A$ with $b \neq 0$, exist $q, r \in A$ such that:

$a = bq + r$ with $r = 0$ or $N(r) < N(b)$

It can be proved that the condition 1) of the definition is superfluous, in the sense that if there is a function satisfying condition 2), then there is a function, not necessarily the same, that satisfies both conditions.

The theorem on the division algorithm of polynomials, asserts that $K[X]$ is Euclidean, taking $N = d$ as Euclidean metric. Taking $N = ||$ the ring $\mathbb{Z}$ is Euclidean. We will prove this again, in a way that will serve us as a model to other cases.

Let $a, b \in Z$ with $b > 0$, if q is the integral part of $|a/b|$, we have: $|a/b| = q + s$ with $0 \leq s < 1$, that is $a = bq + bs$. Since $0 \leq |bs| < |b|$ and $bs = a - bq \in \mathbb{Z}$, since $a, b, q \in \mathbb{Z}$. Putting $r = bs$ we   obtain the result (this form of the algorithm, where we admit negative residues, is slightly different to that of chapter 3).

Consider now the ring of *Gaussian integers*, that is, the set:

$$\mathbb{Z}[i] = \{a + bi/a, b \in \mathbb{Z}\}$$

$\mathbb{Z}[i]$ is a subring of the field $\mathbb{C}$ of complex numbers, and so is a domain. In order to prove that $\mathbb{Z}[i]$ is Euclidean, we could try to proceed as in the proof just seen of the Euclideanity of $\mathbb{Z}$, that is, if $a + bi, c + di \in \mathbb{Z}[i]$ with $c + di \neq 0$, we put:

$$\frac{a+bi}{c+di} = q_1 + q_2 i + s_1 + s_2 i \ (1)$$

with q_1 the integral part of the real part of $|(a + bi)/(c + di)|$ and q_2 the integral part of the imaginary part of that quotient, then:

$$a + bi = (c + di)(q_1 + q_2 i) + (c + di)(s_1 + s_2 i) \ (2)$$

but   as $a + bi, c + di, q_1 + q_2 i \in \mathbb{Z}[i]$ and $\mathbb{Z}[i]$ is a ring, it follows that $(c + di)(s_1 + s_2 i) \in \mathbb{Z}[i]$. Here we can not use the modulus as

Euclidean function, because the modulus of a Gaussian integer need not to be an integer. Instead, the norm N defined for any complex number $x + yi$ as the square of the modulus, i.e.

$$N(x + yi) = x^2 + y^2$$

applied to a Gaussian integer, is an ordinary integer. For this norm to be a Euclidean function in $\mathbb{Z}[i]$ it would suffice having in (2):

$$N((c + di)(s_1 + s_2 i)) < N(c + di)$$

or, since N is multiplicative: $N(zw) = N(z)N(w) \ \forall z, w \in C$; it would be enough to have $N(s_1 + s_2 i) < 1$. However, as $0 \leq s_1, s_2 < 1$, we have: $N(s_1 + s_2 i) = s_1^2 + s_2^2 < 2$, not necessarily < 1. Nevertheless, all works fine with a slight modification, instead of taking q_1 (q_2) as the integral part of the real (imaginary) part of $\left| \frac{a+bi}{c+di} \right|$, we take them as the closer integers to those real or imaginary parts, we obtain $-\frac{1}{2} \leq s_1, s_2 \leq \frac{1}{2}$, and then $N(s_1 + s_2 i) = s_1^2 + s_2^2 \leq \frac{1}{4} + \frac{1}{4} < 1$. Hence we have proved:

Proposition 4.1: The ring $\mathbb{Z}[i]$ of the Gaussian integers is Euclidean with the norm.∎

Example: We will find quotient and residue when dividing $12 + 9i$ by $3 + 2i$. We have:

$$\frac{12 - 9i}{3 + 2i} = \frac{(12 - 9i)(3 - 2i)}{13} = \frac{18}{13} - \frac{51}{13}i = 1 - 4i + \frac{5}{13} + \frac{1}{13}i$$

then: $12 - 9i = (1 - 4i)(3 + 2i) + 1 + i$, with $N(1 + i) = 2 < N(3 + 2i) = 13$.

The set:

$$\mathbb{Z}[\sqrt{-2}] = \{a + b\sqrt{2}i / a, b \in \mathbb{Z}\}$$

is a subring of the field $\mathbb{C}$ of complex numbers. In a similar way to that shown for the Gaussian integers, we have (exercise):

Proposition 4.2: $\mathbb{Z}[\sqrt{-2}]$ is a Euclidean domain with the norm.∎

Let $\theta \in \mathbb{C}$ a primitive cubic root of unity, that is $\theta^3 = 1$ and $\theta \neq 1$. Since $0 = \theta^3 - 1 = (\theta - 1)(\theta^2 + \theta + 1)$, results in $\theta^2 + \theta + 1 = 0$. It is easily verified that, $\mathbb{Z}[\theta] = \{a + b\theta / a, b \in \mathbb{Z}\}$ is a subring of $\mathbb{C}$.

Proposition 4.3: $\mathbb{Z}[\theta]$ is an Euclidean domain with the norm.

Proof: Proceeding as in the case of the Gaussian integers, we got to check that the norm of a complex of the form $s_1 + s_2\theta$, with s_1, s_2 real numbers such that $-(1/2) \leq s_1, s_2 \leq (1/2)$, is lesser than 1 and, in fact, we have:

$$N(s^1 + s^2\theta) = (s^1 + s^2\theta)(s^1 + s^2\theta) = s^{12} - s^1s^2 + s^{22} =$$

$$= (s_1 - \left(\frac{s^2}{2}\right)^2 + \frac{3s^{22}}{4} \leq \frac{9}{16} + \frac{3}{16} < 1. \blacksquare$$

5 - DIVISIBILITY

In this section, the notion of divisibility in $\mathbb{Z}$ is generalized to an arbitrary domain.

Let A be a domain and a, b elements of A. We say that a *divides* b, or that a is a *divisor* of b, and we write $a|b$, iff there exists $c \in A$ such that $b = ac$.

Examples:
  1) $0 \mid a \Rightarrow a = 0$.
  2) $a \mid 0 \forall a \in A$.
  3) $1 \mid a \forall a \in A$.

We have: $a \mid 1 \Leftrightarrow \exists c \in A$ such that $1 = ac$, that is, the divisors of 1 are the invertible elements of A, also called *unities* of A. The set of all unities of A, will be denoted by $U(A)$.

Examples:
  1) $U(\mathbb{Z}) = \{1, -1\}$
  2) $U(A[X]) = U(A)$.
In fact, $f \in U(A[X]) \Leftrightarrow \exists g \in A[X]$ such that $fg = 1 \Rightarrow 0 = gr(f) + gr(g)$, then $gr(f) = 0$ and $gr(g) = 0$, i.e., $f, g \in A$ and as $fg = 1$, $f \in U(A)$. Since clearly $U(A) \subset U(A[X])$, results $U(A[X]) = U(A)$. In particular, $U(Z[X]) = \{1, -1\}$ and if K is a field: $U(K[X]) = U(K) = K - \{0\}$.
  3) $U(Z[i]) = \{1, -1, i, -i\}$.

Indeed, it is clear that $1, -1, i, -i \in U(Z[i])$. Let $z = a + bi \in U(Z[i])$, then exists $w \in Z[i]$ such that $zw = 1$. Taking norms we obtain: $N(z)N(w) = 1$ and as $N(z), N(w) \in N \cup \{0\}$, it results $N(z) = 1 = a^2 + b^2$, but $a, b \in \mathbb{Z}$, $a = \pm 1$ and $b = 0$, or $a = 0$ and $b = \pm 1$, so that $z = \pm 1$ or $z = \pm i$.

4) $U(\mathbb{Z}[\sqrt{-2}]) = \{1, -1\}$.

As in the above example, any unity $a + b\sqrt{(-2)}i \in \mathbb{Z}[\sqrt{-2}]$ has norm 1: $1 = a^2 + 2b^2$, then $b = 0$ and $a = \pm 1$.

5) $U(\mathbb{Z}[\theta]) = \{1, -1, \theta, -\theta, \theta^2, -\theta^2\}$, being θ a primitive cubic root of 1.

Also, in this case, any unity $a + b\theta \in Z[\theta]$ must have norm 1:

$$1 = a^2 - ab + b^2 = \left(a - \frac{b}{2}\right)^2 + \frac{3b^2}{4},$$

that is, $(2a - b)^2 + 3b^2 = 4$, from where: $b = 0, b = 1$ or $b = -1$. If $b = 0$, results in $a = \pm 1$; if $b = 1, a = 0$ or $a = 1$ and if $b = -1, a = 0$ or $a = -1$.

Proposition 5.1: If $a, b, c \in A$, then:

1) $a \mid a$
2) $a \mid b$ and $b \mid c \Rightarrow a \mid c$
3) $a \mid b$ and $a \mid c \Rightarrow a \mid hb + kc$ whichever be $h, k \in A$.

Proof: exercise.■

If two elements $a, b \in A$ divides each other: $a \mid b$ and $b|a$, we say that they are *associates* and we write: $a \sim b$. We have then that $\sim$ is an equivalence relation in A.

Proposition 5.2: a and b are associates $(a \sim b) \Leftrightarrow$ there exists $u \in U(A)$ such that $b = ua$.

Proof: ($\Rightarrow$): $a \sim b \Leftrightarrow a \mid b$ and $b \mid a \Leftrightarrow$ exist $u, v \in A$ such that $b = ua, a = vb$, hence if one of a, b is zero, the other is zero as well, and in such a case $b = 1a$. Assume then that $a \neq 0$ and $b \neq 0$, from $b = ua$ and $a = vb$ we obtain $b = uvb$ and as $b \neq 0, uv = 1$ and $u \in U(A)$.

($\Leftarrow$): From $b = ua$ with $u \in U(A)$, it follows that $\exists v \in A$ such that $uv = 1$, then $a = vb$, and so $a \mid b$ and $b \mid a$. ∎

6 - IRREDUCIBLES AND PRIMES

The concept of prime or irreducible in $\mathbb{Z}$ can be generalized to a domain A, as follows: $a \in A$ is said to be *irreducible* if $a \neq 0, a \notin U(A)$ and satisfies the following condition:

$$b \mid a \Rightarrow b \sim 1 \text{ or } b \sim a$$

Examples: 1) If K is a field, any polynomial of degree 1 in $K[X]$ is irreducible. In fact, let $f \in K[X]$ be of degree 1, then $f \neq 0, f \notin U(K[X]) = K - \{0\}$ since the elements of $K - \{0\}$ are the polynomials of degree 0, and if $g \in K[X]$ and $g \mid f$ then $gr(g) \leq gr(f) = 1$, hence $gr(g) = 0$ or $gr(g) = 1$. If $gr(g) = 0, g \in K - \{0\}$, i. e., $g \sim 1$, while if $gr(g) = 1$, as $f = gh$ with $gr(h) = 0$, results $g \sim f$.

The fields K in which the converse is valid, that is, any irreducible polynomial in $K[X]$ is of degree 1, are said algebraically closed fields. The so-called Fundamental Theorem of Algebra states that $\mathbb{C}$ is algebraically closed.

2) $= (1 + i)(1 - i)$ is not irreducible in $\mathbb{Z}[i]$, since neither $1 + i$ nor $1 - i$ are unities in $\mathbb{Z}[i]$ (example 3 above). $1 + i$ is irreducible in $\mathbb{Z}[i]$, since $N(1 + i) = 2$ and any element whose norm is a prime number is irreducible. Indeed, any element whose norm is a prime number must be irreducible. In fact, if $z \in Z[i], N(z) = p$ prime, then if $w \in \mathbb{Z}[i]$ with $w \mid z$, we have $z = wv$ for some $v \in Z[i]$, and taking norms: $N(w) = 1$ and $N(v) = p$, in which case $w \sim 1$, or $N(w) = p$ and $N(v) = 1$, in which case $v \sim 1$ and $w \sim z$.

3) In the ring $\mathbb{Z}[\theta]$ when θ is a primitive cubic root of 1, $\lambda = 1 - \theta$ is irreducible and $\lambda^2 \sim 3$. In fact, $N(\lambda) = (1 - \theta)(1 - \bar\theta) = 3$, and in a similar way as in the above example, any element of prime norm, must be irreducible. Moreover, $w^2 = (1 - \theta)^2 = 1 - 2\theta + \theta^2 = -3\theta$, then $w^2 \sim 3$ since $-\theta \in U(Z[\theta])$.

Another property of the integers is that of the Fundamental Theorem of Arithmetic: any nonzero integer $\neq \pm 1$ factorizes in a unique way as ±1 multiplied by a product of primes. For a domain A we define:

A is a *Unique Factorization Domain* (U.F.D.) iff:

1) For each $a \in A$ such that $a \neq 0$ and $a \notin U(A)$, there exist irreducibles $p_1, p_2, \ldots, p_n \in A$ such that,

$$a = p_1 p_2 \ldots p_n$$

2) If $p_1 p_2 \ldots p_n = q_1 q_2 \ldots q_m$ with p_i, q_j irreducible elements of A, then $n = m$ and, reordering the q_j if necessary, $p_i \sim q_j \ \forall i = 1, 2, \ldots, n$.

In other words, any nonzero and noninvertible element is a product of irreducibles in an, essentially, unique way.

Example: Let us show an example of a domain which does not satisfy 2) and, as we shall see later, it satisfies 1). Let $\mathbb{Z}[\sqrt{-5}]$ defined by:

$$\mathbb{Z}[\sqrt{-5}] = \{a + \sqrt{5}bi / a, b \in \mathbb{Z}\}$$

$\mathbb{Z}[\sqrt{-5}]$ is a subring of $\mathbb{C}$ and is, therefore, a domain. Let us compute its invertible elements, if $z = a + \sqrt{5}bi \in \mathbb{Z}[\sqrt{-5}]$, we have (being N the norm): $z \in U(\mathbb{Z}\sqrt{-5}] \Leftrightarrow \exists w \in \mathbb{Z}[\sqrt{-5}]$ such that $zw = 1 \Rightarrow N(z)N(w) = 1$, but as $N(a + \sqrt{5}bi) = a^2 + 5b^2 \in \mathbb{N} \cup \{0\}$, results in $N(z) = 1 = a^2 + 5b^2$, then $b = 0$ and $a = \pm1$. It follows that $U(\mathbb{Z}\sqrt{-5}] = \{1, -1\}$. We have:

$$9 = 3 \cdot 3 = (2 + \sqrt{5}i)(2 - \sqrt{5}i) \ (*)$$

If we prove that the elements $3, 2 + \sqrt{5}i$ and $2 - \sqrt{5}i$ are irreducible and that 3 is not an associate neither to $2 + \sqrt{5}i$ nor to $2 - \sqrt{5}i$, it will result by $(*)$ that in $\mathbb{Z}[\sqrt{-5}]$ the condition 2) of the above definition is not valid. It is clear that 3 is not associate neither to $2 + \sqrt{5}i$ nor to $2 - \sqrt{5}i$ by the above proposition and by being $U(Z(\sqrt{-5}) = \{1, -1\}$. Since $3, 2 + \sqrt{5}i$ and $2 - \sqrt{5}i$, all have norm $= 9$ (as is seen from $(*)$), to prove that they are irreducible, will be enough proving that any element of norm $= 9$ is irreducible. In fact, let z such that $N(z) = 9$, then $z \neq 0, z \notin U(\mathbb{Z}[\sqrt{-5})$ and if $w|z$, then $z = wv$ with $v \in \mathbb{Z}[\sqrt{-5}]$, then $9 = N(z) = N(w)N(v)$ so that $N(w) = 1, 3$ or 9; but if it were $N(w) = 3$, putting $w = c + \sqrt{5}di$, it would be $9 = c^2 + 5d^2$ which is impossible for integers c and d. Hence $N(w) = 1$ and $N(v) = 9$ or $N(w) =$

9 and $N(v) = 1$. In the first case we have $w \sim 1$ and in the second $v \sim 1$, so that $w \sim z$.

An element $p \in A$ is said to be prime iff $p \neq 0, p \notin U(A)$ and moreover:

$$p \mid ab \Rightarrow p \mid a \text{ or } p \mid b$$

Proposition 6.1: Prime $\Rightarrow$ irreducible.

Proof: Let p be prime and let $a \mid p$, then $p = ab$ $(a, b \in A)$ so that $p \mid ab$ and, in consequence, $p \mid a$ or $p \mid b$. If $p \mid a$ we have $a \sim p$ and if $p \mid b$ then $a \sim 1$.∎

The converse of the above proposition is valid in $\mathbb{Z}$ (and, as we will see later, in any unique factorization domain), but is not valid in general; for example in $\mathbb{Z}[\sqrt{-5}]$, 3 is irreducible and according to $(*)$: $3 \mid (2 + \sqrt{5}i)(2 - \sqrt{5}i)$, however $3 \mid 2 \pm \sqrt{5}i \Rightarrow 2 \pm \sqrt{5}i = 3(a + \sqrt{5}bi)$ with a and b integers, the it would result $3 \mid 2$ in $\mathbb{Z}$, which is absurd.

7 - GREATEST COMMON DIVISOR

Let a, b be elements of the domain A. A *greatest common divisor* (g.c.d.) of a and b is, by definition, any element $d \in A$ such that:

1) $d \mid a$ and $d \mid b$.
2) $c \in A, c \mid a$ and $c \mid b \Rightarrow c \mid d$.

Proposition 7.1: (Uniqueness) If d and d' are greatest common divisors of a and b, then $d \sim d'$.

Proof: exercise.∎

When we defined a g.c.d. d of two integers, we added the condition that $d \geq 0$, in order to obtain the equality of two of them (because $U(Z) = \{\pm 1\}$). In the case of polynomials over a field F, adding the condition that the g.c.d. is monic (principal coefficient $= 1$) we obtain also the equality, not only the equivalence, because $U(F[X]) = U(F) = F - (0)$.

In a similar way it is defined the greatest common divisor of three or more elements, as a common divisor which is divisible by any common divisor, and its uniqueness, except associates, is proved. For its existence, it suffices the existence for any two elements, as follows from the next proposition.

Proposition 7.2: Let A be a domain such that any two elements have a greatest common divisor. We denote by (a, b) to any greatest common divisor of a and b.

a) $(a, (b, c))$ is a greatest common divisor of a, b, c. Then, there exists a greatest common divisor of three or more elements.
b) $(a, (b, c)) \sim ((a, b), c)$.
c) $(ca, cb) \sim c(a, b)$.
d) $(a, b) \sim 1 \Rightarrow (a, bc) \sim (a, c)$.
e) $(a, b) \sim 1$ and $a \mid bc \Rightarrow a \mid c$.
f) Irreducible $\Rightarrow$ prime.

Proof: a) $(a, (b, c))$ clearly complies the conditions that define a greatest common divisor of a, b, c.

b) $((a, b), c)$ complies them too, then $(a, (b, c)) \sim ((a, b), c)$.

c) Let $d \sim (a, b)$ and $e \sim (ca, cb)$. Since $cd \mid ca$ and $cd \mid cb$, results in $cd \mid e$, then $e = cdu$ for some u. Moreover, as $e \mid ca$ and $e \mid cb$, it follows that $du \mid a$ and $du \mid b$, so then $du \mid d$, hence $u \sim 1$ and $e \sim cd$.

d) $(a, bc) \sim ((a, ac), bc) \sim (a, (a, b)c) \sim (a, c)$.

e) $a \mid bc \Rightarrow (a, bc) = a$, but as $(a, b) \sim 1$, then by d): $(a, bc) \sim (a, c)$, hence $(a, c) \sim a$ and $a \mid c$.

f) Let $p \in A$ irreducible such that $p \mid ab$ and $p \nmid a$, and we will see that $p \mid b$. Let d be a greatest common divisor of p and a; since p is irreducible and $d \mid p$, we must have $d \sim 1$ or $d \sim p$, but as $d \mid a$ and $p \nmid a$, it must be $d \sim 1$, and from e) it follows that $p \mid b$.■

Example: In $\mathbb{Z}[\sqrt{-5}]$ the elements $z = 9$ and $w = 6 + 3\sqrt{5}i$ do not have a greatest common divisor. In fact, if d were a greatest common divisor of them, we would have $2 + \sqrt{5}i \mid d$ since $2 + \sqrt{5}i$ is a common divisor of z and w; and then it would exist $u \in \mathbb{Z}[\sqrt{-5}]$ such that $d = (2 + \sqrt{5}i)u$, from where $u \mid 2 - \sqrt{5}i$ and $u \mid 3$, and hence $u \sim 1$ and $d \sim 2 +$

$\sqrt{5}i$. Since 3 is also a common divisor of z and w, we obtain similarly d~3. Then $2 + \sqrt{5}i \sim 3$ which we have seen is impossible.

Theorem 7.3: If A is an Euclidean domain, then there exists a greatest common divisor of any two elements of A.

Proof: Let N be a Euclidean function on A, and a, b nonzero elements of A. Consider the set:
$$B = \{N(ha + kb)/h, k \in A, ha + kb \neq 0\}$$
this is a subset of $\mathbb{N} \cup \{0\}$, so it has a smallest element $m = N(d), d = sa + tb$. We assert that d is then a greatest common divisor of a, b. Obviously if $c \mid a$ and $c \mid b$, then $c \mid d = sa + tb$. In order to prove that $d \mid a$, put:
$$a = dq + r \text{ with } r = 0 \text{ or } N(r) < N(d)$$
We have: $r = a - dq = (1 - qs)a + (-t)b$, which shows that if it were $r \neq 0$, then it would be $N(r) \in B$, contradicting the minimality of $N(d)$. Then $r = 0$ and d|a. Similarly, $d \mid b$ and d is a greatest common divisor of a, b.∎

Proceeding in a similar way as we did with the integers, we can prove the validity of a Euclidean algorithm in any Euclidean ring, and the last nonzero residue is a g.c.d., the details are left to the exercises at the end of the chapter.

8 - UNIQUE FACTORIZATION DOMAINS

Proposition 8.1: If in a domain A any irreducible is prime, then the condition 2) of the definition of an unique factorization domain is satisfied, that is, if $p_1 p_2 \ldots p_n\} = q_1 q_2 \ldots q_m$ with p_i, q_j irreducibles in A, then $n = m$ and, except a reordering of the $q_j, p_i \sim q_i \forall i = 1, 2, \ldots, n$.

Proof: Let $p_1 p_2 \ldots p_n = q_1 q_2 \ldots q_m$ with the p_i and the q_j irreducible. As p_1 is prime and $p_1 \mid q_1 q_2 \ldots q_m, p_1$ must divide one of the q_j, and reordering them, is necessary, we can assume that: $p_1 \mid q_1$; but q_1 is irreducible and p_1 is noy a unit, then $p_1 \sim q_1$; so there is a unit u such that $q_1 = up_1$, hence:
$$p_2 \ldots p_n = (uq_2) \ldots q_m$$

Proceeding by induction in n, as uq_2 is irreducible, by inductive hypothesis we have $n - 1 = m - 1$ and except for a reordering of the q_j, $p_i \sim q_m$ $\forall i = 2, \ldots, n$.∎

Theorem 8.2: Any Euclidean domain A is an unique factorization domain.

Proof: Let N be an Euclidean function on A. If condition 1) of the definition of unique factorization domain were false, there would exists $a \in A$, $a \neq 0$, $a \notin U(A)$, with smallest $N(a)$ which is not a product of irreducibles, in particular, a would not be irreducible, so there would exists $b \in A$ such that $b \mid a$, $b \nsim 1$, $b \nsim a$, so $a = bc$, $b \nsim 1$, $c \nsim 1$. But $N(a) = N(bc) > N(b)$, and $N(a) > N(c)$. Then, by minimality of $N(a)$, both b and c would be a product of irreducibles and so would be $a = bc$. The condition 2) of the definition of unique factorization domain follows from 7.3, 7.2 f, and 8.1.∎

Proposition 8.3: In a unique factorization domain A, any two elements have a greatest common divisor.

Proof: Let $a, b \in A$. If any of them is 0 or is a unity, the existence of a greatest common divisor is immediate. Assume both nonzero and non unities, then they are the product of irreducibles. Grouping associate irreducibles and irreducibles common to both, we can write:

$$a = p_1^{\alpha_1} p_2^{\alpha_2} \ldots p_n^{\alpha n} q_1^{\beta_1} q_2^{\beta_2} \ldots q_m^{\beta_m}$$
$$b = p_1^{\gamma_1} p_2^{\gamma_2} \ldots p_n^{\gamma n} r_1^{\delta_1} r_2^{\delta_2} \ldots r_l^{\delta l}$$

where the p_i, q_j, r_k are irreducible pairwise non associates and the exponents are natural numbers. If we put:

$$\varepsilon_i = min\{\alpha_i, \gamma_i\} \; (i = 1, \ldots, n)$$

we have that $d = p_1^{\varepsilon_1} p_2^{\varepsilon_2} \ldots p_n^{\varepsilon n}$ satisfies the conditions which define a greatest common divisor of a and b.∎

Proposition 8.4: In an unique factorization domain any irreducible is prime.

Proof: This follows from 8.3 and 7.2 f.∎

9 - THREE THEOREMS OF FERMAT

The following theorem was conjectured by *A.* Girard and surely proved by Fermat, although the first published proof was by Euler.

Theorem 9.1: (of the two squares of Fermat): Every prime of the form $p = 4n + 1$ is a sum of two squares.

Proof: By corollary 7.2 Chap. 5, as $p = 4n + 1$, there exists $x \in \mathbb{Z}$ such that $x^2 \equiv -1 (mod p)$, then $p \mid (x + i)(x - i)$ in the ring $\mathbb{Z}[i]$ of Gaussian integers. If p were prime in $\mathbb{Z}[i]$, we would have $p \mid x + i$ or $p \mid x - i$, i.e., $x \pm i = p(c + di)$ with $c, d \in \mathbb{Z}$ and comparing the imaginary parts we should have $pd = \pm 1$ which is absurd. Then p is not irreducible in $\mathbb{Z}[i]$ so there is $z = a + bi \in \mathbb{Z}[i]$ such that $z \mid p, z \nsim 1$ and $z \nsim p$. Putting $p = zw$ and taking norms, we have $p^2 = N(z)N(w)$, hence we have the following cases:

· $N(z) = 1$ in which case $z \sim 1$, a contradiction.

· $N(z) = p$

· $N(z) = p^2$ in which case $N(w) = 1$, That is $w \sim 1$, then $z \sim p$, a contradiction too.

It follows that $N(z) = p = a^2 + b^2$.∎

The problem of solving the Diophantine equation $y^3 = x^2 + 2$ was possed by Fermat to the English mathematicians Brounker and Wallis. Fermat said that this problem seemed difficult at first sight, but that he founded a beautiful method to solve it. The first published proof was by Euler with some gaps.

Theorem 9.2: $x = \pm 5, y = 3$ are the only solutions of the diophantine equation $y^3 = x^2 + 2$.

Proof: Let (x, y) be a solution of the equation, that is, $x.y \in \mathbb{Z}$ with $y^3 = x^2 + 2$. A common divisor of x and y, must divide 2, but 2 can not be a common factor of x, y as is seen looking at the equation modulus 4, hence x, y must be coprime. In the domain $\mathbb{Z}[\sqrt{-2}]$ we have the factorization:

$$y^3 = (x + \sqrt{2}i)(x - \sqrt{2}i) \ (*)$$

Let us show that 1 is a greatest common divisor of $x + \sqrt{2}i$ and $x - \sqrt{2}i$ in $\mathbb{Z}[\sqrt{-2}]$. If not, it would exists an irreducible common factor z: $z \mid x + \sqrt{2}i$ and $z \mid x - \sqrt{2}i$, then $z \mid 2\sqrt{2}i$ and as z is irreducible, it is also prime in $\mathbb{Z}[\sqrt{-2}]$. It follows that $z \mid \sqrt{2}i$ or $z \mid 2 = (\sqrt{2}i)(-\sqrt{2}i)$, hence $z \mid \sqrt{2}i$ and as $\sqrt{2}i$ is irreducible, since its norm is a prime number, we have $z \sim \sqrt{2}i$. Then we would have $x + \sqrt{2}i = \sqrt{2}i(a + b\sqrt{2}i)$ for some integers a, b, so $x = -2b$ which is absurd, since in such case y would be also even. WE have just proved that $x + \sqrt{2}i$ and $x - \sqrt{2}i$ are coprime in $\mathbb{Z}[\sqrt{-2}]$ and, as this is a unique factorization domain, from $(*)$ it follows that both must be cubes in $\mathbb{Z}[\sqrt{-2}]$ (actually, each must be an unity by a cube, but in this ring all the unities are cubes). Then, must exist $c, d \in \mathbb{Z}$ such that $x + \sqrt{2}i = (c + d\sqrt{2}i)^3$, that is:

$$x + \sqrt{2}i = c(c^2 - 6d^2) + d(3c^2 - 2d^2)\sqrt{2}i$$

from where follows $d(3c^2 - 2d^2) = 1$, then $d = 1$ and $3c^2 - 2d^2 = 1$ or $d = -1$ and $3c^2 - 2d^2 = -1$. In the first case we obtain $c = \pm 1$ and in the second $3c^2 = 1$ and as the last is not possible, we must have $c = \pm 1$, and then $x = c(c^2 - 6d^2) = \pm 5$. ∎

The following theorem is the case $n = 3$ of Fermat's Last Theorem which, with the case $n = 4$, was posed by Fermat as a challenge to his contemporaneous, in particular to Frénicle de Bessy, to Mersenne, to Brounker, and to Wallis. The first published proof of the case $n = 3$ (a proof of the case $n = 4$ was founded in Fermat's papers as an auxiliary result to solve another problem and is one of the two proofs of Fermat, in Number Theory, that remains) was by Euler in his Algebra of 1770. In it, he used divisibility properties of the integers of the form $a^2 + 3b^2$; in particular, he used the following proposition: if s is odd and $s^3 = a^2 + 3b^2$ with a, b coprime, then s is also a number of that form: $s = u^2 + 3v^2$. This proposition, given without sufficient justification, was consider a flaw in Euler's proof, but 200 years later a totally rigorous proof of it was founded among Euler's papers [25].

The following proof is based on the ring $\mathbb{Z}[\theta]$ where θ is a primitive cubis root of unity. Recall some facts already proved in this ring:

1) $\mathbb{Z}[\theta] = \{a + b\theta / a, b \in \mathbb{Z} \text{ and } \theta^3 = 1 \text{ with } \theta \neq 1\}$ is an Euclidean domain (prop.4, 3) and then is an U.F.D.
2) $x \in U(\mathbb{Z}[\theta]) \Leftrightarrow N(x) = 1 \Leftrightarrow x \in \{\pm 1, \pm \theta, \pm \theta^2\}$ where $U(\mathbb{Z}[\theta])$

is the set of unities or inversible elements of $Z[\theta]$ and N denotes the norm:

$$N(a + b\theta) = (a + b\theta)(a + b\theta^2) = a^2 - ab + b^2.$$

3) $\lambda = 1 - \theta$ is prime in $\mathbb{Z}[\theta]$ and $\lambda^2 = -3\theta$.

We extend the congruence notation to the ring $\mathbb{Z}[\theta]$ with the obvious meaning and properties similar to ordinary congruences are valid.

Lemma 9.3: For $x, y \in \mathbb{Z}[\theta]$ we have:

a) $x \equiv 0 (mod\lambda)$ or $x \equiv \pm 1 (mod\lambda)$.

b) $x \equiv \pm 1 (mod\lambda) \Rightarrow x^3 \equiv \pm 1 (mod\lambda^4)$.

c) If x, y are not divisible by λ and if $x^3 \equiv \varepsilon y^3 (mod\lambda^2)$ with ε an unity in $\mathbb{Z}[\theta]$, then $\varepsilon = \pm 1$.

Proof: a) Since $\mathbb{Z}[\theta]$ is Euclidean, there exist $\alpha, \beta \in \mathbb{Z}[\theta]$ such that $x = \lambda\alpha + \beta$ with $N(\beta) < N(\lambda) = 3$, and as in $\mathbb{Z}[\theta]$ there is no element with norm 2, we must have $\beta = 0$ or $N(\beta) = 1$, and in this last case: $\beta = \pm 1$ or $\pm\theta$ or $\pm\theta^2$, but as $1 \equiv \theta \equiv \theta^2 (mod\lambda)$, a) follows.

b) From $x = \lambda\alpha \pm 1$ follows: $x^3 = \lambda^3\alpha^3 \pm 3\lambda^2\alpha^2 + 3\lambda\alpha \pm 1$ and as $3 = -\theta^2\lambda^2$, results: $3\lambda^2\alpha^2 \equiv 0 (mod\lambda^4)$ and $\lambda^3\alpha^3 + 3\lambda\alpha = \lambda^3(\alpha^3 - \theta^2\alpha)$ so it wil be enough to verify that $\alpha^3 - \theta^2\alpha \equiv 0 (mod\lambda)$ to conclude that $x^3 \equiv \pm 1 (mod\lambda^4)$. In fact, by a), $\alpha \equiv 0 (mod\lambda)$ or $\alpha \equiv \pm 1 (mod\lambda)$, if $\alpha \equiv 0 (mod\lambda)$ it is clear that $\alpha^3 - \theta^2\alpha \equiv 0 (mod\lambda)$ and if $\alpha \equiv \pm 1 (mod\lambda)$, results $\alpha^3 \equiv \pm 1 (mod\lambda)$ so that $\alpha^3 \equiv \alpha (mod\lambda)$ and $\alpha^3 - \theta^2\alpha \equiv \alpha(1 - \theta^2) \equiv 0 (mod\lambda)$ because $1 \equiv \theta^2 (mod\lambda)$.

c) Since x, y are not divisible by λ, we have by a) and b): $x^3 \equiv \pm 1 (mod\lambda^4)$ and $y^3 \equiv \pm 1 (mod\lambda^4)$, hence $\varepsilon \equiv \pm 1 (mod\lambda^2)$ and this is only possible if $\varepsilon = \pm 1$ ($\theta \equiv \pm 1$ and $\theta^2 \equiv \pm 1 (mod\lambda^2)$ lead to absurdity).∎

Theorem 9.4: There are no integers x, y, z such that $xyz \neq 0$ and:

$$x^3 + y^3 = z^3$$

Proof: If there exist, we may assume (dividing by its g.c.d.) that they are coprime and, as any factor of two of them must be a factor of the third, we may assume that they are pairwise coprime.

Note that one of them must be divisible by 3. In fact, a cube can only be congruent modulus 9, with 0 or ± 1, hence, to fulfill the condition $x^3 + y^3 \equiv z^3 (mod 9)$, one of those cubes must be $\equiv 0 (mod 9)$, so one of x, y, z must be divisible by 3.

We have then that one, and only one of them, is divisible by 3 and we may assume, by symmetry, that z is divisible by 3 (for example if y were divisible by 3, we could write: $x^3 + (-z)^3 = (-y)^3$). Put $z = 3^k z_0$ with $k \in \mathbb{N}$ and z_0 an integer not divisible by 3.

Recalling that in the ring $\mathbb{Z}[\theta]$, where θ is a primitive cubic root of unity, we have that $\lambda = 1 - \theta$ is prime and that $3 = -\theta^2 \lambda^2$, we obtain:

$$x^3 + y^3 = \varepsilon \lambda^{3n} z_0{}^3 \ (1)$$

where $\varepsilon = (-\theta^2)^{3k} = \pm 1, n = 2k > 1$ is a natural number and x, y, z_0 are integers not divisible by λ (since if an integer is divisible by λ in $\mathbb{Z}[\theta]$, it must be divisible by 3 in $\mathbb{Z}$(exercise). In equation (1), as $\varepsilon = \pm 1$, it can be assumed, changing if necessary z_0 by $-z_0$, that $\varepsilon = 1$. We will prove that the equation:

$$x^3 + y^3 = \lambda^{3n} z_0{}^3 (2)$$

is impossible with $x, y, z_0 \in \mathbb{Z}[\theta]$ nonzero, pairwise coprime and such that $\lambda \nmid xyz_0$ and with $n \in \mathbb{N}$. This will be sufficient to prove the theorem, since two coprime integers remain coprime viewed in $\mathbb{Z}[\theta]$. In order to do so, we will assume that equation (2) is possible, and so it will be an equation of that form with smallest n, which, not to change the notation, we assume is (2) itself, and we will arrive to an equation of the same type with smaller n. Let us show first that (2) es impossible if $n = 1$. Indeed, in such a case, by lemma 9.3 b), we would have:

$$\pm 1 \pm 1 \equiv \pm \lambda^3 (mod \lambda^4)$$

that is $\lambda^3 \equiv 0 (mod \lambda^4)$ or $\lambda^3 \equiv \pm 2 (mod \lambda^4)$, so then $\lambda \mid 1$ or $\lambda \mid 2$ which is absurd. We assume then that $n > 1$.

We have the factorization:

$$\lambda^{3n} z_0{}^3 = x^3 + y^3 = (x + y)(x + \theta y)(x + \theta^2 y) (3)$$

Since λ is prime, it must divide some factor on the right of (3), but then it must divide the other two factors because:

$$x + y \equiv x + \theta y \equiv x + \theta^2 y \ (mod \lambda) \text{ since } 1 \equiv \theta \equiv \theta^2 (mod \lambda)$$

Furthermore, as $n > 1$, λ^2 must divide one of such factors, but only one, since if it divided two of them, it would divided its difference

210

and would result in $\lambda^2 \mid (1 - \theta^i)y$ $(i = 1, 2)$ and so $\lambda \mid y$ which is absurd. We may assume that $\lambda^2 \mid x + y$, for if, for example, $\lambda^2 \mid x + \theta y$, putting $y' = \theta y$ and then $x + \theta y = x + y'$; $x + y = x + \theta^2 y'$; $x + \theta^2 y = x + \theta y'$ and we can proceed as follows, changing y by y'. We have then:

$$x + y = \lambda^{3n-2}u$$
$$x + \theta y = \lambda v$$
$$x + \theta^2 y = \lambda w$$

with $u, v, w \in \mathbb{Z}[\theta]$ pairwise coprime and not divisible by λ. According to (3):

$$z_0{}^3 = uvw$$

and as $\mathbb{Z}[\theta]$ is an U.F.D., there are unities ε_i in $\mathbb{Z}[\theta]$ and elements u_0, v_0, w_0 in $\mathbb{Z}[\theta]$ such that:

$$u = \varepsilon_1 u_0{}^3, v = \varepsilon_2 v_0{}^3, w = \varepsilon_3 w_0{}^3$$

with $\varepsilon_1 \varepsilon_2 \varepsilon_3 = \pm 1$ by lemma 9.3. c). Then:

$$-\varepsilon_1 \lambda^{3n-2} u_0{}^3 = -(x + y) = \theta(x + \theta y) + \theta^2(x + \theta^2 y)$$
$$= \varepsilon_2 \theta \lambda v_0{}^3 + \varepsilon_3 \theta^2 \lambda w_0{}^3$$

from where, multiplying by the inverse of $\varepsilon_3 \theta^2$ which is $\pm \varepsilon_1 \varepsilon_2 \theta$, and simplifying, we obtain:

$$w_0{}^3 + \varepsilon_4 v_0{}^3 = \varepsilon_5 \lambda^{3(n-1)} u_0{}^3 \ (4)$$

where ε_4 and ε_5 are unities in $\mathbb{Z}[\theta]$ such that $\varepsilon_4 \varepsilon_5 = -1$ and u^0, v_0, w_0 are pairwise coprime and not divisible by λ. It will be enough proving that ε_4 and ε_5 are cubes to conclude that (4) is an equation of the same type of (2) but with lesser n, what will be a contradiction. Since the only unities that are cubes in $\mathbb{Z}[\theta]$ are ± 1 (verify), we will have to prove that $\varepsilon_4 = \pm 1$, from where $\varepsilon_5 = \mp 1$. By (4) we have:

$$w_0{}^3 + \varepsilon_4 v_0{}^3 \equiv 0 (mod\lambda^2)$$

and by lemma 9.3. c), results $\varepsilon_4 = \pm 1$.∎

10 - POLYNOMIALS WITH COEFFICIENTS IN A FIELD

Recall that if A is a subring of a ring A' and $x \in A'$, the specialization $\varphi_x \colon A[X] \to A'$ in x, defined by:

$$\varphi_x(\Sigma a_i X^i) = \Sigma a_i x^i$$

is a morphism of rings. We put $\varphi_x(f) = f(x)$ and we say that x is a root of f iff $f(x) = 0$.

Proposition 9.1: Let K be a field, $x \in K$ and $f \in K[X]$,

$$x \text{ is a root of } f \Leftrightarrow X - x \mid f$$

Proof: By the division algorithm, we have $f = (X - \alpha)q + r$ with $q, r \in K[X]$ and $r = 0$ or $d(r) < d(X - \alpha) = 1$, that is $r \in K$. Since the specialization in x is a morphism of rings, we obtain: $f(x) = r$, then:

$$x \text{ is a root of } f \Leftrightarrow r = 0 \Leftrightarrow X - x \mid f . \ \blacksquare$$

Proposition 9.2: If $f \in K[X]$ with $d(f) = n$ ($f \neq 0$), then f has at most n roots in K.

Proof: Let $x_1, \ldots, x_m$ be distinct roots of f in K. Then $X - x_i \mid f$ para $i = 1, \ldots, m$. We have:

$$f = (X - x_1)f_1$$

As $X - x_2 \mid f$ and $(X - x_2, X - x_1) \sim 1$, it follows that $X - x_2 \mid f_1$, hence

$$f = (X - x_1)(X - x_2)f_2$$

Following in the same way, we obtain:

$$f = (X - x_1)(X - x_2)\ldots(X - x_m)f_m$$

from where, taking degrees follows $n \geq m$.$\blacksquare$

The above proposition is also valid for polynomials with coefficients in a domain A, because it is enough to immerse A in a field. Instead, it is not necessarily valid is the ring of coefficients is not a domain, for example, $X^2 - X, \in Z_6[X]$ has in Z_6 the roots: $0, 1, 3$ and 4.

Assuming the Fundamental Theorem of Algebra, which asserts that any polynomial in $C[X]$ of degree ≥ 1 has a root in $\mathbb{C}$, we obtain:

Proposition 9.3: If $f \in \mathbb{C}[X]$ is a polynomial of degree $n \geq 1$, then there exist $x_1, \ldots, x_n \in \mathbb{C}$, not necessarily distinct, such that:

$$f = a(X - x_1)\ldots(X - x_n)$$

with $a \in \mathbb{C}$, being this the factorization of f as a product of irreducibles.

Proof: By the Fundamental Theorem of Algebra, there exists a root $\alpha_1 \in \mathbb{C}$ of f, then

$$f = (X - \alpha_1)f_1$$

If $d(f_1) = 0$ we have already the factorization of f. If $d(f_1) \geq 1$, by the fundamental theorem again, there exists a root $\alpha_2 \in \mathbb{C}$ of f_1, then:

$$f = (X - \alpha_1)(X - \alpha_2)f_2$$

Following in this way, we obtain the result.■

Lemma 9.4: If $z \in \mathbb{C}$ is a root of a polynomial f with real coefficients, then $\bar{z}$ (the conjugate of z) is a root of f too.

Proof: Let $f = \sum_{i=0}^{n} a_i X^i$ with $a_i \in R$. Since z is a root of f, we have $f(z) = \sum_{i=0}^{n} a_i z^i = 0$ and, as conjugation is a morphism of rings: $0 = \sum_{i=0}^{n} \bar{a_i}\bar{z^i} = \sum_{i=0}^{n} a_i \bar{z^i} == f(\bar{z})$.■

Lemma: Let K be a subfield of a field F. If $f, g \in K[X]$ are such that $g \mid f$ in $F[X]$, then $g \mid f$ in $K[X]$.

Proof: By the division algorithm in $K[X]$, we have $f = gq + r$ with $q, r \in K[X]$ and $r = 0$ or $d(r) < d(g)$. In $F[X]$ we have $f = gq'$ with $q' \in F[X]$. Then by the uniqueness of the quotient and residue in $F[X]$, we have $q = q'$ and $r = 0$, so $g \mid f$ in $K[X]$.■

Proposition 9.5: The only irreducible polynomials in $\mathbb{R}[X]$ are those of degree 1 and those of degree 2: $aX^2 + bX + c$ such that $b^2 - 4ac < 0$.

Proof: Let $f \in \mathbb{R}[X]$ be irreducible. As $d(f) \geq 1$, by the Fundamental Theorem of Algebra, f has a root $z \in \mathbb{C}$. If $z \in \mathbb{R}$ results $f = (X - z)g$ with $g \in R[X]$ and as $X - z$ is irreducible in $\mathbb{R}[X]$, we have $g \in \mathbb{R} - \{0\}$, so f has degree 1. If, instead, $z \notin \mathbb{R}$, as $X - z$ and $X - z$ are coprime in $\mathbb{C}[X]$, we have $h = (X - z)(X - \bar{z}) = X^2 - (z + \bar{z})X + z\bar{z} \mid f$ in $\mathbb{C}[X]$, but since $h \in \mathbb{R}[X]$, it follows, by the above lemma, that it divides f in $\mathbb{R}[X]$. Finally, as f is irreducible we must have $g \sim f$, so there exists $a \in \mathbb{R} - \{0\}$ such that:

$$f = a(X^2 - (z + \bar{z})X + z\bar{z})$$

and putting $b = -a(z + \bar{z}), c = az\bar{z}$ we obtain $b^2 - 4ac = a^2(z^2 + 2z\bar{z} + z^2 - 4z\bar{z}) = a^2(z - \bar{z})^2$ and if $z = x + yi$ with $x, y \in R, y \neq 0$; it follows that $b^2 - 4ac = -4a^2y^2 < 0$.∎

The following proposition, clearly exposed by Vieta and Girard, was a crucial step in getting mathematicians to believe in the validity of the Fundamental Theorem of Algebra, and that they tried to prove it.

Proposition 9.6: (relations between coefficients and roots) Let

$$f = X^n + a_{n-1}X^{n-1} + \ldots + a_1X + a_0$$

be a monic polynomial in $\mathbb{C}[X]$, with, possible repeated, roots $x_1, \ldots, x_n$. The following relations between coefficients and roots are valid:

$$a_{n-1} = -\sum_i x_i$$

$$a_{n-2} = \sum_{i<j} x_i x_j$$

$$.$$
$$.$$

$$a_2 = (-1)^{n-1} \sum_{i_1 < i_2 < \ldots < i_{n-1}} x_{i_1} x_{i_2} \ldots x_{i_{n-1}}$$

$$a_1 = (-1)^n x_1 \ldots x_n$$

Proof: We have $f = (X - x_1) \ldots (X - x_n)$ and performing the products and comparing coefficients, using that X is an indeterminate over $\mathbb{C}$, we obtain the result.∎

In finding rational roots of polynomials with integer coefficients, the following is useful (obviously valid, more generally, for roots in the field of quotients of a domain such that any two elements have a gcd).:

Proposition 9.7: If (r/s) with $r, s \in \mathbb{Z}, s \neq 0, r, s$ coprime, is a rational root of the polynomial $f = a_nX^n + a_{n-1}X^{n-1} + \ldots + a_1X + a_0$ with coefficient in $\mathbb{Z}$, then $r \mid a_0$ and $s \mid a_n$.

Proof: exercise.∎

Example: To factorize $f = X^4 - 2X^2 - 3X - 2 \in \mathbb{Z}[X]$, we begin trying the possible rational roots $\pm 1, \pm 2$. We have $f(2) = 0$ and $f(-1) = 0$, then:

$f = (X - 2)(X + 1)(X^2 + X + 1)$

$$= (X - 2)(X + 1)\left(X - \frac{-1 + \sqrt{3}i}{2}\left(X + \frac{1 + \sqrt{3}i}{2}\right).\right)$$

Examples: We consider the factorization of the polynomials of the form $X^5 - 1$, with different fields of coefficients.

1) In $\mathbb{C}[X]$: According to th. 6.4, Chap. 7, the 5th roots of one are $z_k = e^{i\frac{2k\pi}{5}}, k = 0, 1, 2, 3, 4$, Then:
   $X^5 - 1 = (X - z_0)(X - z_1)(X - z_2)(X - z_3)(X - z_4)$

2) In $\mathbb{R}[X]$: Since the conjugate of $e^{i\alpha}$ is $e^{i(-\alpha)}$, we see that $z_4 = z_1$ and $z_3 = z_2$, then:
   $X^5 - 1 = (X - 1)(X^2 - 2Rez_1X + 1)(X^2 - 2Rez_2X + 1)$

If we want to be more specific: put $\varepsilon^5 = 1$, with $\varepsilon \neq 1$, so then $\varepsilon^4 + \varepsilon^3 + \varepsilon^2 + \varepsilon + 1 = 0$, or:

$$\left(\varepsilon^2 + \frac{1}{\varepsilon^2}\right) + \left(\varepsilon + \frac{1}{\varepsilon}\right) + 1 = 0$$

if we put $z = \varepsilon + \frac{1}{\varepsilon}$, we obtain $z^2 + z - 1 = 0$, hence $z = -\frac{1}{2} \pm \frac{\sqrt{5}}{2}$. As $z_1 = \frac{1}{z_1}$, we have $2Rez_1 = z_1 + \frac{1}{z^1} = -\frac{1}{2} + \frac{\sqrt{5}}{2}$, and similarly $2Rez_2 = -\frac{1}{2} - \frac{\sqrt{5}}{2}$ and then:

$$X^5 - 1 = (X - 1)\left(X^2 + \frac{1 - \sqrt{5}}{2}X + 1\right)\left(X^2 + \frac{1 + \sqrt{5}}{2}X + 1\right)$$

3) In $\mathbb{Q}[X]$:
$$X^5 - 1 = (X - 1)(X^4 + X^3 + X^2 + X + 1)$$
since the polynomial $f = X^4 + X^3 + X^2 + X + 1$ is irreducible in $\mathbb{Q}[X]$. This can be seen verifying that f has no rational roots, and so no factors of degree 1, and that f can not be a product of two second degree polynomial in $Q[X]$, since such decomposition would contradict the uniqueness of the factorization in 2).

More generally, we will prove the irreducibility over $\mathbb{Q}$ of the $p-$ cyclotomic polynomial (p prime) in section 11.

4) In $\mathbb{Z}_5[X]$:

$$X^5 - 1 = (X - 1)^5$$

(easily seen expanding the binomial).

11 - UNIQUE FACTORIZATION IN POLYNOMIAL RINGS

In this section, we will prove that if A is an U.F.D., then the polynomial ring $A[X]$ is also an U.F.D..

In order to do that, we consider the embeddings: $A \subset A[X] \subset K[X]$ where K is the field of quotients of A. Note that a prime p in A, remains, as we will see, prime in $A[X]$, but becomes a unit in $K[X]$, or that a polynomial, as pX, is irreducible in $K[X]$, but it is reducible in $A[X]$, so we must be careful about the ring in which we are considering an element or a relationship.

Let A be a domain. A polynomial $f \in A[X]$ is called *primitive* iff no prime of A divides f, or what is the same, no prime of A divides to all the coefficients of f.

Lemma 11.1: Let A be a domain:

a) p prime in $A \Rightarrow p$ prime in $A[X]$.

b) (Gauss's lemma) $f, g \in A[X], f, g$ primitivos $\Leftrightarrow fg$ primitivo.

Proof: a) Let p be a prime in A, and let $p \mid fg$ with $f, g \in A[X]$. We write explicity: $f = \sum_i a_i X^i, g = \sum_j b_j X^j$, so then

$$fg = \sum_k \left(\sum_{i+j=k} a_i b_j \right) X^k$$

Assume by contradiction, that $p \nmid f$ and $p \nmid g$. Let r be the smallest index such that $p \nmid a_r$, and s the smallest index such that $p \nmid b_s$. The coefficient of X^{r+s} in fg, is:

$$a_0 b_{r+s} + \ldots + a_{r-1} b_{s+1} + a_r b_s + a_{r+1} b_{s-1} + \ldots + a_{r+s} b_0$$

p divides this coefficient, also divides any summand on the left of $a_r b_s$ by the election of r, and divides any summand on the right of $a_r b_s$, by the election of s. Then p must divide $a_r b_s$, and therefore $p \mid a_r$ or $p \mid b_s$, a contradiction.

b) p prime in A and $p \mid fg \Rightarrow$ by a), $p \mid f$ or $p \mid g$. ∎

Note that if A is an unique factorization domain, it is clear that a polynomial is primitive if, and only if, 1 is a greatest common divisors of its coefficients.

Proposition 11.2: Let A an unique factorization domain and K its quotient field.

a) Any nonzero element of $A[X]$, can be written in the form: af with $a \in A$ and $f \in A[X]$ primitive. Moreover, there is uniqueness in such writing, as follows:

$$af = bg;\ a, b \in A - (0);\ f, g \in A[X] \text{ primitive} \Rightarrow a \sim b \text{ in } A, \text{and } f \sim g \text{ in } A[X]$$

b) If $f, g \in A[X]$ are primitive, then:

$$g \mid f \text{ in } K[X] \Rightarrow g \mid f \text{ in } A[X]$$

c) If $f \in A[X]$ is primitive, then:

$$f \text{ is irreducible in } K[X] \Leftrightarrow f \text{ is irreducible in } A[X].$$

Proof: a)   Let $h \in A[X], h \neq 0$, if a is a greatest common divisor of the coefficients of h, we can write $h = af$ with $f \in A[X]$ primitive. Let now $af = bg$ with $a, b \in A, f, g \in A[X]$ both primitive. Since a is a greatest common divisor of the coefficients of bg, we have $a \sim b$ in A, then $b = ua$ with $u \in U(A)$, so then $f = ug$ and $f \sim g$ in $A[X]$.

b) We have $f = gh$ with $h \in K[X]$. If $b \in A$ is a common multiple of the denominators of the coefficients of h, we have $bh \in A[X]$ and, according to a) we can write $bh = ah'$ with $a \in A, h' \in A[X]$ primitive. Hence $bf = agh'$. Since gh' is primitive (lemma 11.1), by the uniqueness in a) we obtain $f \sim gh'$ in $A[X]$, so $g \mid f$ in $A[X]$.

c) ($\Rightarrow$) If g|f in A[X],then g is primitive and g|f in K[X],hence g~1 or g~f in K[X], but since f and g are primitive, it follows by b) that g~1 or g~f in A[X].

c) ($\Leftarrow$) Let $g \mid f$ in $K[X]$, then there exists $g' \in A[X]$, primitive such that $g' \sim g$ in $K[X]$, then   by b), $g' \mid f$ in $A[X]$, so $g' \sim 1$ or $g' \sim f$ in $A[X]$, and $g \sim 1$ or $g \sim f$ in $K[X]$. ∎

Theorem 11.3: A U.F.D. $\Rightarrow A[X]$ U.F.D.

Proof: 1) Assume the first condition of the definition is false, and take a polynomial $f \in A[X]$ of smallest degre among those that, not being zero or a unit, do not admit a decomposition as a product of irreducibles. According to the above proposition, we can write: $f = af'$ with $a \in A$ and $f' \in A[X]$ primitive. Since a is a product of irreducibles in A, which are also irreducible in $A[X]$ by lemma 11.1 and prop.6.1, we have that f' can not be a product of irreducibles in $A[X]$. In particular f' is not irreducible in $A[X]$, so then, there exist $g, h \in A[X]$ non unities, such that $f' = gh$. Since g and h are primitive and non unities, then both g and h must have degree $< d(f)$ (a primitive polynomial of degree zero is a unit) and, by the election of f, both g and h are a product of irreducibles, so f' too, which is absurd.

2) The uniqueness of the decomposition will result from prop. 8.1, once we prove that any irreducible is prime in $A[X]$. Let then f be irreducible in $A[X]$, if $f \in A$, f is irreducible in A and, as A is an U. F. D., f is prime in A and, by lemma 8.1, f is also prime in $A[X]$. If, instead, $f \notin A$ it must be primitive and if $f \mid gh$ with $g, h \in A[X]$, putting $g = ag', h = bh'$ (g', h' primitives, $a, b \in A$), by the above proposition f is irreducible in $K[X]$, so $f \mid g'$ or $f \mid h'$ in $K[X]$ and by prop. 11.2.b), $f \mid g'$ or $f \mid h'$ in $A[X]$, hence $f \mid g$ or $f \mid h$ in $A[X]$.∎

From the ring $A[X]$ we can form the ring $A[X, Y] = (A[X])[Y]$ of polynomials in an indeterminate Y with coefficients in $A[X]$. In such a case, it follows that X is trascendental over $A[Y]$. In fact, if

$$\sum_i \left(\sum_j a_{ij} Y^j \right) X^i = 0$$

We obtain $\sum_j (\sum_i a_{ij} X^i) Y^j = 0$, from where, by the transcendence of Y over $A[X]$, results $\sum_i a_{ij} X^i = 0 \ \forall j$, then $a_{ij} = 0 \ \forall i, j$ and, so $\sum_j a_{ij} Y^j = 0 \ \forall i$.

Inductively we define the polynomial ring in n indeterminates: $X_1, \ldots, X_n$ by:

$$A[X_1, \ldots, X_n] = (A[X_1, \ldots, X_{n-1}])[X_n]$$

Corollary 11.4: If A is an U.F.D., then the polynomial ring in n indeterminates: $A[X_1, \ldots, X_n]$ is an U.F.D. as well.

Proof: It follows by induction from theorem 11.3.∎

We now turn to examine some irreducibility criteria. For each prime $p \in \mathbb{Z}$, there is a ring epimorphism $\theta_p : \mathbb{Z} \to \mathbb{Z}_p$, that sends a in its class $\bar{a} \bmod p$. This morphism can be extended to an epimorphism, also denoted $\theta_p : \mathbb{Z}[X] \to \mathbb{Z}_p[X]$ in the obvious way, that is:

$$\theta_p \left(\sum_i a_i X^i \right) = \sum_i a_i X^i$$

Strictly speaking, X is an indeterminate over $\mathbb{Z}$, so we must take another indeterminate over $\mathbb{Z}_p$, but is customary to use the same letter to both. Let $f \in \mathbb{Z}[X]$be a primitive polinomial of degree ≥ 1: $f = a_n X^n + \ldots + a_1 X + a_0$ with $a_n \neq 0$. If f is reducible in $\mathbb{Z}[X]$, we must have $f = gh$ with g and h of degree ≥ 1 (because f is primitive), so for any prime $p \in \mathbb{Z}$ such that $p \nmid a_n$ we have, denoting $\theta_p(f)$ by f, in $\mathbb{Z}[X]$: $f = gh$ with $1 \leq grg = grg, grh = grh \leq n - 1$, so then f is reducible in $\mathbb{Z}_p[X]$ too. He have proved then, that if there exists a prime $p \in \mathbb{Z}$, such that $p \nmid a_n$ and f is irreducible in $\mathbb{Z}_p[X]$, then f is irreducible in $\mathbb{Z}[X]$. For example, $f = X^3 + 345X^2 + 42X + 105$ is irreducible in $\mathbb{Z}[X]$ since modulus 2 we have: $f = X^3 + X^2 + 1$ is irreducible in $\mathbb{Z}_2[X]$ because it does not have roots in $\mathbb{Z}_2$.

The following irreducibility criterion is a generalization of one proved by T. Schönemann (Journal de Crelle, 1846) for the case of the ring $\mathbb{Z}$ and proved by F. Eisenstein in 1850 for the ring $\mathbb{Z}[i]$.

Proposition 11.5: (Eisenstein's criterion) Let A be an U.F.D and $f \in A[X]$ of degree $n \geq 1$, given by:

$$f = a_n X^n + \ldots + a_1 X + a_0$$

If there exists a prime p in A such that $p \nmid a_n, p \mid a_i \ \forall i = n - 1, \ldots, 0$ and $p^2 \nmid a_0$, then:

a) f primitive $\Rightarrow f$ irreducible in $A[X]$.
b) If K is the field of quotiente of A, f is irreducible in $K[X]$.

Proof: a) Assume f primitive and let $f = gh$ with $g, h \in A[X]$ necessarily primitive. Put:

$$g = Xr + \ldots + b_1 X + b_0$$
$$h = c_s Xs + \ldots + c_1 X + c_0$$

with $b_r \neq 0, c_s \neq 0$. If $r = 0$, g is an unity or if $s = 0$, h is a unit, so we assume $r, s \geq 1$ (and then $n \geq 2$). Since $p \mid a_0 = b_0 c_0$ and $p^2 \nmid a_0$, we have (say) $p \mid b_0$ and $p \nmid c_0$. Since $p \mid a_1 = b_0 c_1 + b_1 c_0$, it follows that $p \mid b_1$. Following in this way, we obtain $p \mid b_r$, then $p \mid a_n = b_r c_s$ which is a contradiction. Hence it must be $r = 0$ or $s = 0$, so f is irreducible in $A[X]$.

b) We have $f = af'$ with $a \in A$ and $f' \in A[X]$ primitive. As $p \nmid a$, for if not we must have $p \mid a_n$, it is clear that the coefficients of f' satisfy the conditions that allow us to conclude by a), that f' is irreducible in $A[X]$, and then f is irreducible in $K[X]$.∎

Examples: 1) Let $f = X^n - 2 \in \mathbb{Z}[X]$. Taking $p = 2$, the hypothesis of Eisenstein's criterion are satisfied, hence $X^n - 2$ is irreducible in $\mathbb{Q}[X]$. In $\mathbb{Q}[X]$ there are then irreducible polynomials af any degree, in contrast with $\mathbb{R}[X]$ or $\mathbb{C}[X]$.

2) The polynomial $f = Y^3 - X^2 - 2$ is irreducible in $\mathbb{R}[X, Y]$, since $p = X - \sqrt{2}$ es irreducible, and then prime, in $A = \mathbb{R}[X]$ and the hypothesis of Eisenstein's criterion are verified.

In certain cases, it is convenient to make a change of variable in order to apply Eisenstein's criterion. The following lemma, whose proof we leave as an exercise, can be useful for that.

Lemma 8.6: Let A be a ring. The function $\theta: A[X] \to A[Y]$ defined by $\theta(X) = Y + a$ with $a \in A$ is an isomorphism of rings.∎

Example: Both, Schönemann and Eisenstein, used the criterion to prove the irreducibility in $\mathbb{Q}[X]$. of the so-called cyclotomic polynomial: $f = X^{p-1} + X^{p-2} + \ldots + X + 1$ for $p \in \mathbb{N}$ prime. Gauss had given a more involved proof earlier. In fact, $(X - 1)f = X^p - 1$ and applying $\theta: X \mapsto Y + 1$, we obtain $Yf(Y + 1) = (Y + 1)^p - 1 = \sum_{i=1}^{p} \frac{p}{i} Y^i$ and then:

$$\theta(f) = \sum_{i=1}^{p} \frac{p}{i} Y^{i-1}$$

As $p \mid \frac{p}{i} \; \forall i = 1, \ldots, p-1$; $p^2 \nmid \frac{p}{1}$ and $p \nmid (p/p) = 1$, it follows by Eisenstein's criterion, that $\theta(f)$ is irreducible in $\mathbb{Q}[Y]$, hence f is irreducible in $\mathbb{Q}[X]$.

EXERCISES

Ex. 1: If $n, m \in \mathbb{N}$ are coprime, and $f: \mathbb{Z}_n \to \mathbb{Z}_m$ a ring homomorphism, then $f = 0$

Ex. 2: a) Let $f: A \to B$ be a morphism of rings. f is monomorphism if, and only if, $f(a) = 0 \Rightarrow a = 0$.

b) Any nonzero morphism of fields is a monomorphism.

Ex. 3: Let $\mathbb{R}_{>0}$ denote the set of strictly positive real numbers. In this set we consider two operations, an "addition" given by the ordinary product $\cdot$, and a "multiplication" ∇ defined by:

$$x \nabla y = x^{\log_2 y}$$

Prove that the function $f: (\mathbb{R}, +, \cdot) \to (\mathbb{R}_{>0}, \cdot, \nabla)$ defined by $f(x) = 2^x$ is a bijection that preserves the operations and so, or otherwise, $(\mathbb{R}_{>0}, \cdot, \nabla)$ is a ring. Hence f is a ring isomorphism.

Ex. 4: Let A be a domain and $a, b \in A$. If there exist $n, m \in \mathbb{N}$ coprime and such that $a^n = b^n$ and $a^m = b^m$, then $a = b$.

Ex. 5: Any morphism $f: A \to B$ of domains, can be extended to a morphism of its quotient fields.

Ex. 6: Use Fermat's little theorem to prove that for any prime p there are distinct polynomials that agree as polynomial functions on $\mathbb{Z}_p$.

a) Prove that the set $\mathbb{Z}[\sqrt{-2}] = \{a + b\sqrt{2}i / a, b \in \mathbb{Z}\}$ is a subring of $\mathbb{C}$ and that it is an Euclidean domain.

b) Which of the following are subrings of $\mathbb{C}$?.

c) $\{a + b\theta \; / \; a, b \in \mathbb{Z}; \; \theta^3 = 1, \theta \neq 1\}$.

d) $\{a + b[3]\sqrt{2} \; / a, b \in \mathbb{Z}\}$.

e) Give examples in $\mathbb{Z}[i]$ such that when dividing two elements, there isn't uniqueness of quotient and residue.

f) If A, is an Euclidean domain with an Euclidean function N which is multiplicative: $N(ab) = N(a)N(b)$, then:

    i. $U(A) = \{a \in A / N(a) = N(1)\}$.

    ii. $N(a) = p$ prime $\Rightarrow a$ is irreducible.

Ex. 7: In $\mathbb{Z}[\theta]$ with $\theta^3 = 1, \theta \neq 1$, prove that:

a) $1 - \theta$ and $2 + 3\theta$ are not associated, while $1 - \theta$ and $1 - \theta^2$ are associated.

b)   The only unities that are cubes are ± 1.

Ex. 8:   Let $\mathbb{Z}[\sqrt{2}] = \{a + b\sqrt{2} \, / a, b \in \mathbb{Z}\}$.

a) $\mathbb{Z}[\sqrt{2}]$ is a subring of $\mathbb{R}$.

b)   We define the norm $N(x)$ of an element $x = a + b\sqrt{2} \in \mathbb{Z}[\sqrt{2}]$ by $N(x) = (a + b\sqrt{2})(a - b\sqrt{2}) = a^2 - 2b^2$. Verify that the norm is multiplicative: $N(xy) = N(x)N(y)$.

c) $x \in U(\mathbb{Z}[\sqrt{2}]) \Leftrightarrow N(x) = \pm 1$.

d)   All the powers with an integral exponent of $3 + 2\sqrt{2}$ are unities of $\mathbb{Z}[\sqrt{2}]$, then there is an infinity of unities in this ring.

Ex. 9:   If K is a field and $a \in K, a \neq 0$ prove:

a) $X - a \mid X^n - a^n \, \forall \, n \in \mathbb{N}$.

b) $X - a \nmid X^n + a^n \, \forall \, n \in \mathbb{N}$.

c) $X + a \mid X^n - a^n \Leftrightarrow n$ is even.

d) $X + a \mid X^n + a^n \Leftrightarrow n$ is odd.

Ex. 10:   Prove that in the ring $\mathbb{Z}[i]$:

a)   5 is not irreducible.

b) $1 + 2i$ is irreducible.

Ex. 11:   Prove that in $A = \mathbb{Z}\sqrt{-5}$ the following are valid:

a) $\alpha \in U(a) \Leftrightarrow N(\alpha) = 1$ (where $N = ||\,|^2$).

b) $N(\alpha)$ prime in $\mathbb{Z} \Rightarrow \alpha$ irreducible in A.

c)   In A is valid the condition 1) of the definition of D.F.U.

Ex. 12:   If A is an Euclidean domain and $\mathbb{N}$ an Euclidean function, in order to find a g.c.d. of two elements $a, b \in A$ and to express it in the form $sa + tb$, we can use an Euclidean algorithm, to obtain $gcd(a, b)$ as the last nonzero residue. Prove it, and find in this way a g.c.d. and express it in the form $sa + tb$ in the following cases:

a) $X^3 - X^2 + X - 1$; $.X^3 + 2X^2 + X + 2$ in $\mathbb{R}[X]$.

b) $1 + 5i$; $-1 + 5i$ in $\mathbb{Z}[i]$.

c) $4 + 5\theta$; $1 - 4\theta$ in $\mathbb{Z}[\theta]$.

Ex. 13:   In $K[X]$ where K is a field, the following properties are valid:

a) r is the residue when dividing n by $m \Leftrightarrow X^r - 1$ is the residue when dividing $X^n - 1$ by $X^m - 1$. In particular: $X^m - 1 \mid X^n - 1 \Leftrightarrow m \mid n$.

b) $gcd(X^n - 1, X^m - 1) = (X^d - 1)$, where $d = gcd(n, m)$.

c) $gcd(X^{n-1}+\ldots+X+1, X^{m-1}+\ldots+X+1) = X^{d-1}+\ldots+X+1$, where $d = gcd(n. m)$.

Ex. 14: In $\mathbb{Z}[i]$ write $z = 3 + 11i$ as a product of irreducibles.

Ex. 15: In $\mathbb{Z}[i]$ the following are valid:

a) z invertible $\Rightarrow \bar{z}$ invertible.

b) z irreducible $\Rightarrow \bar{z}$ irreducible.

Ex. 16: Let $a \in \mathbb{Z}$ such that $\lambda \mid a$ in $\mathbb{Z}[\theta]$, then $3 \mid a$ in $\mathbb{Z}$.

Ex. 17: If z is irreducible in $\mathbb{Z}[i]$, then $N(z) = 2$ or $N(z) = p$ with $p \in \mathbb{Z}$ a prime of the form $4n + 1$ or $N(z) = p^2$ with $p \in \mathbb{Z}$ a prime of the form $4n + 3$.

Ex. 18: Let $p \in \mathbb{Z}$ prime,

a) If p is of the form $4n + 3, p$ remains prime in $\mathbb{Z}[i]$.

b) If $p = 2$ or p is of the form $4n + 1, p$ is a product of two conjugate primes in $\mathbb{Z}[i]$.

Ex. 19: $a \in \mathbb{Z}$ is a sum of two squares $\Leftrightarrow$ in the descomposition of a as a product of primes (in $\mathbb{Z}$), the prime factors of the form $4n + 3$ occurs with an even exponent.

Consider the Diophantine equation:

$$x^2 + 1 = 2y^3 \ (*)$$

prove:

a) $1 + i \mid 1 + xi, \forall\, x \in \mathbb{Z}$ odd.

b) $2 \nmid x + i \,\forall\, x \in \mathbb{Z}$.

c) If x is odd: $gcd(x + i, x - i) \sim 1 + i$.

d) If x, y satisfy $(*)$, then $\dfrac{x+i}{1+i}$ must be a cube $(a + bi)^3$ in $\mathbb{Z}[i]$, and we have:

$(a - b)(a^2 + b^2 + 4ab) = 1$, and: $(a + b)(a^2 + b^2 - 4ab) = x$

Conclude that $x = \pm 1, y = 1$ are the only solutions of the equation.

Ex. 20: Consider the Diophantine equation:

$$x^2 + 4 = y^3 \ (**)$$

a) If x is odd, then $gcd(x + 2i, x - 2i) \sim 1$ in $\mathbb{Z}[i]$.

b) If x, y satisfy (**) and x is odd, then $x +$ 2i must be a cube $(a + bi)^3$ in $\mathbb{Z}[i]$ and then $x = \pm 11, y = 5$.

c) If x, y satisfy (**) and $x = 2z$ is even, then $z^2 + 1 = 2w^3$ and by the exercise above, we have $x = \pm 2, y = 2$.

Ex. 21: Knowing that $1 + i$ is a root of $X^5 + 2X^4 + 2X^3 - X^2 - 2X - 2$, factorize it in $\mathbb{R}[X]$, b) $\mathbb{C}[X]$

Ex. 22: Factorize $X^5 - X$ over:
a) $\mathbb{R}[X]$, b) $\mathbb{C}[X]$, c) $\mathbb{Z}_5[X]$, d) $\mathbb{Z}_3[X]$

Ex. 23: (Lagrange's interpolation formula) Let K be any field. Given $n + 1$ distinct elements $a_0, a_1, \ldots, a_n \in K$, and given $b_0, b_1, \ldots, b_n$ elements of K not necessarily distinct, the polynomial:

$$b_n \frac{\prod_{j \neq i} X - a_j}{\prod_{j \neq i} a_i - a_j}$$

takes the value b_i when X is specialized in a_i, and takes the value 0 when X is specialized in a_j with $j \neq i$. Hence the polynomial:

$$f = \sum_{i=0}^{n} b_i \frac{\prod_{j \neq i} X - a_j}{\prod_{j \neq i} a_i - a_j}_i$$

is a polynomial of degree $\leq n$ such that $f(a_i) = b_i \; \forall \, i = 0, 1, \ldots, n$.

Appendix 1
CONSTRUCTION OF REAL NUMBERS

The results obtained in chapter 6 "suggest" how to build up $\mathbb{R}$ from $\mathbb{Q}$. Each real number is determined by a non-empty, bounded from above, set of rational numbers; or by a Cauchy sequence of rational numbers; or by an increasing, bounded from above, sequence of rational numbers; or by a sequence of nested intervals of rational numbers whose lengths tend to zero. Two sequences of rational numbers determine the same real number if its difference is a null sequence (a sequence that tends to zero) and as we saw this condition define an equivalence relation in the set of all sequences

In general, when the existence of a mathematical object is suspected or desired, it is appropriate, if not indispensable, to assume it, derive some consequences and examining them, finding an idea to prove or refute it. In the case at hand, that of the real numbers, we have assumed its existence, that is the existence of an ordered field that satisfies the least upper bound property, from where we can derive some consequences such as the following:

1) Any real number is the least upper bound of some set of rational numbers. In fact, if $x \in \mathbb{R}$, taking:

$$A = \{q \in \mathbb{Q}/q < x\}$$

by the density of $\mathbb{Q}$ in $\mathbb{R}$, it follows that x is the least upper bound of A.

Then any real number may be thinking as a non-empty and bounded from above set of rational numbers, of which it is the supremum. Therefore, some criteria should be available, in order to decide when two of such sets of rational numbers have the same supremum, without assuming its existence.

Note that according to exercise 22, Chap. 3, the sets: $A = \{(1 + (1/n))^n / n \in \mathbb{N}\}$ and $B = \{\sum_{i=0}^{n} \frac{1}{i!} / n \in \mathbb{N}\}$ are both bounded from above (3 is an upper bound of both of them). By the same exercise, for each $n \in \mathbb{N}$ we have:

$$\left(1 + \frac{1}{n}\right)^n \leq \sum_{i=1}^{n} \frac{1}{i!}$$

Then, for each $a \in A$, there is $b \in B$ such that $a \leq b$.

Let us show that this condition is sufficient to have: $s = supA \leq t = supB$. In fact, t is an upper bound of A, since $a \in A \Rightarrow \exists b \in B$ such that $a \leq b \leq t$. Moreover, in case that $s \notin A$ it is also a necessary condition, since if $s \leq t, a \in A$ and if for each $b \in B$ it were $b < a$, we would have $t \leq a$ and then $s \leq a$, but as $a < s$ because $s \notin A$, results in a contradiction. We have proved:

2) If A and B have supremum: $s = supA, t = supB$ with $s \notin A$, we have:

$s \leq t \Leftrightarrow$ for each $a \in A$ exists $b \in B$ such that $a \leq b$

hence, if also $t \notin B$:

$$s = t \Leftrightarrow \begin{cases} \text{for each } a \in A \text{ exists } b \in B \text{ such that } a \leq b \text{ and} \\ \text{for each } b \in B \text{ exists } a \in A \text{ such that } b \leq a) \end{cases}$$

Then if we restrict ourselves to consider non-empty, bounded from above, sets of rational numbers such that if they have a supremum, i.e., if it belongs to $\mathbb{Q}$, the supremum does not belong to the set, the above is a criterion to find if two such sets have the same supremum without determining it. The restriction to those sets is not essential, but it simplifies some proofs and is not an inconvenience since it is enough to remove the supremum. For example, if $A = \{\frac{1}{n} / n \in \mathbb{N}\}$ we replace it by $A - \{1\}$ and if $A = \{2\}$ we replace A by a set that determines the same real number (for example $\{q \in \mathbb{Q} / q < 2\}$).

In what follows we will prove the existence of $\mathbb{R}$ from that of $\mathbb{Q}$, which, in turn, can be reduced to that of $\mathbb{Z}$, this to that of $\mathbb{N}$, and, finally, the last to Set Theory.

According to 1), each real number is determined by a set of rational numbers and according to 2) it is convenient to consider only those sets whose supremum, in case of being rational, does not belong to the set.

We start then with the family S of all non-empty, bounded from above, sets of rational numbers A, such that its supremum, if exists, does not belong to A. By 2) it is advisable to define in S a relation $\sim$ as follows:

$$A \sim B \text{ iff} \begin{cases} \text{for each } a \in A \text{ } exists \text{ } b \in B \text{ such that } a \leq b \text{ and} \\ \text{for each } b \in B \text{ exists } a \in A \text{ such that } b \leq a. \end{cases}$$

or, else, to define a relation $\lesssim$ in S by:

$A \lesssim B$ iff for each $a \in A$ exists $b \in B$ such that $a \leq b$, and then define $A \sim B$ by: $A \lesssim B$ and $B \lesssim A$.

The relation $\lesssim$ is clearly reflexive and transitive, from where it follows that $\sim$ is an equivalence relation in S.

We think a real number as an equivalence class under the relation $\sim$ and in the quotient set $\dfrac{S}{\sim}$ of those classes, we will define sum and product operations and an order relation, that will do it an ordered field satisfying the least upper bound property.

In order to do that, if $A, B \in S$ we define:

$$A + B = \{a + b / a \in A , b \in B\}$$

It follows that $A + B \in S$, since clearly $A + B$ is non-empty and bounded from above and, moreover, if $s = sup(A + B)$ exists, that is belongs to $\mathbb{Q}$, it can not belongs to $A + B$, for if not we would have $s = t + u$ with $t \in A$ and $u \in B$, hence, if $a \in A$ then $a + u \in A + B$ and so $a + u \leq s$, that is $a \leq t$ and it would follow that t would be the supremum of A belonging to A, a contradiction, then $A + B \in S$.

We obviously have:

$$A \lesssim C \text{ and } B \lesssim D \Rightarrow A + B \lesssim C + D$$

from which follows:

$$A \sim C \text{ and } B \sim D \Rightarrow A + B \sim C + D$$

which shows that, if we define the sum of two classes $\bar{A}, \bar{B}$ with $A, B \in S$ by:

$$\bar{A} + \bar{B} = \overline{A + B} \text{ (1)}$$

, the sum is well defined.

This sum is clearly associative and commutative and if we define:

$$O = \{q \in \mathbb{Q} / q < 0\} \text{ (2)}$$

results that $\bar{O}$ is its neutral element, i.e., $A + O \sim A$ for any $A \in S$. In fact, $A + O \lesssim A$ since for each $a \in A$ we have $a + q \leq a$ whichever be $q \in O$. Furthermore $A \lesssim A+O$, since if $a \in A$, as A does not contain its supremum, it must exists $a' \in A$ with $a < a'$, then taking $q = a - a'$, we have $q \in Q, q < 0$ and $a \leq a' + q$ with $q \in O$.

For each $A \in S$, let:

$-A = \{-c/c$ is an upper bound, but not the supremum, of $A\}$ (3)

$-A$ is non-empty because A is bounded from above; $-A$ is bounded from above because A is non-empty (if $a \in A$, then $-a$ is an upper bound of $-A$). If the supremum $-s$ of $-A$ exists and belongs to $-A$, we wold have that s is an upper bound of A, but then it would be the supremum of, since if t is an upper bound of A but not the supremum, we would have $-t \leq -s$ hence $t \geq s$, and so s would be the supremum of A. It follows that $-A \in S$.

Let us see that $\overline{-A}$ is the additive inverse of $\bar{A}$, that is $A + (-A) \sim 0$. We have $A + (-A) \lesssim 0$, since if $a \in A$ and $-c \in -A$, then $a - c < 0$.

$0 \lesssim A + (-A)$, since given $q \in O$ (i. e., $q \in Q$ with q<0) exists $a \in A$ such that $a - q$ is an upper bound of A (corollary 1.4) then, as $a - q$ is not the supremum of A, it follows that $q - a \in -A$, then $q = a + q - a \in A + (-A)$.

We have proved:

a) In $\frac{S}{\sim}$ with the sum defined by (1) the properties S.1 to S.4, defined in chapter 2, are valid, that is the sum is well defined, is associative, commutative, has a neutral element $\bar{0}$, defined by (2) and each element $\bar{A}$ has an additive inverse $\overline{-A}$ with $-A$ given by (3).

In $\frac{S}{\sim}$ a relation $\leq$ can be defined as is suggested 2):

$$\bar{A} \leq \bar{B} \; iff \; A \lesssim B \; (4)$$

Let us show that this definition does not depend on the election of the representative of the classes. Indeed, let $A \sim C, B \sim D$, and $A \lesssim B$. We have $C \lesssim A, A \lesssim B$, and $B \lesssim D$, from where by the transitivity of $\lesssim$ results $C \lesssim D$.

The transitivity of $\lesssim$ also proves that $\leq$ is transitive.

Given $A, B \in S$, we have $A \lesssim B$ that is: for each $a \in A$ exists $b \in B$ such that $a \leq b$, or exists $a \in A$ such that $a > b \; \forall b \in B$, from where $B \lesssim A$.

Also, clearly $A \lesssim B \Rightarrow A + C \leq B + C$.

Defining $\bar{A} < \bar{B}$ as $\bar{A} \leq \bar{B}$ and $\bar{A} \neq \bar{B}$, as by definition of $\sim$ we have $\bar{A} \leq \bar{B}$ and $\bar{B} \leq \bar{A} \Rightarrow \bar{A} = \bar{B}$, results:

b) The relation $<$ just defined, verify trichotomy, transitivity, and compatibility the sum.

It is not possible to define, as we did with the sum, the product of two classes A, B as the class of the set AB of all products ab with $a \in$

A and $b \in B$, since this set may be not bounded from above. For example, if $A = B = \{q \in Q/q < 3\}$, the set of such products contains $\mathbb{N}$. But in case that both A and B consist only of positive numbers, the described process works, as it is suggested by lemma 6.1. We then will proceed defining first a product for classes that admit a representative with positive elements, which, as will see, are the classes $\bar{A}$ such that $\bar{A} > \bar{0}$ and then generalizing the definition by the "rule of signs".

Note that for $A \in S$:

$$\bar{0} < \bar{A} \Leftrightarrow \exists a \in A \text{ with } a > 0 \ (5)$$

In fact, if $\bar{0} < \bar{A}$ we have $\bar{0} \leqslant \bar{A}$ and $\bar{0} \nsim \bar{A}$. If it were $a \leq 0 \ \forall a \in A$, it would be $a < 0 \ \forall a \in A$ since if not the supremum of A would belong to A. but then $\bar{A} \leqslant \bar{0}$ and, as $\bar{0} \leqslant \bar{A}$ by hypothesis, we would have $\bar{A} = \bar{0}$. The converse is clear, so (5) is valid.

If $\bar{0} < \bar{A}$ and $\bar{0} < \bar{B}$ we define:

$$AB = \{ab/a \in A, b \in B \text{ with } a > 0 \text{ and } b > 0\}$$

Also, we define:

$$\bar{A}\bar{B} = \overline{AB} \ (6)$$

Note that $AB \in S$ since AB is clearly non-empty and bounded from above and, if its supremum, if exists, would belong to AB, we would have $u = st$ with $s \in A$ and $t \in B$, from   where,   for   each $a \in A$ with $a > 0$, as $at \in AB$ we   would   have $at \leq u = st$, then $a \leq s$ and $s \in A$ would result in the supremum of A, so $AB \in S$.

It is clear that the product given by (6) is well defined, is associative and commutative. To see that it has a neutral element, we consider:

$$I = \{q \in Q/0 < q < 1\} \ (7)$$

and verify that $AI = A$, or $AI \leqslant A$ and $A \leqslant AI$. In fact, if $a \in A, a > 0$ and $q \in I$ we have $aq < a$ then $AI \leqslant A$. In addition, if $a \in A, a > 0$, exists $b \in A$ such that $a < b$ (because sup A, if exists, does not belong to A) and taking $q = \frac{a}{b}$, we have $q \in I$ and $a = bq$ so that $A \leqslant AI$.

Let us show that each $\bar{A}$ with $\bar{0} < \bar{A}$ has a multiplicative inverse. Let:

$$A^{-1} = \{b^{-1}/b \text{ is an upper bound but not the supremum of } A\} \ (8)$$

A^{-1} is obviously non-empty, and is bounded from above, since by (5) exists $a \in A$ such that $a > 0$, then, if b is an upper bound of A, we have $a \leq b$, so then $b^{-1} \leq a^{-1}$, i.e., a^{-1} is an upper bound of A^{-1}.

We have $AA^{-1} \leqslant I$, since if $a \in A, a > 0$ and if b is an upper bound of $A, a < b$ then $ab^{-1} \in I$.

Also, we have $I \lesssim AA^{-1}$, since if $0 < q < 1, t = q^{-1} > 1$ and by prop. 1.6, exists $a \in A$ such that ta is an upper bound of A, that is $a^{-1}t^{-1} = a^{-1}q \in A^{-1}$, then $q = aa^{-1}q \in AA^{-1}$.

It follows then that $\bar{A}\bar{A}^{-1} = I$, that is $\bar{A}^{-1}$ is the multiplicative inverse of $\bar{A}$.

If $\bar{A}, \bar{B}, \bar{C}$ are positive, i.e., $O < \bar{A}, \bar{B}, \bar{C}$; distributivity is valid:

$$\bar{A}(\bar{B} + \bar{C}) = \bar{A}\bar{B} + \bar{A}\bar{C} \ (9)$$

In fact, to prove $A(B + C) \lesssim AB + AC$, take $a \in A, d \in B + C$ with $a, d > 0$ and $d = b + c, b \in B, c \in C$. At least one of b, c must be positive, let c>0. Let $b' = b$ if $b > 0$ or $b' = -b$ if $b < 0$ or b' any strictly positive element of B if $b = 0$. In any case, we have $ad \leq ab' + ac$, then $A(B + C) \lesssim AB + AC$.

On the other hand, if $a, a' \in A, b \in B$ and $c \in C$ all of them strictly positive and with, say, $a \leq a'$, then $ab + a'c \leq a'(b + c)$ and hence $AB + AC \lesssim A(B + C)$.

If, moreover, $\bar{C} < \bar{B}$ then:

$$\bar{A}(\bar{B} - \bar{C}) = \bar{A}\bar{B} - \bar{A}\bar{C} \ (10)$$

where $B - C$ denotes $B + (-C) = B + (-C)$. In fact, $AB + (-(AC)) \lesssim A(B + (-C))$ because if $a \in A, a > 0, b \in B, b > 0$ and $-x \in -AC$ (i.e., x is an upper bound but not the least upper bound of AC), $ab - x = a(b - a^{-1}x)$ and $a^{-1}x$ is an upper bound of C and is not its supremum (since A does not contains its supreme, there is $a' \in A$ such that $a' > a$, and so $a'^{-1}x$ is an upper bound of C such that $a'^{-1}x < a^{-1}x$), then $- a^{-1}x \in -C$.

Moreover, $A(B + (-C)) \lesssim AB + (-(AC))$. Since if $a \in A, a > 0, b \in B, -c \in -C$ (that is, c is an upper bound but not the least upper bound of C) con $b > c$, then exists $t > 1$ such as $t^{-1}c$ is an upper bound but not the least upper bound of C (if is not the supremum of C, there exist c' an upper bound of C which is not the supremum of C, such that $c > c'$ and putting $t = \frac{c}{c'}$ we have $t > 1$ and $t^{-1}c = c'$ is an upper bound but not the supremum of C) and taking $a' \in A$ such as $a' \geq a$ and that $a't$ be upper bound of A (prop.1.6), we obtain: $a(b - c) \leq a'(b - c) = a'b - a'c$ with $a'c = a'tt^{-1}c$ an upper bound but not the supremum of AC.

Summing up this results, we have proved:

c) For positive classes, the product defined by (6), is well defined, it is associative, commutative, has a neutral element given by (7), for each class there is a multiplicative inverse given by (8), and the distributive laws (9) and (10) are valid.

In order to extend the product to arbitrary classes, we define:

$$AB = \begin{cases} 0 \ if \ A = 0 \ or \ B = 0 \\ -A(-B) \ if \ 0 < A \ and \ B < 0 \\ -(-A)B \ if \ A < 0 \ and \ 0 < B \\ (-A)(-B) \ if \ A < 0 \ and \ B < 0 \end{cases}$$

It is clear that this new product is associative, commutative, $\bar{I}$ is its neutral element and each nonzero class $\bar{A}$, has a multiplicative inverse (if $\bar{A} < \bar{0}$, then by compatibility with the sum, $\bar{0} < -\bar{A}$ and if $-\bar{B}$ is the multiplicative inverse of $-\bar{A}$, then $\bar{B}$ is the multiplicative inverse of $\bar{A}$, since $\overline{AB} = (-\bar{A})(-\bar{B}) = I$. Note that in the above argument, we have used that $-(-\bar{C}) = C$, which allow us to write any class as an opposite of some of them, a property that follows easily from those enounced in a)).

Let us show that the product is distributive:

$$\bar{A}(\bar{B} + \bar{C}) = \bar{A}\bar{B} + \bar{A}\bar{C}$$

This can be proved considering each case. Let us see, for example, the case in which: $\bar{0} < \bar{A}, \bar{0} < \bar{B}, \bar{C} < \bar{0}$ and $\bar{B} + \bar{C} < \bar{0}$. We have:

$$\bar{A}(B + C) = -\bar{A}(-(\bar{B} + \bar{C})) = -\bar{A}((-\bar{B}) + (-\bar{C})) = -\bar{A}(-\bar{C} - \bar{B})$$
$$= -[\bar{A}(-\bar{C}) - \overline{AB}] = -\bar{A}(-\bar{C}) + \overline{AB} = \overline{AB} + \overline{AC}$$

We now will prove that in $\frac{S}{\sim}$ the least upper bound property is valid. In order to do that, let $\Omega = (\overline{A_k})_{k \in K}$ a non-empty, bounded from above, subset of $\frac{S}{\sim}$. Put:

$$T = \bigcup_{k \in K} A_k$$

and we shall prove that $\bar{T}$ is the supremum of Ω.

First, let us show that $T \in S$. T is clearly non-empty. To see that it is bounded from above, let $\bar{C}$ an upper bound of Ω, that is $A_k \lesssim C \ \forall k \in K$, so then $T \lesssim C$. If the supremum s of T, exists and belongs to T, we would have $s \in A_k$ for some $k \in K$ and it would result that s would be the supremum of A_k, contradicting that $A_k \in S$. Hence $T \in S$.

Second, T is an upper bound of Ω, since $A_k \lesssim T \ \forall k \in K$.

Finally, if $\bar{U}$ is an upper bound of Ω, we have $A_k \lesssim U \ \forall k \in K$, so then if $a \in T$, exists $k \in K$ such that $a \in A_k$, hence exists $b \in U$ such that $a \leq b$, and so $T \lesssim U$.

We have proved:

Theorem: If there exists an ordered field, then also exists an ordered field that satisfies the least upper bound property. ∎

As we already said, the existence of an ordered field can be proved from Set Theory. Also, it can be proved that, essentially, there is only one ordered field that satisfies the least upper bound property. In contrast there an infinity of ordered fields essentially distinct.

Appendix 2

CONSTRUCTIONS WITH RULER AND COMPASS

Constructions with ruler and compass arise in ancient Greece. They consist in, starting from given segments or angles, determining, using only ruler and compass, one or more segments or angles with specific properties, using the ruler or straight edge only to join two points already constructed, not to measure.

The most famous, which have been, for more than two thousand years, the object of the efforts of countless mathematicians, both professionals and amateurs, are the following:

1) The construction of regular polygons.
2) The trisection of an angle.
3) The duplication of the cube.
4) The quadrature of the circle.

In Euclid's Elements, we find the construction of the equilateral triangle and the regular pentagon. As the square is easily constructed and the bisection of an angle is possible with ruler and compass, the regular polygons of 2^m sides with $m \geq 2$ are constructible too. Moreover, the Greeks had methods available to prove that if two regular polygons of r and s sides are contructibles and r and s are coprime, then the regular polygon of rs sides is contructible. So the regular polygon of 15 sides is constructible. More than two thousand years passed before making further progress in this problem. In 1796 $a - 19$ −old Gauss, proves that the regular polygon of 17 sides is constructible and later, he generalized this result to any polygon of p sides, where $p = 2^n + 1$ is a Fermat prime. Wantzel in 1837 complements these results, by coming to the definitive conclusion: the regular polygon of n sides is constructible by ruler and compass if, and only if n takes one of the following forms:

$$n = 2^m \text{ or } n = 2^{m-2}p_1 \dots p_k$$

where $m \geq 2$ and $p_1 \ldots p_k$ are distinct Fermat primes.

The problem of the trisection of the angle, refers to an arbitrary angle. Some triangles are trisectable with ruler and compass, but what it is about, is the possibility of trisecting any angle in that way. The impossibility of the trisection of the angle with ruler and compass (that is the demonstration that some angles are not trisectable) was proved by Wantzel in 1837.

The problem of the duplication of the cube, also called Delos's problem, consists in, given a cube (its edge), construct with ruler and compass, the edge of a cube with double volume. Wantzel in 1837 proved that such a construction is not possible.

The quadrature of the circle consists in, given a circle (its radius), construct with ruler and compass a square (its side) with the same area as the circle. This construction is impossible, which was proved by Lindemann in 1882 by demonstrating that the number π is transcendent, since any constructible number must be algebraic.

The mathematics necessary to prove the transcendence of π are beyond the scope of this book, so we will not deal with the quadrature of the circle. Instead, we will prove the impossibility of the trisection of the angle and of the duplication of the cube. We will also see the constructability of certain polygons and the impossibility of constructing others.

We will pose the problems on the complex plane. If the data is a segment, which is the case in 1), 2) and 3), we arrange the coordinate system in such a way that the segment coincides with the interval $[0, 1]$. If the data is an angle we arrange it with a side coincident with the x −axis, and if z is the intersection of the unit circunference with the other side, we describe the angle as the three points $0, 1, z$. If there are more segments or angles we proceed in a similar way (for example if there is another segment as data, we arrange it to coincide with the interval $[0, a]$ for some real and positive a). In this way the data are complex numbers: $0, 1, z, a \ldots$ and departing from these, we construct new points with ruler and compass, that is, intersecting two lines, or a line and a circunference, or two circunferences, with the condition that these lines or circunferences are *admissible*, that is the lines determined by points data or already constructed and circunferences of center a number data or already constructed and radius determined by two of such points. For our purpose will be sufficient that the data are $0, 1$.

If $z, w \in \mathbb{C}$ with $z \neq w$, the line determined by z and w is the set:

$$L(z, w) = \{u \in \mathbb{C}/\exists \lambda \in \mathbb{R} \text{ such that } u = z + \lambda(z - w)\}$$

and the circumference with center z and radius r $(r \in \mathbb{R}_{>0})$ is the set:

$$C(z, r) = \{u \in \mathbb{C}/|u - z| = r\}$$

The set C (not to confund with the emphazised $\mathbb{C}$ of the set of complex numbers) of the constructible complex numbers, is determined by the conditions:

1) $0, 1 \in C$.
2) If $z, w, u, v, x, y \in C$, then the elements of the following intersections belong to C, with the condition that the intersected figures do not coincide:

$$L(z, w) \cap L(u, v) \; L(z, w) \cap C(u, |v - x|) \; C(z, |w - y|) \cap C(u, |v - x|)$$

We will now prove a series of properties of C which correspond with classical geometric constructions. It is advisable to make the graphs in each case, to visualize the geometric ideas.

(A) $z, w \in C \Rightarrow z + w \in C$.

If $z \neq w$ we have: $z + w \in C(z, |w|) \cap C(w, |z|)$

If $z = w$ we have: $z + w = 2z \in C(z, |z|) \cap L(0, z)$

(B) $z \in C \Rightarrow -z \in C$.

If $z = 0$ is clear. If $z \neq 0$, we have: $-z \in C(0, |z|) \cap L(0, z)$

(C) $a \in \mathbb{R} \cap C \Rightarrow ai \in C$

We have successively: $(2 \in C$ by (A) and $-1 \in C$ by (B))

$$\sqrt{3}i \in C(1, 2) \cap C(-1, 2)$$
$$i \in C(0, 1) \cap L(0, \sqrt{3}i)$$
$$ai \in L(0, i) \cap C(0, |a|)$$

(D) If $z, w, u \in C$ with $z \neq w$, then the parallel to $L(z, w)$ passing through u, is admissible.

In fact, by (A) and (B) we have, $u + w - z \in C$ and $L(u, u + w - z)$ is the parallel to $L(z, w)$ passing through u.

(E) $z \in C \Leftrightarrow Re(z)$ and $Im(z) \in C$.

$(\Rightarrow)$: $Re(z) \in L(0, 1) \cap$ parallel to $L(0, i)$ passing through z.

$Im(z) \in L(0, i) \cap$ parallel to $L(0, 1)$ passing through z.

$(\Leftarrow)$: $z \in L \cap L'$ where L is the parallel to $L(0, i)$ passing through $Re(z)$ and L' is the parallel to $L(0, 1)$ passing through $Im(z)$.

(F) $a, b \in \mathbb{R} \cap C \Rightarrow ab \in C$ and, if $a \neq 0, \frac{b}{a} \in C$.

Let $a \neq 0$, we have:

$ab \in L(0,1) \cap$ parallel to $L(i,b)$ passing through ai. $\frac{b}{a} \in$
$L(0,1) \cap$ parallel to $L(ai,b)$ passing through i.

(G) $e^{i\alpha}, e^{i\beta} \in C (\alpha, \beta \in \mathbb{R}) \Rightarrow e^{i(\alpha+\beta)} \in C$.

$e^{i(\alpha+\beta)} \in \mathbb{R}(e^{i\beta}, |1 - e^{i\alpha}|) \cap C(0,1)$.

(H) Let $\alpha, r \in \mathbb{R}$ with $r \neq 0$, we have: $r \in C$ and $e^{i\alpha} \in C \Leftrightarrow re^{i\alpha} \in C$

$(\Rightarrow)$: $re^{i\alpha} \in L(0, e^{i\alpha}) \cap C(0, |r|)$

$(\Leftarrow)$: $r \in L(0,1) \cap C(0, |re^{i\alpha}|)$, $e^{i\alpha} \in C(0,1) \cap L(0, re^{i\alpha})$

(I) $z \in C, z \neq 0 \Rightarrow z^{-1} \in C$.

Putting $z = |z|e^{i\alpha}$ with $\alpha \in \mathbb{R}$, we have by (H): $|z| \in$
C and by (F): $|z|^{-1} \in C$. Furthermore, $e^{i(-\alpha)} \in C(0,1) \cap$ parallel to $L(0,i)$ passing through $e^{i\alpha}$ then $e^{i(-\alpha)} \in C$, and
by (H) it follows, $|z|^{-1}e^{i(-\alpha)} \in C$

(J) $z, w \in C \Rightarrow zw \in C$.

Putting $z = |z|e^{i\alpha}, w = |w|e^{i\beta}$ by (H): $|z|, |w|, e^{i\alpha}, e^{i\beta} \in$
C; then by (F) $|zw| \in C$ and by (G) $e^{i(\alpha+\beta)} \in C$ and, finally,
by (H): $zw \in C$.

Proposition: C is a subfield of $\mathbb{C}$.

Proof: It follows readily from the results just proved.∎

Proposition: $r \in \mathbb{R}_{>0} \cap C \Rightarrow \sqrt{r} \in C$. More generally, if $w \in$
C and $z^2 = w$, then $z \in C$.

Proof: We have, $\sqrt{ri} \in \mathbb{C}\left(\frac{r-1}{2}, \frac{r+1}{2}\right) \cap L(0, i)$ since

$$\left|\sqrt{ri} - \left(\frac{r-1}{2}\right)\right|^2 = \left(\frac{r-1}{2}\right)^2 + r = \left(\frac{r+1}{2}\right)^2$$

then $\sqrt{ri} \in C$ and so $\sqrt{r} = \sqrt{ri}(-i) \in C$.

Moreover, if $z^2 = w$ with $w \in C$ and $w \neq 0$, putting $w = |w|e^{i\alpha}, z = |z|e^{i\beta}$, we have $|z| = \sqrt{|w|} \in C$ by the first part and $e^{i\beta} \in C$ by the classic bisection of the angle, that is, the intersection: $C(1, |e^{i\alpha} -$

236

1|) $\cap\, C(e^{i\alpha}, |e^{i\alpha} - 1|)$ is not empty and if u is one of its elements, we have that both $e^{i\frac{\alpha}{2}}$ and $e^{i\frac{\alpha}{2}} + \pi$ belong to the intersection: $L(0, u) \cap C(0, 1)$.∎

In what follows, if F is a subfield of R, we will use the notation:
$$F(i) = \{a + bi \,/\, a, b \in F\}$$
and it is easily verified that $F(i)$ is a subfield of $\mathbb{C}$.

Recall that if $\alpha \in \mathbb{R}$ is positive and $\sqrt{\alpha} \notin F$, we denote $F(\sqrt{\alpha}) = \{a + b\sqrt{\alpha}\}$ (Chap. 6, sec. 3, example) which is a subfield of $\mathbb{R}$.

Lemma: If F is subfiled of $\mathbb{R}$ and $x, y, u, v \in F(i)$, we have, if the figures do not coincide:

a) $L(x, y) \cap L(u, v) \subset F(i)$.

b) $\exists \alpha$ real and positive such that $L(x, y) \cap C(u, |v|) \subset F(\sqrt{\alpha})(i)$.

c) $\exists \alpha$ real and positive such that $C(x, |y|) \cap \mathbb{C}(u, |v|) \subset F(\sqrt{\alpha})(i)$.

Proof: a) Consider $L(x, y) \cap L(u, v)$ with $x, y, u, v \in F(i)$. We have,
$$z \in L(x, y) \cap L(u, v) \Leftrightarrow \text{exist } \lambda, \mu \in \mathbb{R} \text{ such that } z = x + \lambda(y - x)$$
$$= u + \mu(v - u)$$
For each complex t, we shall denote by t_1 to its real part and by t_2 to its imaginary part. Then the equality $x + \lambda(y - x) = u + \mu(v - u)$ with $\lambda, \mu \in \mathbb{R}$ becomes,
$$x_1 + \lambda(y_1 - x_1) = u_1 + \mu(v_1 - u_1)$$
$$x_2 + \lambda(y_2 - x_2) = u_2 + \mu(v_2 - u_2)$$
This is a linear system with coefficients in F; hence, either there is no solution (the lines do not intersect), there are infinitely many solutions (the lines coincide), or there is a unique solution $\lambda, \mu \in F$. Then if $z \in L(x, y) \cap L(u, v)$ and the lines do not coincide, result in $z_1, z_2 \in F$ and $z \in F(i)$.

b) We analyze the intersection of a line $L(u, v)$ and a circumference $\mathbb{C}(w, r)$ with $u, v, w, r \in F(i)$ and $r \in \mathbb{R} > 0$ (so that $r \in F$). We have:
$$z \in L(u, v) \cap \mathbb{C}(w, r) \Leftrightarrow \exists \lambda \in \mathbb{R} \text{ such that } z$$
$$= u + \lambda(v - u) \text{ and } |z - w| = r$$

that is, exists $\lambda \in \mathbb{R}$ such that $|u - w + \lambda(v - u)|^2 = r^2$, i.e.:

$$(u_1 - w_1 + \lambda(v_1 - u_1))^2 + (u_2 - w_2 + \lambda(v_2 - u_2))^2 = r^2$$

This is an equation of degree ≤ 2 on λ with coefficients in F, i.e., of the type:

$$a\lambda^2 + b\lambda + c = 0$$

with $a, b, c \in F$, then   if $L(u, v) \cap C(w, r) \neq \emptyset$ it   must   exists $\alpha \in F, \alpha > 0$ such that $\lambda \in F(\sqrt{\alpha})$, hence $z \in F(\sqrt{\alpha})(i)$.

c) In the case of two circumferences $C(w, r), C(u, s)$ with $w, u \in F(i), r, s \in F, r, s > 0$, we have: $z \in C(w, r) \cap C(u, s) \Leftrightarrow$

$$(z_1 - w_1)^2 + (z_2 - w_2)^2 = r^2 \ (1)$$
$$(z_1 - u_1)^2 + (z_2 - u_2)^2 = s^2 \ (2)$$

Subtracting these we obtain,

$$2(u_1 - w_1)z_1 + 2(u_2 - w_2)z_2 + u_1{}^2 - w_1{}^2 + u_2{}^2 - w_2{}^2 = r^2 - s^2$$

If $C(w, r)$ and $C(u, s)$ do not coincide and they intersect, it must be $u \neq w$, that is $u_1 \neq w_1$ or $u_2 \neq w_2$, from where:

-If $u_1 \neq w_1$ and $u_2 = w_2$, we have $z_1 \in F$.

-If $u_2 \neq w_2$ and $u_1 = w_1$, we have $z_2 \in F$.

-If $u_1 \neq w_1$ and $u_2 \neq w_2$, we have $z_1 = az_2 + b$ con $a, b \in F$.

Replacing z_1 or z_2 in (1) or (2), we obtain an equation of degree ≤ 2 in z_1 or z_2, and so $z_1, z_2 \in F(\sqrt{\alpha})$ for some $\alpha \in F, \alpha > 0$. Hence $z \in F(\sqrt{\alpha})(i)$.∎

To avoid cumbersome notations, we will write $\mathbb{Q}(\sqrt{\alpha^1}, \sqrt{\alpha^2})$ instead of $(\mathbb{Q}(\sqrt{\alpha_1}))(\sqrt{\alpha_2})$ (see example at the end of sec. 3 chap.6) and similarly in other cases.

Theorem: $z \in \mathbb{C}$ is constructible $\Leftrightarrow$ there are positive real numbers $\alpha_1, \alpha_2, \ldots, \alpha_n$ such that $z \in \mathbb{Q}(\sqrt{\alpha_1}, \sqrt{\alpha_2}, \ldots, \sqrt{\alpha_n})(i)$.

Proof: ($\Rightarrow$): If z is constructible, it can be obtained, starting from 0 and 1 by a finite number n of the steps described in 2). As $0, 1 \in \mathbb{Q}(i)$ it follows from the above lemma that in one step we only can reach $\mathbb{Q}(\sqrt{\alpha_1})(i)$ for some α_1 real and positive, from there, in another step, by the same lemma, we only can reach $\mathbb{Q}(\sqrt{\alpha_1}, \sqrt{\alpha_2})(i)$

with α_2 real and positive. In n steps, we can only reach then an extension of the type: $\mathbb{Q}(\sqrt{\alpha_1}, \sqrt{\alpha_2}, \ldots, \sqrt{\alpha_n})(i)$ with $\alpha_1, \alpha_2, \ldots, \alpha_n$ positive real numbers.

($\Leftarrow$): We will prove that $\mathbb{Q}(\sqrt{\alpha_1}, \sqrt{\alpha_2}, \ldots, \sqrt{\alpha_n})(i) \subset \mathbb{C}$. In fact, as $1 \in \mathbb{C}$ by (A) follows $\mathbb{N} \subset \mathbb{C}$, then by (B) $\mathbb{Z} \subset \mathbb{C}$ and by (I) $\mathbb{Q} \subset \mathbb{C}$; by prop. 7.2 $\sqrt{\alpha_1} \in \mathbb{C}$ and using again (A), (B), (I) we obtain $\mathbb{Q}(\sqrt{\alpha_1}) \subset \mathbb{C}$, and continuing in that way we arrive at $\mathbb{Q}(\sqrt{\alpha_1}, \sqrt{\alpha_2}, \ldots, \sqrt{\alpha_n})(i) \subset \mathbb{C}$. $\blacksquare$

Proposition: (Construction of the regular pentagon) $\varepsilon = e^{i\frac{2\pi}{5}}$ is constructible.

Proof: We have:
$$0 = \varepsilon^5 - 1 = (\varepsilon - 1)(\varepsilon^4 + \varepsilon^3 + \varepsilon^2 + \varepsilon + 1)$$
and as $\varepsilon \neq 1$, then: $\varepsilon^4 + \varepsilon^3 + \varepsilon^2 + \varepsilon + 1 = 0$, that is:
$$\left(\varepsilon^2 + \frac{1}{\varepsilon^2}\right) + \left(\varepsilon + \frac{1}{\varepsilon}\right) + 1 = 0$$
or putting,
$$z = \varepsilon + \frac{1}{\varepsilon} \quad (*)$$
we have, $z^2 + z - 1 = 0$, hence $z = -\frac{1}{2} \pm \frac{\sqrt{5}}{2}$, so that $z \in \mathbb{Q}(\sqrt{5})$. Since by $(*)$ $\varepsilon^2 - z\varepsilon + 1 = 0$, it follows that $\varepsilon \in \mathbb{Q}(\sqrt{5})(\sqrt{\alpha})(i)$ with $\alpha \in \mathbb{Q}(\sqrt{5}), \alpha > 0$, and by the previous theorem ε is constructible. $\blacksquare$

Proposition: The regular polygonal of 17 sides is constructible.

Proof: If ε is a primitive $p -$ root of unity (p prime), then the others are $\varepsilon^k, k = 1, \ldots, p - 1$, and all of them satisfy the equation:
$$x^{p-1} + x^{p-2} + \ldots + x + 1 = 0 \ (1)$$
Reordering and grouping these roots is an essential step in Gauss's treatment of cyclotomy.

First of all, there exists (as Gauss proved) $g \in \mathbb{N}$, such that the powers of g: $g^0, g^1, \ldots, g^{p-2}$ are congruent in some order to $1, 2, \ldots, p - 1$. For example, if $p = 17$, it is enough to take $g = 3$. In fact, we have the following table of congruences modulus 17:

g^0	g^1	g^2	g^3	g^4	g^5	g^6	g^7	g^8	g^9	g^{10}	g^{11}	g^{12}	g^{13}	g^{14}	g^{15}
1	3	9	10	13	5	15	11	16	14	8	7	4	12	2	6

(in modern words we say that g is a *cyclic generator of the group of unities* of $\mathbb{Z}_{17}$).

The primitive p − roots of unity are then: $\varepsilon^{g^h}, h = 0, \ldots, p - 2$. To simplify notation, it is convenient to write (as Gauss did) $[k] = \varepsilon^k$. Note that with this notation, we have $[k][h] = [k + h]$.

Next, we regroup the roots as follows: for each divisor m of $p - 1$, so $mr = p - 1$, and each primitive root $[k]$, we consider the $m - orbit$ of $[k]$:

$$O = \{[k], [kg^r\}], [kg^{2r}\}], \ldots, [kg^{(m-1)r}]\}$$

which is formed, starting from $[k]$, "multiplying" each $[t]$ by g^r to obtain $[tg^r]$, or recalling that $[t] = \varepsilon^t$, raising each $[t]$ to the g^r power: $(\varepsilon^t)^{g^r} = \varepsilon^{tg^r} = [tg^r]$. Note that the element that follows the last in O, is $[kg^{m-1}g^r] = [kg^{mr}] = [kg^{p-1}] = [k]$, so this process permute cyclically the roots in O.

Adding the elements of the orbit, we obtain what Gauss called $m -$ period of $[k]$:

$$(m, k) = [k] + [kg^r] + [kg^{2r}] + \ldots + [kg^{(m-1)r}]$$

there are r distinct m- periods when k varies, since if $[k]$ and $[k']$ are in the same $m -$ orbit, if and only if, $(m, k) = (m, k')$.

Note that to multiply two $m -$ periods $(m, k)(m, h)$, it is convenient to group the products as follows:

$$(m, k)(m, h) = ([k] + [kg^r] + [kg^{2r}] + \ldots + [kg^{(m-1)r}])([h] + [hg^r]$$
$$+ [hg^{2r}] + \ldots + [hg^{(m-1)r}]) =$$
$$= [k + h] + [(k + h)g^r] + \ldots + [(k + h)g^{(m-1)r}] +$$
$$+ [k + hg^r][(k + hg^r)g^r] + \ldots + [(k + hg^r)g^{(m-1)r}] + \ldots +$$
$$+ [k + hg^{(m-1)r}][(k + hg^{(m-1)r})g^r] + \ldots + [(k$$
$$+ hg^{(m-1)r})g^{(m-1)r}] =$$
$$= (m, k + h) + (m, k + hg^r) + \ldots + (m, k + hg^{(m-1)r})$$

In case $p = 17$, there are $r = 2$, $8 -$ periods, namely:

$$(8, 1) = [1] + [9] + [13] + [15] + [16] + [8] + [4] + [2]$$
$$(8, 3) = [3] + [10] + [5] + [11] + [14] + [7] + [12] + [6]$$

We have $(8, 1) + (8.3) = -1$, because $(8, 1) + (8, 3)$ is the sum $\varepsilon + \varepsilon^2 + \ldots + \varepsilon^{16}$ which, by (1), equals -1.

240

Also, $(8,1)(8,3) = -4$, for, by the above expression of the product of periods, we have:

$$(8,1)(8,3) = (8,4) + (8,11) + (8,6) + (8,12) + (8,15) + (8,8)$$
$$+ (8,13) + (8,7)$$

but $(8,4) = (8,15) = (8,8) = (8,13) = (8,1)$, and $(8,11) = (8,6) = (8,12) = (8,7) = (8,3)$, so then $(8,1)(8,3) = 4\{(8,1) + (8,3)\} = -4$.

Then the $8-$ periods are roots of the equation: $x^2 + x - 4 = 0$, so they belong to the field $\mathbb{Q}(\sqrt{17})$.

The $4-$ periods are:

$$(4,1) = [1] + [13] + [16] + [4] \; ; \; (4,2) = [2] + [9] + [15] + [8]$$
$$(4,3) = [3] + [5] + [14] + [12] \; ; \; (4,6) = [6] + [10] + [11] + [7]$$

We have $(4,1) + (4,2) = (8,1)$ and $(4,1)(4,2) = -1$, then there are roots of the equation: $x^2 - (8,1)x - 1 = 0$. Since the discriminant $\alpha = (8,1)^2 + 4$ is positive and belongs to $\mathbb{Q}(\sqrt{17})$, we have $(4,1), (4,2) \in \mathbb{Q}(\sqrt{17}, \sqrt{\alpha})$ with $\alpha \in \mathbb{Q}(\sqrt{17}), \alpha > 0$. Similarly $(4,3), (4,6) \in \mathbb{Q}(\sqrt{17}, \sqrt{\beta})$, with $\beta \in \mathbb{Q}(\sqrt{17}), \beta > 0$.

The 2- periods are: $(2,1) = [1] + [16]$; $(2,4) = [4] + [13]$, etc. and we have $(2,1) + (2,4) = (4,1)$; $(2,1)(2,4) = (4,3)$, so $(2,1), (2,4)$ are roots of the equation $x^2 - (4,1)x + (4,3) = 0$, and they are real numbers (because both are sum of a complex number and its conjugate), then there exists $\gamma \in \mathbb{Q}(\sqrt{17}, \sqrt{\alpha}, \sqrt{\beta}), \gamma > 0$, such that:

$$(2,1) \in \mathbb{Q}(\sqrt{17}, \sqrt{\alpha}, \sqrt{\beta}, \sqrt{\gamma}),$$

Finally, as $(2,1) = \varepsilon + \frac{1}{\varepsilon}$, we have $\varepsilon^2 - (2,1)\varepsilon + 1 = 0$, hence there exists $\delta \in \mathbb{Q}(\sqrt{17}, \sqrt{\alpha}, \sqrt{\beta}, \sqrt{\gamma}), \delta > 0$ such that:

$$\varepsilon \in \mathbb{Q}(\sqrt{17}, \sqrt{\alpha}, \sqrt{\beta}, \sqrt{\gamma}, \sqrt{\delta}) (i)$$

Then by theorem above, ε is contructible.∎

In order to treat other classical constructions we shall need the following:

Lemma: Let f be the polynomial function with rational coefficients defined by:

$$f(z) = z^3 + pz^2 + qz + r$$

If f has a constructible root, then it has a rational root.

Proof: Let z be a constructible root of f. By the above theorem, we will have $z \in$ $\mathbb{Q}(\sqrt{\alpha^1}, ..., \sqrt{\alpha^n})(i)$ where $\alpha_1, ..., \alpha_n$ are real numbers > 0 and $\alpha^i \in$ $\mathbb{Q}(\sqrt{\alpha^1}, ..., \sqrt{\alpha^{i-1}})$. Putting $K = \mathbb{Q}(\sqrt{(\alpha^{\wedge}1}))...(\sqrt{(\alpha^{\wedge}(n-1)}))$ and $\alpha = \alpha^{\wedge}n$, we have $z \in K(\sqrt{\alpha})(i)$, then $z = a + bi$ with $a, b \in K(\sqrt{\alpha})$. Since $\bar{z} = a - bi$ is also a root of f and as the sum of the roots is rational (by the relations between the roots and the coefficients: Chap. 5, sec. 7) and $z \in$ $K(\sqrt{\alpha})(i)$ it follows that f must have a root $w \in K(\sqrt{\alpha})$, because $z + \bar{z} \in K(\sqrt{\alpha})$. Putting $w = c + d\sqrt{\alpha}$ with $c, d \in K$ we have, similarly, that $c - d\sqrt{\alpha}$ is also a root of f, and as the sum of the roots must be rational, results that f must have a root in K. Following in that way, we find that f must have a rational root.■

We will apply this lemma to obtain the proof of the impossibility of certain constructions.

Proposition: The duplication of the cube is not possible with ruler and compass.

Proof: This is the famous Delos Problem; given a cube, construct a cube with double volume; that is, given the edge of a cube (which we can take as 1), construct, with ruler and compass, the edge of a cube whose volume is twice of the given. In other words, construct $\sqrt[3]{2}$. Since $\sqrt[3]{2}$ satisfies:

$$z^3 - 2 = 0$$

and this equation has no rational roots, $\sqrt[3]{2}$ is not constructible.■

Proposition: The regular heptadecagon is not constructible.

Proof: If it were, the number $\varepsilon = e^{i\frac{2\pi}{7}}$ would be constructible. Since ε satisfies:

$$0 = \varepsilon^7 - 1 = (\varepsilon - 1)(\varepsilon^6 + \varepsilon^5 + \varepsilon^4 + \varepsilon^3 + \varepsilon^2 + \varepsilon + 1)$$

and as $\varepsilon \neq 1$, we have:

$$\left(\varepsilon^3 + \frac{1}{\varepsilon^3}\right) + \left(\varepsilon^2 + \frac{1}{\varepsilon^2}\right) + \left(\varepsilon + \frac{1}{\varepsilon}\right) + 1 = 0$$

Putting $z = \varepsilon + \frac{1}{\varepsilon}$, it follows that z would be also constructible. But z satisfies the equation:

$$z^3 + z^2 - 2z - 1 = 0$$

As this equation has no rational roots, z and, so, ε are not constructible.∎

Proposition: The trisection of the angle is impossible with ruler and compass.

Proof: There are trisecable angles, for example those of $90°, 180°$, etc. The problem refers to the trisection of an arbitrary angle. We will show that the angle of $120°$ can not be trisectable, and at the same time that the regular eneagon is not constructible. In fact, if any of those constructions were possible, then the number $\varepsilon = e^{i\frac{2\pi}{9}}$ would be constructible. Since ε satisfies: $\varepsilon^9 = 1$, we have $0 = (\varepsilon^3 - 1)(\varepsilon^6 + \varepsilon^3 + 1)$ and as $\varepsilon^3 \neq 1$, it follows:

$$\left(\varepsilon^3 + \frac{1}{\varepsilon^3}\right) + 1 = 0$$

from where, putting $z = \varepsilon + \frac{1}{\varepsilon}$, follows:

$$z^3 - 3z + 1 = 0$$

and this equation has no rational roots, so z and then ε are not constructible.∎

EXERCISES

Ex. 1: Which of the following are constructible numbers?:

$$a)\sqrt{(2 + \sqrt{3})}, \; b)3\sqrt{2} + \frac{\sqrt{3+\sqrt{2}}}{1+\sqrt{3}}i, \; c)\sqrt[3]{7}, \; d)\sqrt[6]{2}$$

Ex. 2: If the regular polygons of n and m sides are constructible and n and m are coprime, then the regular polygon of nm sides is also contructible.

Ex. 3: The regular polygon of 2^n sides, with $n \geq 2$, is constructible.

Ex. 4: From $\left(e^{i\alpha}\right)^3 = e^{i3\alpha}$ deduce: $\cos(3\alpha) = -3\cos\alpha + 4\cos^3\alpha$ and conclude that $\cos 20°$ is not contructible.

Ex. 5: From the impossibility of the trisection of the angle of 120°, prove that the angles of 60°, 30°, 15°, etc. are not trisectable. Then there is an infinite number of non-trisectable angles.

Appendix 3
QUATERNIONS
AND OCTONIONS

1 - INTRODUCTION

The Italian algebraists of the 16th century introduced complex numbers when considering third-degree equations with real roots, that in order to find them, they must solve an auxiliary second-degree equation with no real roots. However, the majority of mathematicians, continued to consider complex numbers as imaginary or impossible entities until at the beginning of the 18th century, Gauss, Argand, and Wessell, interpreted them geometrically as points of the plane. Hamilton in 1833 formalizes this idea algebraically, presenting them as ordered pairs of real numbers, with operations defined by:

$$(a,b) + (c,d) = (a+c, b+d)$$
$$(a,b)(c,d) = (ac - bd, ad + bc)$$

and he tried to extend the nice properties (which today we describe as those of a division algebra) of these pairs (or vectors as he called them) to three dimensions, that is, to ordered triplets or vectors $(a,b,c) = a + bi + cj$, where the sum is the ordinary addition of vectors, that is, coordinate to coordinate, and the problem is to define a product that preserves those properties. For several years the idea did not settle, until 1843 when, taking a walk around Dublin, a sudden inspiration reveals him that if, instead of triplets, considering quadruples: $(a,b,c,d) = a + bi + cj + dk$ it can be defined a product between them, which preserves the desired properties except for the commutativity of the product, and engraves on Brougham bridge, the fundamental relations:

$$i^2 = j^2 = k^2 = ijk = -1.$$

At the end of the same year, John Graves, friend and connoisseur of Hamilton work, extends the construction to eight dimensions, having this time, to abandon not only the commutativity but also the associativity of the product, obtaining the division algebra of octonions or "Cayley numbers", called so because they were rediscovered by Cayley in 1845. This octonions, despite not being associative, verify some laws of weak associativity, called alternative laws:

$$x(xy) = x^2y; \; (xy)y = xy^2$$

and they form an algebra called alternative. So that, alternative algebras began with the octonions, but also ended with them, or rather, with a generalization of them to an arbitrary field: Cayley-Dickson algebras. In fact, a theorem of Bruck and Kleinfeld states that an alternative division ring is, or a Cayley-Dickson division algebra or an associative division ring. For a proof of this result we refer to [35]. Here we will prove a particular case, already proved by Zorn in 1930: an alternative, division $\mathbb{R}$ − álgebra of finite dimension, must be either the field $\mathbb{R}$ of real numbers, the field $\mathbb{C}$ of complex numbers, the associative but not commutative division ring H of quaternions, or the alternative but not associative, neither commutative, division ring O of octonions. The search of the alternative division $\mathbb{R}$ − algebras of finite dimension, will take us in a natural way, to define quaternions and octonions. As particular cases, will be proved the, sometimes called "Final Theorem of Arithmetic" attributed to Wieierstrass: the only associative, commutative division $\mathbb{R}$ − algebras of finite dimension, except for isomorphisms, are $\mathbb{R}$ and $\mathbb{C}$; and the theorem of Frobenius (1877): the associative, division $\mathbb{R}$ − álgebras of finite dimension are, except for isomorphisms, $\mathbb{R}, \mathbb{C}$ or H.

We also will prove a theorem of Hurwitz on "composition algebras": the only real algebras that admit composition are of dimension 1, 2, 4 or 8, and are the four mentioned above. Additionally, we generalize the definition of cross product and prove that such a product can only exist in dimensions 1, 3 or 7.

Finally, we will prove the Four Square Theorem using Hurwitz Quaternions.

The word ring, coined by Hilbert in the context of rings of cyclotomic integers, is used generally in Mathematics, requiring an associative product. For some applications, this requirement is very restrictive and it is convenient enlarging its meaning to design a set A with two (internal, binary) operations sum and product, such that:

1) The sum is associative, commutative, it has a neutral element 0, and, for each $x \in A$, exists an additive inverse: $-x$.
2) The product is distributive, on both sides, respect to the addition:

$$x(y + z) = xy + xz$$
$$(x + y)z = xz + yz$$

A ring is said to be associative if the associativity of the product is valid, commutative is the product is commutative, with identity if there exists a neutral element of the product.

The concepts of subring, morphism of rings, among others, do not depend on the properties of the product, so they transfer without change to this more general situation.

A, perhaps, familiar example of a non-associative and noncommutative ring is the Euclidean space of three dimensions with the vectorial or cross product. Recall that if $\{x_1, x_2, x_3\}$ is the canonical basis, the vectorial product $x \times y$ of two vectors $x = a_1 x_1 + a_2 x_2 + a_3 x_3$ and $y = b_1 x_1 + b_2 x_2 + b_3 x_3$ (a_i, b_i real numbers) is defined by:

$$x \times y = (a_2 b_3 - a_3 b_2)x_1 + (a_1 b_3 - a_3 b_1)x_2 + (a_1 b_2 - a_2 b_1)x_3$$

This space with the ordinary sum of vectors and the product above is a ring in which the following identities are verified:

$$x \times y = -y \times x \ (1)$$
$$(x \times y) \times z = x \times (y \times z) + y \times (z \times x) \ (2)$$

From the first follows the non commutativity of the product, for it is easy to find vectors x, y such that $x \times y \neq 0$ (example: $x_1 \times x_2 = x_3$) and from the second follows that the product is neither associative, as is equally easy to find x, y, z such that $y \times (z \times x) \neq 0$ (example: $x = y = x_1, z = x_2$).

We are interested in a special kind of rings, the, so-called, algebras. If K is a field, a $K - algebra$ is a $K -$ vector space with a product between vectors such that, with this product and the sum already existent in the vector space structure, becomes a ring in a way that the product between vectors and the product by scalars verify the following compatibility condition:

$$a(xy) = (ax)y = x(ay)$$

where the first letters of the alphabet denote scalars and the last, vectors.

The dimension of a K − algebra is that of the corresponding vector space. A K − algebra is said to be associative, commutative, with identity, if the corresponding ring has such properties. A subalgebra is a subset which is both a subspace and a subring and a morphism of algebras is a linear morphism of rings.

Familiar examples of algebras are:

- $K[X]$ the polynomial ring in an indeterminate X, with coefficients in the field K. It has infinite dimension because $\{1, X, \ldots, X^n, \ldots\}$ is a base.
- $\mathbb{R}$ is a $\mathbb{Q}$ − algebra of infinite dimension.
- $\mathbb{C}$ is an $\mathbb{R}$ − algebra of dimension two.
- $M_n(K)$, that is, the square matrices $n \times n$ over a field K, is an associative K − algebra of dimension n^2, but is not commutative if $n \geq 2$.

An associative K − algebra A with identity 1, is said to be a division algebra iff for any $x \in A, x \neq 0$, there exists a (two sided) multiplicative inverse x^{-1}. We remark that this definition is applicable only to associative algebras, later we shall give a wider definition.

The first example of a not commutative division algebra was given by Hamilton with quaternions.

To construct them, take an $\mathbb{R}$ − vector space $\mathbb{H}$ of dimensión 4 and a base $\{x_1, x_2, x_3, x_4\}$. Among the vectors of the choosen base, we define a product, taking x_1 as a neutral element, that is $x_1 x_r = x_r x_1 = x_r$ for $r = 1, 2, 3, 4$, so then we will write $x_1 = 1$ (this is a vector, not to be confused with the scalar 1). We also will write, as usual, $x_2 = i, x_3 = j, x_4 = k$ and we define:

$$i^2 = j^2 = k^2 = -1, ij = k, ji = -k, jk = i, kj = -i, ki = j, ik = -j$$

and extend the definition to a product of two arbitrary vectors: $x = \sum_r a_r x_r$, $y = \sum_s b_s x_s$ $(a_r, b_s \in \mathbb{R}$ where the index varied between 1 and 4, unless otherwise specified), by bilinearity, that is:

$$xy = \sum_{r,s} a_r b_s (x_r x_s)$$

where $x_r x_s$ must be replace by its value defined above.

It is routine to verify that $\mathbb{H}$ with this product and the sum and product by scalars of the underlying vector space structure is an $\mathbb{R}$ − algebra with identity, associative but not commutative. Let us show that it is a division algebra:.

For $x = \sum_r a_r x_r$, we define its conjugate $\bar{x}$ by:

$$\bar{x} = a_1 x_1 - \sum_{r=2}^{4} a_r x_r$$

Doing the products we obtain:

$$x\bar{x} = \bar{x}x = (a_1^2 + a_2^2 + a_3^2 + a_4^2) \cdot 1$$

If we define the norm $N(x)$ of x as the real number $a_1^2 + a_2^2 + a_3^2 + a_4^2$, then we have clearly:

$$N(x) = 0 \Leftrightarrow x = 0$$

hence, if $x \neq 0$ we have $N(x) \neq 0$ and:

$$x \cdot (\frac{1}{N(x)})\bar{x} = (\frac{1}{N(x)})\bar{x} \cdot x = 1$$

then $\mathbb{H}$ is a division algebra.

But a question emerges: why choose the relationships given in the definition of multiplication and no others?. As we shall see, in trying to obtain all associative, division $\mathbb{R}$ − algebras of finite dimension, the relationships above are the only possible in four dimensions except for isomorphisms. Furthermore, besides H the only such algebras are (Frobenius's theorem) $\mathbb{R}$ and $\mathbb{C}$.

The example above of the Euclidean space with the cross product is not only a ring but an $\mathbb{R}$ − álgebra, where in addition the identities (1) and (2) are verified. This is a particular case of the so-called Lie algebras.

A K − algebra is a *Lie algebra* iff the following identities are verified:

$$x^2 = 0 \text{ (which impplies } xy = -yx)$$
$$(xy)z + (yz)x + (zx)y = 0$$

The last one is called *Jacobi identity*.

Parallel to Lie algebras are the so-called *Jordan algebras*. These are K − algebras such that the following identities are valid:

$$xy = yx$$
$$x^2(yx) = (x^2 y)x$$

and were introduced by P. Jordan in an attempt to formalize the mathematics of Quantum Mechanics.

From an associative $K -$ álgebra, with the product denoted by juxtaposition, can be derived both, a Lie algebra, defining a new product $[,]$ by the commutator:

$$[x, y] = xy - yx$$

and also a Jordan algebra with the new product:

$$x \cdot y = \frac{1}{2}(xy + yx)$$

wherein this last case we assume that $2 \neq 0$ in K.

The imposition of additional identities, as in the case of Lie and Jordan algebras, is typical in the theory of non (not necessarily) associative algebras. For a systematic exposition, we refer to [37]. The additional identities relevant here, are the, so-called, alternative laws, which are a kind of weak associativity.

3 - ALTERNATIVE ALGEBRAS

A ring or a $K -$ algebra A are said to be *alternative* for any $x, y \in A$ the following *alternative laws* are verified:

$$x^2 y = x(xy) \text{ and } xy^2 = (xy)y$$

Alternative rings and algebras, play an important role in the classification of both, Lie and Jordan algebras and also in the study of the foundations of Geometry.

Defining the *associator* B of $x, y, z \in A$ by:

$$B(x, y, z) = x(yz) - (xy)z$$

the associativity becomes $B(x, y, z) = 0$ for any $x, y, z \in A$ and the alternative laws become the identities: $B(x, x, y) = B(x, y, y) = 0$.

Lemma 3.1: In an alternative $K -$ algebra A:

a) The associator is an alternate trilinear mapping. That is, it is linear in each variable and change sign when we permute any two arguments, for example:

$$B(x, y, z) = -B(y, x, z)$$

b) $x(yx) = (xy)x$ (which, then, can be simply denoted by xyx), (*flexible law*).

c) If x, y anticommute, that is $xy = -yx$, then: $x(yz) = -y(xz)$ and $(zx)y = -(zy)x$

d) The *Moufang* identity: $(zx)(yz) = z(xy)z$ is valid.

e) A is *power associative*, i.e., if we define inductively: $x^{n+1} = x^n x$ with $n \in \mathbb{N}$, then we have $x^{n+m} = x^n x^m$ for any $x \in A$; $n, m \in \mathbb{N}$.

f) For any $x \in \epsilon A$, the mapping $\phi: K[X] \to A$, defined by $\phi(\sum a_i X^i) = \sum a_i x^i$ is a morphism of $K -$ algebras.

Proof: a) To verify that the associator is linear in each variable is left as a straightforward exercise. By the first alternative law we have:

$$0 = B(x + y, x + y, z) =$$
$$= B(x, x, z) + B(x, y, z) + B(y, x, z) + B(y, y, z)$$
$$= B(x, y, z) + B(y, x, z)$$

then $B(x, y, z) = -B(y, x, z)$. The other verifications to obtain the alternativity are similar.

b) $B(x, y, x) = -B(y, x, x) = 0$, or $x(yx) = (xy)x$.

c) By a) we have $B(x, y, z) + B(y, x, z) = 0$, then if $xy + yx = 0$ results:

$$0 = B(x, y, z) + B(y, x, z) + (xy + yx)z = x(yz) + y(xz)$$

For the other identity, the proof is similar.

d) We have:

$$(zx)(yz) - ((zx)y)z = B(zx, y, z) = B(y, z, zx)$$
$$= y(z^2 x) - (yz)(zx) =$$
$$= y(z^2 x) - B(yz, z, x) - (yz^2)x = B(y, z^2, x) - B(yz, z, x) =$$
$$= B(y, z^2, x) - B(x, yz, z) = B(y, z^2, x) - x(yz^2) + (x(yz))z =$$
$$= B(y.z^2, x) + B(x, y, z)z - B(x, y, z^2) = B(x, y, z)z$$

Then:

$$(zx)(yz) = B(x, y, z)z + ((zx)y)z$$
$$= B(x, y, z)z - B(z, x, yz) + z(xy)z = z(xy)z.$$

e) By induction on n, we obtain readily, using the flexible law, that $xx^n = x^{n+1}$. Then, by induction on m we will see that $x^n x^m = x^{n+m}$, where we may assume that $n > 1$. We have:

$$x^n x^{m+1} = (xx^{n-1})(x^m x) = x(x^{n-1}x^m)x = xx^{n-1+m}x = x^{n+m+1},$$

where we have used Moufang's identity and the flexible law.

f) This follows from e), since

$$\left(\sum_i a_i x^j\right)\left(\sum_j b_j x^j\right) = \sum_k \sum_{i+j=k} a_i b_j x^i x^j = \sum_k \left(\sum_{i+j=k} a_i b_j\right) x^k$$

Where in the last equality we use the power associativity.∎

4 - DIVISION ALGEBRAS

To avoid trivialities, we assume that A is a nonzero $K-$ algebra: $A \neq (0)$.

If A is an associative algebra with identity 1, we say that A is a division algebra, iff any nonzero element $x \in A$ has a multiplicative inverse, that is for any $x \neq 0$, there exists $x^{-1} \in A$ such that:

$$xx^{-1} = x^{-1}x = 1$$

as usual (using associativity!) one proves that such inverse is unique.

When considering *non-associative*, that is not necessarily associative, algebras, it is suitable to adopt the following definition:

A $K-$ álgebra A (with or without identity) is said to be a *division algebra* iff for any $x, y \in A$, with $x \neq 0$, the equations:

$$xz = y, wx = y$$

have unique solutions.

Defining $L_x: A \to A$ and $R_x: A \to A$ as the multiplications by x on left and on right respectively, that is: $L_x(z) = xz$ and $R_x(w) = wx$; the previous definition is equivalent to say that those mappings, which are K- linear, are bijective.

If A is associative and has an identity, both definitions of division algebra are equivalent. More generally:

Lemma 4.1: If A is a $K-$ álgebra with identity 1, such that, for each $x \neq 0$ exists x' such that:

  a) $xx' = x'x = 1$
  b) $B(x', x, y) = 0 \; \forall \, y \in A$,

then x' is unique, and A is a división algebra.

Proof: x' is unique, since from $xy = 1$, follows $x' = x'(xy) = (x'x)y = y$. Moreover, given x, y with $x \neq 0$, $x'y$ is a solution of $xz = y$. In fact:

$$x(x'y) = (xx')y = y$$

and is the only one, since if $xz = y$, then $x'y = x'(xz) = (x'x)z = z$, and similarly yx' is the only solution of $wx = y$. ∎

In the case of finite dimension, we have the following equivalence:

Lemma 4.2: For a finite dimensional K − algebra A, we have: A is a division algebra $\Leftrightarrow$ A is *integral* (i.e.: $xy = 0 \Rightarrow x = 0$ or $y = 0$).

Proof: ($\Rightarrow$): This implication is valid whether or not the dimension is finite, since, if $x \neq 0$, 0 is the only solution of $xy = 0$.

($\Leftarrow$): The integrality implies that if $x \neq 0$, the linear mappings L_x and R_x are injective, but as the dimension is finite, both of them must be onto.

Lemma 4.3: If A is an alternative, division algebra, then:
a) A has an identity 1.
b) Any $x \neq 0$ has an unique multiplicative inverse (on both sides, denoted then x^{-1}).
c) If $x \neq 0$, then $B(x, x^{-1}, y) = 0 \; \forall \; y$.

Proof: a) Let $x \neq 0$ and let y the only solution of $xy = x$ (so then $y \neq 0$). Since $xy^2 = xy = x$, it follows that $y^2 = y$. Then for any $z \in A$, we have $0 = (y^2 - y)z = y(yz - z)$ and, hence $yz = z \; \forall z \in A$. Similarly, $zy = z \; \forall z \in A$. then $y = 1$ is the identity of A.

b) From $wx = 1$, follows $x = x(wx) = (xw)x$, then $xw = 1$ (1 and xw is the only solution of $ux = x$).

c) By Moufang's identity, we have: $x(x^{-1}y)x = yx$, then $B(x, x^{-1}, y)x = 0$ and then $B(x, x^{-1}, y) = 0$. ∎

5 - ℝ-ÁLGEBRAS

By lemma 4.3 above, an alternative division K − algebra has an identity 1. In what follows, usually we will identify the field K with the subspace $K \cdot 1$ generated by the identity vector and, for example, if

253

we write $x + a$ where x is a vector and a scalar, we mean, of course, $x + a \cdot 1$.

Lemma 5.1: If A is an alternative division $\mathbb{R}$ − algebra of finite dimension, with identity 1 (lemma4.3), then:

a) For each $x \in A$ exist $a, b \in \mathbb{R}$ such that:
$$x^2 + ax + b = 0$$
If, moreover, $x \notin \mathbb{R} \cdot 1$, then there exists $c \in \mathbb{R}, c \neq 0$ such that:
$$(x + \frac{a}{2})^2 = -c^2$$

b) If $A \neq \mathbb{R} \cdot 1$, then there exists $u \in A$ such that $u^2 = -1$.

c) If $u \in A$ verifies: $u^2 = -1$, then we have:
$$xu = ux \Leftrightarrow x \in \mathbb{R} \cdot 1 + \mathbb{R} \cdot u$$

d) For any $y \in A$ if there exists $u \in A, u \neq 0$ such that it anticommutes with y, that is $yu = -uy$, then $y^2 \in \mathbb{R} \cdot 1$.

e) If u, y are such that $u^2 = -1, y^2 \in \mathbb{R} \cdot 1$ and $y \notin \mathbb{R} \cdot 1 + \mathbb{R} \cdot u$, then:
$$yu + uy \in \mathbb{R} \cdot 1$$

Proof: a) As A is finite dimensional, the set of all the powers of x must be linearly dependent, so there must be a relation:
$$\sum_{i=0}^{n} a_i x^i = 0$$
with the a_i real numbers not all of them zero. That is, x is a root of a nonzero polynomial $f = \sum_{i=0}^{n} a_i X^i$. As any polynomial in $\mathbb{R}[X]$ is a product of polynomials of degree 1 or 2 and since the specialization $\mathbb{R}[X] \to A: X \to x$ is a morphism of algebras (lemma 3.1 f) and A is an integral algebra (lemma 4.2) it follows that x must satisfy a polynomial of the second-degree, i.e., there exist $a, b \in \mathbb{R}$ such that:
$$x^2 + ax + b = 0$$
Completing the square we obtain (by compatibility of scalars and vectors): $(x + (\frac{a}{2})^2 = \frac{a^2}{4} - b$, then if $x \notin \mathbb{R} \cdot 1$ we must have $\frac{a^2}{4} - b < 0$ and putting $\frac{a^2}{4} - b = -c^2$ results a).

b) It is enough to take $u = \frac{1}{c}(x + \frac{a}{2})$ in a).

254

c) Let $u^2 = -1$ and $xu = ux$. If $x \notin \mathbb{R} \cdot$

1 there exist real numbers a, c such that $\left(x + \dfrac{a}{2}\right)^2 = -c^2 = (cu)^2$ then:

$$0 = \left(x + \frac{a}{2}\right)^2 - (cu)^2 = (x + \frac{a}{2} + cu)(x + \frac{a}{2} - cu)$$

since $xu = ux$. As A is integral, it follows that $x \in \mathbb{R} \cdot 1 + \mathbb{R} \cdot u$.
The other implication is clear.

d) From $yu = -uy$ follows: $y^2u = y(yu) = -y(uy) = -(yu)y = (uy)y = uy^2$, that is $y^2u = uy^2$. On the other side, by a), y satisfies an equation with real coefficients: $y^2 + ay + b = 0$ and as u commutes with y^2 and with b, it must commute also with ay: $(ay)u = u(ay)$. If $a \neq 0$ follows $yu = uy = -uy$ then $y = 0$ since $u \neq 0$. In case that $a = 0$ results $y^2 = -b \in \mathbb{R} \cdot 1$.

e) By a) there are real numbers a, b, c, d such that:
$$(y + u)^2 + a(y + u) + b = 0 \ (1)$$
$$(y - u)^2 + c(y - u) + d = 0 \ (2)$$

Adding this equations:
$$2(y^2 - 1) + (a + c)y + (a - c)u + b + d = 0$$

Since $y^2 \in \mathbb{R} \cdot 1$, if it were $a + c \neq 0$ it would be $y \in \mathbb{R} \cdot 1 + \mathbb{R} \cdot u$ contrary to the hypothesis. Then we must have $a + c = 0$. then we must have $a - c = 0$ since $u \notin R \cdot 1$. We have then: $a = c = 0$ and substracting (2) from (1), results $yu + uy \in \mathbb{R} \cdot 1$.■

Theorem 5.2: If A an alternative, division $\mathbb{R}$ − álgebra of finite dimension, then ($\simeq$ means $\mathbb{R}$ − algebra isomorphism):
$$A \simeq \mathbb{R}, A \simeq \mathbb{C}, A \simeq \mathbb{H} \text{ or } A \simeq \mathbb{O}$$
where $\mathbb{O}$ is an $\mathbb{R}$ − álgebra of dimension 8 that will be defined later.

Proof: If $A = \mathbb{R} \cdot 1$ there is nothing to prove ($A \simeq \mathbb{R}$).

Let $A \neq \mathbb{R} \cdot 1$. By lemma 5.1. b), exists $i \in A$ such that $i^2 = -1$. It follows that $C = \mathbb{R} \cdot 1 + \mathbb{R} \cdot i$ is a subalgebra of A isomorphic to $\mathbb{C}$.

Any element of A that commute with i, must belong to C by lemma 5.1. c). Then, if A es commutative, we must have $A \simeq \mathbb{R}$ or $A \simeq \mathbb{C}$, so we have proved the Final Theorem of Arithmetic.

Assume then that exists $x \in$
A such that does not commute with i. Putting:

$$z = xi - ix \neq 0$$

we have that z anticommutes with i. In fact, from:

$$zi = (xi - ix)i = -x - ixi$$
$$iz = i(xi - ix) = ixi + x$$

we obtain $zi = -iz$. From lemma 5.1 d), follows that $z^2 \in \mathbb{R} \cdot 1$. Since $z \notin \mathbb{R} \cdot 1$ because it anticommutes with i, it follows that $z^2 = -c^2$ with $c \in \mathbb{R}, c \neq 0$. Putting $j = c^{-1}z$ we obtain:

$$ij = -ji, j^2 = -1$$

Defining k by $k = ij$ we obtain the relations:

$$k^2 = -1; \ ik = -j; \ ki = j; \ ij = k; \ ji = -k; \ jk = i; \ kj = -i$$

(for example: $jk = j(ij) = (ji)j = (-ij)j = -ij^2 = i$).

Furthermore, the set $\{1, i, j, k\}$ is linearly independent over $\mathbb{R}$, since if $a + bi + cj + dk = 0$, that is $a + bi + (c + di)j = 0$, then $c + di = 0$ and $a + bi = 0$ since $j \notin \mathbb{C}$ (anti-commutes with i), hence $a = b = c = d = 0$. We have then that

$$H = \mathbb{R} \cdot 1 + \mathbb{R} \cdot i + \mathbb{R} \cdot j + \mathbb{R} \cdot k$$

is a subalgebra of A isomorphic to the algebra $\mathbb{H}$ of quaternions.

Let us show that if an element $x \in A$ associates with i, j, then it belongs to H. That is:

$$B(x, i, j) = 0 \Rightarrow x \in H \ (*)$$

In fact, let $B(x, i, j) = 0$ and assume $x \notin \mathbb{R} \cdot 1$ (for if not there would be nothing to prove). By lemma. 5.1, exist $a, c \in \mathbb{R}$ such that $\left(x + \frac{a}{2}\right)^2 = -c^2$. Putting:

$$y = x + \frac{a}{2} \ (1)$$

we have $y^2 \in \mathbb{R} \cdot 1$ and as we can assume that $y \notin \mathbb{R} \cdot 1 + \mathbb{R} \cdot u$ for $u = i, j, k$ (for if not, y and in consequence x, would belong to H), applying lemma 5.1 e) for $u = i, j, k$ results:

$$yi + iy = a_1$$
$$yj + jy = a_2$$
$$yk + ky = a_3$$

where a_1, a_2, a_3 are real numbers. Putting:

$$w = y + ((a_1)/2)i + ((a_2)/2)j + ((a_3)/2)k \ (2)$$

we have,

$$wi + iw = yi + iy - a_1 = 0$$

that is, w anticommutes with i. In a similar way, it is proved that w anticommutes with j and with k. As $B(x, i, j) = 0$, from (1) and (2) it follows that $B(w, i, j) = 0$, then:

$$wk = -kw = -(ij)w = -i(jw) = i(wj) = (iw)j = -(wi)j$$
$$= -w(ij) = -wk$$

then $2wk = 0$ and so $w = 0$. From (1) and (2) we obtain:

$$x = -(a/2) - ((a_1)/2)i - ((a_2)/2)j - ((a_3)/2)k$$

which proves (*). From (*) we obtain Frobenius's Theorem: if A is associative, then it is isomorphic to $\mathbb{R}, \mathbb{C}$ or H.

Assume now that there is some $x \in A$ which does not associate with i, j. Let

$$v = B(x, i, j) \neq 0$$

and we will show that v anticommutes with i. In fact:

$$vi = -B(x, j, i)i = -(x(ji))i + (xj)i^2 = (xk)i - xj$$
$$iv = -iB(i, x, j) = -i^2(xj) + i((ix)j) = xj + i((ix)j)$$

and will be enough to prove that $(xk)i = -i((ix)j)$ or that $i(xk)i = (ix)j$ but this follows from Moufang identity: $i(xk)i = (ix)(ki) = (ix)j$.

Similarly one verifies that v anticommutes with j and with k.

From lemma 5.1.d), it follows that $v^2 \in R \cdot 1$, then $v^2 = -c^2$ with $c \in R, c \neq 0$ and putting $h = c^{-1}v$ we obtain the relations:

$$h^2 = -1; \ ih = -hi; \ jh = -hj; \ kh = -hk$$

Let O be the subspace of A defined by:

$$O = \mathbb{R} \cdot 1 + \mathbb{R} \cdot i + \mathbb{R} \cdot j + \mathbb{R} \cdot k + \mathbb{R} \cdot h + \mathbb{R} \cdot (ih) + \mathbb{R} \cdot (jh) + \mathbb{R} \cdot (kh)$$

O has dimension 8 since if we have,

$$a_0 + a_1 i + a_2 j + a_3 k + (a_4 + a_5 i + a_6 j + a_7 k)h = 0$$

as h does not associate with i, j because $i(jh) = -i(hj) = h(ij) = hk = -kh = -(ij)h$, where in the second equality we use lemma 1, c). it follows that $h \notin H$ (H is associative) and, in consequence: $a_4 + a_5 i + a_6 j + a_7 k = 0$ (if it were $\neq 0$, from lemma 4.3 would follow $h \in H$) then $a_i = 0 \ \forall i = 0, 1, \ldots, 7$.

We have:

$$k(jh) = -k(hj) = h(kj) = -hi = ih$$
$$h(ih) = -h(hi) = -h^2i = i$$
$$(ih)(ih) = -(ih)(hi) = -ih^2i = -1$$

wherein the first two we use lemma 3.1, c) and in the third Moufang identity. In a similar way, we obtain all entries of the following table, using the notation:

$$u_1 = i, u_2 = j, u_3 = k, u_4 = h, u_5 = ih, u_6 = jh, u_7 = kh$$

	u_1	u_2	u_3	u_4	u_5	u_6	u_7
u_1	-1	u_3	$-u_2$	u_5	$-u_4$	$-u_7$	u_6
u_2	$-u_3$	-1	u_1	u_6	u_7	$-u_4$	$-u_5$
u_3	u_2	$-u_1$	-1	u_7	$-u_6$	u_5	$-u_4$
u_4	$-u_5$	$-u_6$	$-u_7$	-1	u_1	u_2	u_3
u_5	u_4	$-u_7$	u_6	$-u_1$	-1	$-u_3$	u_2
u_6	u_7	u_4	$-u_5$	$-u_2$	u_3	-1	$-u_1$
u_7	$-u_6$	u_5	u_4	$-u_3$	$-u_2$	u_1	-1

From this table, it follows that O is a subalgebra of A.

We will finally show that $A = O$. In order to do so, let $x \in A$. We can assume $x \notin \mathbb{R} \cdot 1$, so then exist $a, c \in \mathbb{R}, c \neq 0$ such that $\left(x + \frac{a}{2}\right)^2 = -c^2$. Putting

$$y = x + (a/2) \quad (3)$$

we have $y^2 \in R \cdot 1$ and, as we can assume $y \notin \mathbb{R} \cdot 1 + \mathbb{R} \cdot u_s$ for $s = 1, \dots, 7$, by lemma 5.1 e), there are real numbers a_s such that:

$$y u_s + u_s y = a_s$$

Let

$$w = y + \sum_{s=1}^{7} \frac{a_s}{2} u_s \quad (4)$$

w anticommutes with each u_s ($s = 1, \dots, 7$), since:

$$w u_s + u_s w = y u_s + u_s y + \sum_{s=1}^{7} \frac{a_s}{2} (u_s u_t + u_t u_s) = a_s - a_s = 0$$

Hence we have:

$$wu_7 = -u_7 w = (u_4 u_3)w = -(u_4 w)u_3 = (wu_4)u_3 = (wu_4)(u_1 u_2)$$
$$= -u_1((wu_4)u_2) =$$
$$= u_1((u_4 w)u_2 = -u_1((u_4 u_2)w) = u_1(u_6 w) = -u_1(wu_6)$$
$$= w(u_1 u_6) = -wu_7$$

then $w = 0$ and from (3) and (4):

258

$$x = -\frac{a}{2} - \sum_{s=1}^{7} \frac{a_s}{2} u_s \in O$$

and the theorem is proved.∎

It would remain to define the octonion algebra $\mathbb{O}$, but now it is clear how to do that. It suffices to take an 8 − dimensional $\mathbb{R}$ − vector space and a base $\{1, u_1, \ldots, u_7\}$, then defining products of the base elements by the preceding table and by the requirement that 1 is a neutral element of the product and extending the product by bilinearity, that is:

$$\left(\sum_{s=0}^{7} a_s u_s \right) \left(\sum_{t=0}^{7} b_t u_t \right) = \sum_{r=0}^{7} \left(\sum_{s+t=ra_s u_s} a_s b_t u_s u_t \right)$$

where $u_0 = 1$ and where $u_i u_j$ must be replaced by the product defined in terms of the base.

An alternative $\mathbb{R}$ − álgebra is obtained. In fact, the distributivity and the alternativity are valid for the elements of the base, and they are valid in general. Moreover, O results a division algebra as we will show next. We can proceed as we did with the quaternions, that is defining the conjugate of $x = a_0 + \sum_{s=1}^{7} a_s u_s$ by:

$$\bar{x} = a_0 - \sum_{s=1}^{7} a_s u_s$$

from where, since $u_s u_t = -u_s u_t$ for any $s, t = 1, \ldots 7$ with $s \neq t$, and $u_s^2 = -1$ if $s \neq 0$, we obtain:

$$x\bar{x} = \bar{x}x = \left(\sum_{s=0}^{7} a_i^2 \right) \cdot 1$$

Hence, putting $N(x) = \sum_{s=0}^{7} a_i^2$, it follows that $N(x) = 0 \Leftrightarrow x = 0$ and if $x \neq 0$, it has an inverse: $N(x)^{-1} x$. As $B(x, \bar{x}, y) = 0$ since as $x + \bar{x} = a \in \mathbb{R}$ we have:

$$B(x, x, y) = B(a - x, x, y) = B(a, x, y) - B(x, x, y) = 0$$

Then by lemma 2, $\mathbb{O}$ es a división algebra.

Another way of verifying that $\mathbb{O}$ is a division algebra is, first proving $\overline{xy} = \bar{y}\bar{x}$ and then proving the multiplicativity of N:

$$N(xy) = N(x)N(y)$$

using Moufang identity, which we leave as an exercise. Then that $\mathbb{O}$ is a division algebra follows as in the next proposition.

A *normed* $\mathbb{R}$ −algebra is an $\mathbb{R}$ −algebra with identity 1, with an inner product $<,>$, such that the corresponding norm: $N(x) = < x, x >$(usually the norm is defined by $\|x\| = \sqrt{< x, x >}$, but to be coherent with the use that we have given to it until now, we will continue calling norm to an N such that $N = \|\,\|^2$) is multiplicative:

$$N(xy) = N(x)N(y)$$

For example, if in the octonion algebra $\mathbb{O}$, we take as orthonormal the definitory base, we have an inner product whose norm coincide with the norm defined in the above section.

Normed algebras are also called algebras with composition or Hurwitz Euclidean algebras.

Proposition 6.1: A finite dimensional normed algebra is a division algebra.

Proof: By definition of an inner product we have $< x, x >= 0 \Rightarrow x = 0$, then : $xy = 0 \Rightarrow N(xy) = N(x)N(y) = 0 \Rightarrow N(x) = 0$ or $N(y) = 0 \Rightarrow x = 0$ or $y = 0$, and the result follows from lemma 4.2.■

Ley A be a normed $\mathbb{R}$ −algebra. Each $x \in A$ can be written univocally as: $x = a \cdot 1 + x'$ with $a \in R$ and $x' \in S$ where S is the ortogonal complement of $\mathbb{R} \cdot 1$. We define the conjugate $\bar{x}$ of x by $\bar{x} = a \cdot 1 - x'$, then it is clear that $x\bar{x} = \bar{x}x$.

Lemma 6.2: For any x, u, v in a normed $\mathbb{R}$ −algebra A, we have:

$$< xu, v >=< u, \bar{x}v >$$

Proof: As the inner product can be expressed in function of the norm as:

$$< x, y >= (1/2)\{N(x + y) - N(x) - N(y)\}$$

and the norm is multiplicative, we have:

$$< xy, y >= N(y) < x, 1 >$$

and replacing $y = u + v$, we obtain

$$< xu, v > +< xv, u >= 2 < u, v >< x, 1 > (1)$$

Putting $x = a \cdot 1 + x'$ with $a \in$
$\mathbb{R}$ and x' in the ortogonal complement of $\mathbb{R} \cdot 1$. By (1) it follows that:
$$< x'u, v >= -< x'v, u >$$
hence: $< xu, v >= a < u, v > +< x'u, v >=< u, av > -< x'v, u >$
$=< u, \bar{x}v >.$ ∎

Proposition 6.3: Any normed real algebra is alternative.

Proof: For any u, x we have by lemma 6:
$$< x\bar{x}, u >=< x, xu >= N(x)\frac{1}{2}\{N(1 + u) - N(1) - N(u)\} =$$
$$< N(x) \cdot 1, u >$$
and as the inner product is not degenerate:
$$N(x) \cdot 1 = x\bar{x}$$
Considering $< xz, xy >$, we have on one side:
$$< xz, xy >= (1/2)\{N(x(z + y)) - N(xz) - N(xy)\} = N(x) < z, y$$
$$>=< z, (x\bar{x})y >$$
and on the other by lemma 6:
$$< xz, xy >=< z, \bar{x}(xy) >$$
then: $\bar{x}(xy) = (x\bar{x})y$, changing here x by $\bar{x}$, we have $x(\bar{x}y) = (\bar{x}x)y$
and as $x\bar{x} = \bar{x}x$:
$$x(\bar{x}y) = (\bar{x}x)y \ (2)$$
Since $x + \bar{x} \in R \cdot 1$, we have: $[(x + \bar{x})x]y = x[(x + \bar{x})y]$,   from
where by (2):
$$x^2y = x(xy)$$
Similarly follows the other alternative law. ∎

By propositions 6.1 and 6.3, a finite-dimensional normed real algebra, is a division alternative algebra, so by theorem 5.2 we obtain
Hurwitz Theorem:

Theorem 6.4: A normed finite dimensional $\mathbb{R}$ −álgebra is necessarily isomorphic to $\mathbb{R}, \mathbb{C}, \mathbb{H}$ or $\mathbb{O}$. ∎

Hurwitz first proved, in 1898, that a real normed finite dimensional algebra, occurs only in dimensions 1, 2, 4 or 8, and later, published posthumously in 1923, that those algebras are the specified above.

7 - CROSS PRODUCT

A few decades after Hamilton invented them, Maxwell used quaternions to formulate his Theory of Electromagnetism. Heaviside and Gibbs noted that the theory becomes simpler if the "real part" of quaternions were discarded; they introduced a product of vectors that later was called Gibbs vectorial product or cross product, by the notation used by Gibbs to denote it: $\times$.

Consider the subspace V of $\mathbb{H}$ formed by the so-called "*pure quaternions*", that is the subspace generated by $\{i, j, k\}$. If $u, v \in V$, we have:

$$uv = -<u, v> 1 + w \quad (1)$$

where $<, >$ is the inner product such that the base $\{i, j, k\}$ is ortogonal, this notion of inner product or dot product was also introduced by Gibbs, and where $w \in V$ is a pure quaternion which is precisely, by definition, the *cross product* of u by v: $w = u \times v$. Since $vu = -<u, v> 1 - w$, we can set alternatively:

$$u \times v = (1/2)(uv - vu) \quad (2)$$

This product $\times : V \times V \rightarrow V$ has the following properties:

a) Bilinearity: that is, is distributive on both sides and is compatible with scalars, so V with $\times$ is an $\mathbb{R}$-álgebra.
b) Orthogonality: $<u, u \times v> = <u \times v, v> = 0$.
c) Magnitude: $N(u \times v) = N(u)N(v) - <u, v>^2$.

The first two follow readily from (2) and the last one from (1), noting that w is in the ortogonal complement of $\mathbb{R} \cdot 1$.

The equivalent definitions (1) and (2), as well as its consequences a), b), c) are valid not only on quaternions but, taking as V the ortogonal   complement   of $R \cdot$ 1, also in any real normed algebra of finite dimension. They are valid then in case of complex numbers, where the cross product becomes trivial: $u \times v = 0$, and also in the case of octonions. Although some properties, as Jacobi identity, valid for the cross product arisen from quaternions are no longer valid on octonions.

We see then, that if we define a product $\times : V \times V \to V$ in an $R-$ vector space V of finite dimension m, as an operation such that a), b), c) are valid, we have the existence of such a product in dimensions $1, 3,$ and 7. In what follows we shall prove that only in those cases there is such a cross product.

Theorem 7.1: If V is an $\mathbb{R}-$ Euclidean space of dimension $m > 0$ with a cross product, then $m = 1, 3$ or 7.

Proof: Consider the $\mathbb{R}-$ vector space $A = \mathbb{R} \times V$ (the cartesian product). Its dimension is $m + 1$, and we define an inner product in A by:
$$< (a, u), (b, v) >= ab + < u, v > (3)$$
where $a, b \in \mathbb{R}, u, v \in$
V, and where both inner products, that already existent in V and the one just defined, are distinguished by the elements to which they apply (similar observation for the corresponding norms).

In A we define a product as follows:
$$(a, u)(b, v) = (a, b - < u, v >, av + bu + u \times v)$$
It is routine, to verify that with this product, A becomes an $\mathbb{R}$ −álgebra with identity $(1, 0)$. We will prove that A is normed, that is, that:
$$N((a, u)(b, v)) = N(a, u)N(b, v)$$
In fact, from (3) follows: $N(a, u) = a^2 + N(u)$, then:
$$N((a, u)(b, v)) = (ab - < u, v >, av + bu + u \times v) =$$
$$= a^2b^2 - 2ab < u, v > + < u, v >^2 + N(av + bu + u \times v) \ (4)$$
but, since by b), u and v are ortogonal to $u \times v$, and by condition c), we obtain:
$$N(av + bu + u \times v) = a^2N(v) + b^2N(u) + 2ab < u, v$$
$$> + N(u)N(v) - < u, v >^2$$
and replacing in (4), we have:
$$N((a, u)(b, v)) = (a^2 + N(u))(b^2 + N(v)) = N(a, u)N(b, v)$$
Finally, since A is a real normed algebra of dimension $m + 1$, by Hurwitz theorem we must have: $m + 1 = 1, 2, 4$ or 8.∎

By analogy with the ring $\mathbb{Z}_i$ of Gaussian integers, we will study the Euclideanity, with the norm as Euclidean metric, of the quaternions with integral coordinates, that is of:

$$L = \{a_0 + a_1 i + a_2 j + a_3 k \in \mathbb{H} \mid a_t \in \mathbb{Z}, t = 0,1,2,3\}.$$

This is clearly a subring of $\mathbb{H}$ called the ring of Lipschitz quaternions. Since it is not commutative, we must distinguish between Euclideanity on the left and on the right. We will proceed on the right.

Given $x, y \in L$ with $y \neq 0$, we want to find out if there exist $q, r \in L$ such that: $x = yq + r$, and $N(r) < N(y)$

As in the case of Gaussian integers, we can consider $y^{-1}x$ and take q as a quaternion, such that its coordinates are "the" most near integers to the respective coordinates of $y^{-1}x$, and we have:

$$y^{-1}x = qs$$

Where $s = s_0 + s_1 i + s_2 j + s_3 k$ verifies $-\frac{1}{2} \le s_t < \frac{1}{2}$,, then $x = yq + ys$ and, putting $ys = r$, we have that $r \in L$ since $r = x - yq, x, y, q \in L$ and

$$N(r) = N(y)N(s) \le N(y)$$

Since $N(s) = s_0^2 + s_1^2 + s_2^2 + s_3^2 \le 1$. However, to have Euclideanity, it is neccesary to have $N(s) < 1$, being the unfavorable case that in which: $|s_t| = \frac{1}{2}$ for all t. In such case, if we take:

$$\varepsilon = \frac{1}{2}(1 + i + j + k)$$

It follows that $s + \varepsilon \in L$, These considerations lead us to consider instead of L, the set:

$$H = L \cup (L + \varepsilon)$$

Where $L + \varepsilon = \{x + \varepsilon \mid x \in L\}$

It is straightforward to verify that H is a subring of $\mathbb{H}$ (for example: $i\varepsilon = (1 + j) + \varepsilon = \varepsilon k$; $\varepsilon^2 = -1 + \varepsilon$, (note that any quaternion x satisfies $x^2 - (x + \bar{x})x + x\bar{x}$, and obviously $\varepsilon + \bar{\varepsilon} = 1, \varepsilon\bar{\varepsilon} = N(\varepsilon) = 1$), etc.) called the ring of Hurwitz quaternions.

We have:

Proposition: a) $x \in H \Rightarrow \exists y \in L$ such that $N(y) = N(x)$.

b) $N(x) \in \mathbb{Z}$ for any $x \in H$.

c) If $x, y \in H$ with $y \neq 0$, then there exist $q, r \in H$ such that:
$$x = yq + r \text{ and } N(r) < N(y)$$
d) $x \in H \Rightarrow \bar{x} \in H$.

Proof: a) Assume $x \in L + \varepsilon$, for if $x \in L$ there is nothing to prove. If $x\varepsilon \in L$, as $N(\varepsilon) = 1$, we have $N(x) = N(x\varepsilon)$ and we are done. If $x\varepsilon \in L + \varepsilon$, then $x\varepsilon^2 = x\varepsilon - x \in L$, and $N(x\varepsilon^2) = N(x)$.

$x\varepsilon^2 = x\varepsilon - x \in L$, and $N(x\varepsilon^2) = N(x)$.

b) Follows immediately from a).

c) Let, as before, q a quaternion such that its coordinates are most near integers to the respective coordinates of $y^{-1}x$, so that $y^{-1}x = q + s$, where $s = s_0 + s_1 i + s_2 j + s_3 k$, verifies: $|s_t| \leq \frac{1}{2}$ for any $t = 0,1,2,3$, so that $N(s) \leq 1$. If for some t we have $|s_t| \leq \frac{1}{2}$, then putting $r = ys$ we obtain c). If, on the contrary, we have $|s_t| \leq \frac{1}{2}$ for $t = 0,1,2,3$, then $s + \varepsilon \in L$ and it follows that $q + s = q - \varepsilon + s - \varepsilon \in H$, and $x = y(q + s)$.

d) The result is obvious if $x \in L$, so we assume that $x \in L + \varepsilon$, that is $x - \varepsilon \in L$, then $\bar{x} - \bar{\varepsilon} \in L$, but $\bar{\varepsilon} + \varepsilon = 1$, so that $\bar{x} - \varepsilon = \bar{x} - \bar{\varepsilon} + 2\varepsilon - 1 \in L$, that is $\bar{x} \in L + \varepsilon$. ∎

In any ring A, The unities are the elements of the ring that have a two sided inverse, that is $x \in A$ is a unit iff there exists $y \in A$ such that $xy = yx = 1$. The set of unities of A is denoted by $U(A)$.

Proposition: a) In both L and H, the unities are the elements of norm 1.

b) $U(L) = \{\pm 1, \pm i \pm j \pm k\}$ and so there are 8 unities in L.

c) $U(H) = \{u, \varepsilon u, \varepsilon^2 | u \in L\}$ and so there are 24 unities in H.

Proof: a) If u is a unity in L (respectively in H), there exists $v \in L$ (respectively $v \in H$), such that $uv = vu = 1$, then $1 = N(uv) = N(u)N(v)$ and as, in both cases, the norms are integers, we must have $N(u) = 1$. Conversely, if $N(u) = 1$, since $N(u) = u\bar{u}$, by d) of the proposition above, $\bar{u}$ is, in each case, the inverse of u.

b) Let $u = a_0 + a_1 i + a_2 j + a_3 k \in L$ such that $N(u) = 1$, that is $a_0{}^2 + a_1{}^2 + a_2{}^2 + a_3{}^2 + 1$, and as the a_t are integers the result follows readily.

c) Let $x \in H$ with $N(x) = 1$. If $x \in L$, then $x \in U(L)$. If $x \in L + \varepsilon$, we consider two cases: $\bar\varepsilon x \in L$ or $\bar\varepsilon x \in L + \varepsilon$. If $\bar\varepsilon x \in L$, taking $u = \bar\varepsilon x$, we have $x = \varepsilon u$ with $u \in U(L)$. If $\bar\varepsilon x \in L + \varepsilon$, then $\bar\varepsilon x - x = \bar\varepsilon^2 x \in L$, and taking $u = \bar\varepsilon^2 x$ we obtain $x = \varepsilon^2 u$ u with $u \in L$. So that result the 24 unities enounced above (they are distinct, because $\varepsilon \notin L$).∎

Lemma: If $p \in \mathbb{Z}$ is a prime, then there are integers a, b, m such that:
$$1 + a^2 + b^2 = mp \ \ with \ \ 0 < m < p$$

Proof: For $p = 2$ the result is clear. Let $p = 2n + 1$ an odd prime and consider the following subsets of $\mathbb{Z}_p$:
$$A = \{\bar{a}^2 \,|a = 0, 1, ..., n\}$$
$$B = \{-1 - \bar{b}^2 | b = 0, 1, ..., n\}$$
A has $n + 1$ elements, for if $\bar{a}^2 = \bar{b}^2$, with $0 \leq a, c \leq n = \frac{p-1}{2}$ then $\bar{a} = \bar{c}$ or $\bar{a} = -\bar{c}$, but as $-n \leq a - c \leq n$ and $0 \leq a + c \leq 2n = p - 1$, it follow that $a = c$ or $a = -c$, and in the last case $a = c = 0$. It follows that B also has $n + 1$ elements. Since $\mathbb{Z}_p$ has $p = 2n + 1$ elements, A and B must have an element in common, so there exist a, b with $0 \leq a, b \leq n$ such that: $a^2 + 1 + b^2 = mp$, but: $mp = 1 + a^2 + b^2 \leq 1 + 2n^2 < p^2$. So $0 < m < p$.∎

Taking $x = 1 + ai + bj$, we have $N(x) = 1 + a^2 + b^2$, and by the lemma above, we obtain:

Corollary: If $p \in \mathbb{Z}$ is a prime,then there exist a positive integer m, and an $x \in H$ such that:
$N(x) = mp$ and $0 < m < p$.∎

Theorem: (Four squares theorem) Any positive integer is a sum of four squares.

Proof: Since a sum of four squares is the norm of a Lipschitz quaternion, and the norm is multiplicative, it is enough to prove that any prime is the norm of a Lipschitz quaternion or, by a proposition above, that it is the norm of a Hurwitz quaternion.

Let $p \in \mathbb{Z}$ a prime, by the corollary there exist $x \in H$, and $m \in \mathbb{Z}$, such that:

$$N(x) = x\bar{x} = mp. \text{ And } 0 < m < p$$

Between all of such m there is a smallest one, assume it is the one above. We will we prove that we must have $m = 1$, and then $N(x) = p$.

By Euclideanity, there exist $q, r \in H$ such that:

$$p = xq + r \text{ and } N(r) < N(x) = mp,$$

Then $\bar{x}p = mpq + \bar{x}r$, So then $pu = \bar{x}r$ for some $u \in H$ ($u = \bar{x} - mq$). Taking norms, we obtain:

$$pN(u) = mN(r)$$

Then, as m and p are coprime, we obtain that p divides N(r), so then:

$$N(r) = m'p \text{ with } 0 \leq m' < m$$

But $N(r) = 0$, implies $r = 0$ so that $p = xq$ and, taking norms, $p^2 = mpq$, then $m = 1$. On the contrary, if $N(r) \neq 0$ we obtain $m' > 0$, wich contradicts the election of m. ∎

The Four Square Theorem was stated by Bachet in 1621 in his commented translation of Diophant's Arithmetic. Fermat asserted that he had a proof, but the first published proof was by Lagrange in 1770, improving early results obtained by Euler.

Legendre in 1798, proved that a positive integer is a sum of three squares if, and only if, it is of the form $4^k(8m + 7)$, although his proof had some gaps, saved later by Gauss.

Waring generalized the theorem to a sum of higher exponents. In 1770 he stated: every positive integer is the sum of 4 squares, 9 cubes, 19 fourth powers, and so on. Waring arrived to those conclusions by empirical methods.

His "and so on" can be interpreted as: For every k there is an s such that any natural number can be written as a sum of at most s powers of exponent k. Stated like thus, was proved by Hilbert in 1909. Another problem is to determine an s for each k.

APPENDIX 4

EXPONENTIAL AND TRIG-ONOMETRICAL FUNC-TIONS

1 - MEANS

The Pythagoreans introduced the notions of arithmetic, geometric and harmonic means, related to their study of a vibrating string and then they gave a general definition of a mean as follows (with current notations):given two "magnitudes" (positive real numbers) a and b with $a > b$, we say that a magnitude x is a *mean of* a and b if we have:

$$\frac{a - x}{x - b} = \frac{u}{v}$$

where u, v independently take the values a, b, x. So we have the means:

$$\frac{a-x}{x-b} = \frac{a}{a} \quad \text{(arthmetic) (1)}$$

$$= \frac{a}{x} \quad \text{(geometric)(2)}$$

$$= \frac{a}{b} \quad \text{(harmonic) (3)}$$

From (1) we obtain: $x = \frac{a+b}{2}$, from (2): $x = \sqrt{ab}$ and from (3): $x = \frac{2ab}{a+b}$

The following inequalities between these means are valid (exercise):

$$\frac{2ab}{a + b} \leq \sqrt{ab} \leq \frac{a + b}{2}$$

The means of two magnitudes can be generalized, in an obvious way, to means of n positive (> 0) real numbers $x_1, x_2,...,x_n$ as follows:

Arithmetic: $\dfrac{x_1 + \cdots + x_n}{n}$

Geometric: $\sqrt[n]{x_1 \ldots x_n}$

Harmonic: $\dfrac{n x_1 \ldots x_n}{x_1 + \cdots + x_n}$

All of them are used in statistics as concentration measures.

The inequalities between the means can also be generalized. We will start with the fundamental inequality between the geometric and arithmetic means, already proved in exercise 18 chap. 6, but from which we will give here other proofs.

Theorem: Let $x_1, x_2, ..., x_n$ be n positive real numbers. The following inequality is valid:

$$\sqrt[n]{x_1 \ldots x_n} \leq \frac{x_1 + \cdots + x_n}{n}$$

And the equality is valid only in case that $x_i = x_j \ \forall \, i, j$.

This theorem will be a corollary of the following

Proposition: Let $x_1, x_2,..., x_n$ be n positive real numbers. If the x_i vary with a fixed sum $S = x_1 + \cdots + x_n$, then the product $x_1 \ldots x_n$ reaches its máximum only in case $x_i = x_j \ \forall \, i, j$ (and, in such case: $x_i = \dfrac{S}{n} \ \forall \, i$).

Proof: If there were any $x_i > \dfrac{S}{n}$, it must exist some $x_j < \dfrac{S}{n}$. Let $x_1 > \dfrac{S}{n}$ and $x_2 < \dfrac{S}{n}$.

Putting:

$$y_1 = \frac{S}{n} \quad \text{and} \quad y_2 = x_1 + x_2 - \frac{S}{n}$$

The sum of the n numbers $y_1, y_2, x_3, \ldots, x_n$ remains S, but

$$y_1 y_2 x_3 \ldots x_n > x_1 x_2 x_3 \ldots x_n$$

Since from $x_1 > \dfrac{S}{n}$ and $x_2 < \dfrac{S}{n}$ follow:

$$x_1 \left(\frac{S}{n} - x_2 \right) > \frac{S}{n} \left(\frac{S}{n} - x_2 \right)$$

Then:

$$y_1 y_2 = \frac{S}{n} \left(x_1 + x_2 - \frac{S}{n} \right) > x_1 x_2. \blacksquare$$

The theorem follows from the proposition, since given n positive numbers $x_1, \ldots, x_n$, if we call S to its sum, by the above proposition, if they are not all equal, then:

$$x_1 \ldots x_n < \frac{S^n}{n^n}$$

Which is equivalent to the inequality of the means. ∎

It is clear that the theorem implies the proposition, so they are equivalent. The following proposition is also equivalent to them.

Proposition: Let $x_1, \ldots, x_n$ be positive real numbers números reales positivos. If the product $P = x_1 \ldots x_n$ is fixed, then the sum $x_1 + \cdots + x_n$ reaches its minimum if and only if $x_i = x_j = \sqrt[n]{P} \ \forall \ i, j$.

Prop.: As we have said, this proposition is equivalent to the theorem. However we will give a proof dual to the proof given in the proposition above. If they are not all equal, we must have for example: $x_1 > \sqrt[n]{P}$ and $x_2 < \sqrt[n]{P}$ and putting $y_1 = \sqrt[n]{P}$, $y_2 = \frac{x_1 x_2}{\sqrt[n]{P}}$, we have $y_1 y_2 = x_1 x_2$, but since:

$$x_1 \left(1 - \frac{x_2}{\sqrt[n]{P}} \right) < \sqrt[n]{P} - x_2$$

We obtain: $y_1 + y_2 = \sqrt[n]{P} + \frac{x_2}{\sqrt[n]{P}} < x_1 + x_2$. ∎

Note that if $n = 2$, from prop.1 follows that between all rectangles with given perimeter, the square is the one with maximun área; and from prop.2 follows that between all rectangles with given area, the square has minimum perimeter.

Similar observations are valid for $n = 3$ and, with the relevant definitions, for any dimension.

Corollary: (Inequality between harmonic and geometric means) If $x_1, \ldots, x_n$ are real numbers > 0, then:

$$\frac{n x_1 \ldots x_n}{x_1 + \cdots + x_n} \le \sqrt[n]{x_1 \ldots x_n}$$

And the equality is valid only if $x_i = x_j \ \forall \ i, j$.

Proof: Just apply the theorem to the n numbers $\frac{1}{x_1}, \ldots, \frac{1}{x_n}$. ∎

The application of the inequality between the means is a recurring theme in the mathematical olimpiads, here is an example:

Example: (British Mathematical Olimpiad 1996) Find all solutions in positive real numbers a, b, c, d of the system of equations:

$$12 = a + b + c + d$$

$$abcd = 27 + ab + ac + ad + bc + bd + cd.$$

By the inequality between the geometric and arithmetic means of the numbers ab, ac, ad, bc, bd, cd and by the second equation, we have:

$$6\sqrt{abcd} \le abcd - 27$$

From where $\sqrt{abcd} \ge 9$, then:

$$3 \le \sqrt[4]{abcd} \le \frac{a + b + c + d}{4} = 3$$

So that $a = b = c = d = 3$ is the only solution of the system.

Exercise: a) The *quadratic mean* of n positive numbers is the square root of the arithmetic mean of the squares of those numbers. Prove the following inequality between the arithmetic and quadratic means of the positive real numbers $x_1, \dots, x_n$:

$$\frac{x_1 + \cdots + x_n}{n} \le \sqrt{\frac{x_1^2 + \cdots + x_n^2}{n}}$$

(hint: use Cauchy's inequality, ex. 14 chap 3)

b) If a, b are positive numbers such that $a + b = 1$, prove:

$$\left(a + \frac{1}{a}\right)^2 + \left(b + \frac{1}{b}\right)^2 \ge \frac{25}{2}$$

And generalize it.

2 - BERNOULLI'S INEQUALITY

In exercise 5, b, chap. 3, we have seen that if $x \in \mathbb{R}$ and $x \ge 0$, then for any $n \in \mathbb{N}$, the following inequality is valid:

$$x^n \ge 1 + n(x - 1)$$

The generalizations of that result to a real positive exponent, are the, so called, Bernoulli's inequalities.

Theorem: (Bernoulli's inequality) If $x > 0$, $c \ge 0$ are real numbers with $x \ne 1$, then:

$$c < 1 \Rightarrow x^c < 1 + c(x - 1) \qquad (1)$$
$$c > 1 \Rightarrow x^c > 1 + c(x - 1) \qquad (2)$$

Proof: We will prove first (1) assuming c rational. Let then $c = \frac{m}{n}$ with m, n natural numbers and $m < n$. By the inequality between arithmetic and geometric means applied to the n numbers: $x, x, \dots, x, 1, 1, \dots, 1$ with m xs and $n - m$ ones, we have:

$$x^{\frac{m}{n}} = \sqrt[n]{x^m} < \frac{mx + n - m}{n} = 1 + \frac{m}{n}(x - 1)$$

We have then that (1) is valid with c rational. Assume now that that c is real with $0 < c < 1$ and consider two subcases: $x < 1$ and $x > 1$.

If $x < 1$ we take $r \in \mathbb{Q}$ such that $0 < r < c$, then:

$$x^c < x^r < 1 + r(x - 1) < 1 + c(x - 1)$$

Where we have used the validity of (1) for a rational r, and the fact that the funtion $f(t) = x^t$ is strictly decreasing if $x < 1$ (theorem 6.3 chap. 6).

Let now $x > 1$. If If it were $x^c > 1 + c(x - 1)$, by theorem 5.3. chap. 6, there would be a rational r such that:

$$x^c > x^r > 1 + c(x - 1)$$

Since the function $f(t) = x^t$ is strictly increasing as $x > 1$ we would have: $c > r > 0$, but then, as $1 + c(x - 1) > 1 + r(x - 1)$ it would result $x^r > 1 + r(x - 1)$ which contradicts the rational case proved above. We must have then:

$$x^c \leq 1 + c(x - 1) \ \forall c\colon 0 < c < 1 \quad (3)$$

Take $\varepsilon > 0$ such that the following two conditions are verified: $0 < c - \varepsilon$ and $c + \varepsilon < 1$. By the inequality between the geometric and arithmetic means of $x^{c-\varepsilon}$ and $x^{c+\varepsilon}$, we have:

$$x^c < \frac{x^{c-\varepsilon} + x^{c+\varepsilon}}{2}$$

But by (3) we have:

$$x^{c-\varepsilon} \leq 1 + (c - \varepsilon)(x - 1)$$
$$x^{c+\varepsilon} \leq 1 + (c + \varepsilon)(x - 1)$$

Then $x^c < 1 + c(x - 1)$.

So that the case (1) is proved. For the case (2), assumethat $c > 1$ and $x > 0$. Putting $y = 1 + c(x - 1)$, we have $y > 0$ and by case (1):

$$y^{\frac{1}{c}} < 1 + \frac{1}{c}(y - 1) = x$$

Then $x^c > y = 1 + c(x - 1)$. ∎

Bernoulli's inequalities will be used in the following sections to base the properties of the exponential function.

3 - NUMBER e

We have defined the number e in chap. 6 as the supremum of the set:

$$\left\{\left(1+\frac{1}{n}\right)^n / n \in \mathbb{N}\right\}$$

We are going to see that such supremum is also the supremum of the set:

$$\left\{\left(1+\frac{1}{x}\right)^x / x \in \mathbb{R}, x > 0\right\}$$

In order to do that, we will prove:

Proposition: For all real number $x > 0$, let f, g be the functions defined by:

$$f(x) = \left(1+\frac{1}{x}\right)^x \quad , \quad g(x) = \left(1+\frac{1}{x}\right)^{x+1}$$

Then: f is strictly increasing, g is strictly decreasing, the supremum of the first equals the infimum of the second and it is the number e.

Dem.: Let y such that $0 < x < y$. By one of Bernoulli's inequalities, we have:

$$\left(1+\frac{1}{x}\right)^{\frac{x}{y}} < 1+\frac{x}{y}\frac{1}{x} = 1+\frac{1}{y}$$

And, since the exponetial function of base > 1 is strictly increasing, we obtain:

$$\left(1+\frac{1}{x}\right)^x < \left(1+\frac{1}{y}\right)^y \qquad (y > x > 0)$$

That is f is strictly increasing.

Being, as before, $y > x > 0$, by the same Bernoulli's inequality, we have:

$$\left(\frac{x}{x+1}\right)^{\frac{x+1}{y+1}} < 1+\frac{x+1}{y+1}\left(\frac{x}{x+1}-1\right) = 1 - \frac{1}{y+1} = \frac{y}{y+1}$$

And as the exponential function of base < 1 is strictly decreasing:

$$\left(\frac{x}{x+1}\right)^{x+1} > \left(\frac{y}{y+1}\right)^{y+1} \quad \text{or} \quad \left(1+\frac{1}{x}\right)^{x+1} < \left(1+\frac{1}{y}\right)^{y+1}$$

That is: g is strictly decreasing. We have then that f is strictly increasing, that g is strictly decreasing, and that $f(x) \le g(x)$ $\forall\, x$ (by definitions of f and g). Let:

$$A = \left\{\left(1 + \tfrac{1}{n}\right)^n / n \in \mathbb{N}\right\};\ B = \{f(x)\,/x \in \mathbb{R}, x > 0\};\ C = \{g(x)/x \in \mathbb{R}, x > 0\}$$

By definition $e = supA$. Moreover, B is bounded from above, since by Arquimedianity for any $x \in \mathbb{R}$, with $x > 0$ there is an $n \in N$ such that $n > x$, then $f(x) < f(n) \le e$. Put then $e' = supB$, we have just proved that $e' \le e$ and as $A \subset B$, it follows that $e \le e'$. We conclude that: $e = e'$.

Furthermore, C is bounded from below, as any element $f(x)$ of B is a lower bound of C, in fact if $g(x') \in C$ we have: if $x \le x'$: $f(x) \le f(x') \le g(x')$ and if $x > x'$: $f(x) \le g(x) \le g(x')$. There is then $e'' = infC$, and the reasoning above proves that $e' \le e''$. If it were $e'' > e'$, taking $x > \frac{e'}{e'' - e'}$, it would result:

$$e'' - e' > \frac{e'}{x} \ge \left(1 + \frac{1}{x}\right)^x \frac{1}{x} = g(x) - f(x)$$

What is in contradiction with:

$$f(x) \le e' < e'' \le g(x)$$

then $e' = e''$. ∎

4 - REAL EXPONENTIAL FUNCTION

We refer here to the exponential funtion of base e.

Proposition: (Inequality of the exponential function) $\forall x \in \mathbb{R}$ with $x \ne 0$, we have:

$$e^x > 1 + x$$

Proof: 1) $x > 0$: put $y = \frac{1}{x}$, then from $\left(1 + \frac{1}{y}\right)^y < e$, follows $(1 + x)^{\frac{1}{x}} < e$ from where

$$1 + x < e^x$$

2) $-1 < x < 0$: putting $y = -\frac{x+1}{x}$, we have $y > 0$ and $y + 1 = -\frac{1}{x}$ then:

$$e < \left(1 + \frac{1}{y}\right)^{y+1} = \left(\frac{y+1}{y}\right)^{y+1} = \left(\frac{1}{x+1}\right)^{-\frac{1}{x}}$$

And as $-x > 0$ it results $e^{-x} < \frac{1}{x+1}$ so then $e^x > x + 1$.

3) $x \le -1$: as $1 + x \le 0$ and $e^x > 0$ we have $e^x > x + 1$. ∎

Let us show the usefulness of the above inequality, and at the same time the singularity of the number e, when solving a problem of extremes, posed by Steiner:

Example: The function $x^{\frac{1}{x}}$ $(x > 0)$ takes its máximum value for $x = e$. In other words: $e^{\frac{1}{e}} > x^{\frac{1}{x}}$ $\forall x > 0$.

In fact, by the inequality of the exponential function:

$$e^{\frac{x-e}{e}} > 1 + \frac{x-e}{e} = \frac{x}{e} \quad \forall x \ne e$$

So then $e^{\frac{x}{e}}\frac{1}{e} > \frac{x}{e}$ or $e^{\frac{x}{e}} > x^{\frac{1}{x}}$.

From now on we will use notions learned in a first course of Calculus: functions of a real variable, limits, continuous and derivable functions, extreme values, Taylor development, sequences and series.

Theorem: The exponential function defined by $f(x) = e^x$ is derivable, it is its own derivative, and admits the following expansion series $\forall x \in \mathbb{R}$:

$$e^x = 1 + x + \frac{x^2}{2!} + \cdots + \frac{x^n}{n!} + \cdots$$

Dem.: By the inequality of the exponential function, if $x \ne 0$ we have $e^{-x} > 1 - x$, then $1 > (1 - x)e^x$, i.e. $e^x < 1 + xe^x$. We have then:

$$1 + x < e^x < 1 + xe^x \quad \forall x \ne 0 \quad (1)$$

From (1) follows that $e^x \to 1$ when $x \to 0$ anl also:

$$1 < \frac{e^x - 1}{x} < e^x \quad \forall x \ne 0$$

Then $\frac{e^x - 1}{x} \to 1$ when $x \to 0$. So then:

$$\frac{e^{x+h} - e^x}{h} = e^x \frac{e^h - 1}{h} \to e^x \quad \text{when } h \to 0$$

It follows that f is derivable $\forall x$ and it is its own derivative, so it is indefinitely derivable and has a Taylor-MacLaurin expansion:

$$e^x = 1 + x + \cdots + \frac{x^{n-1}}{(n-1)!} + \frac{x^n}{n!}e^{\theta x}, \quad (0 < \theta < 1) \quad (2)$$

Since $(n - 1)!^2 \ge n^{n-2}$ for $n \ge 2$ (this can be obtained, for example, by induction since from $(n - 1)!^2 \ge n^{n-2}$ it follows that $n!^2 \ge n^n$

and it remains to prove that $n^n \geq (n+1)^{n-1}$ and this is is is equivalent to $n+1 \geq \left(1 + \frac{1}{n}\right)^n$ we have:

$$\left|\frac{x^n}{n!}e^{\theta x}\right| \leq \left(\frac{|x|}{\sqrt{n}}\right)^n e^{|x|} \leq \frac{1}{2^n}e^{|x|} \quad (3)$$

Where the last inequality is valid taking n such that $\sqrt{n} \geq 2|x|$. So then $\frac{x^n}{n!}e^{\theta x} \to 0$ when $n \to \infty$ and we obtain the expansion of the statement. ∎

Let us show another way to solve the problem posed in the example above. Let $f(x) = x^{\frac{1}{x}} = e^{\frac{1}{x}\log x}$, where log is the Neperian logarithm of base e. We have $f'(x) = e^{\frac{1}{x}\log x}\left(\frac{1}{x^2} - \frac{1}{x^2}\log x\right)$ and this derivative is zero only if $x = e$, moreover $f'(x) > 0$ if $0 < x < e$ and $f'(x) < 0$ if $x > e$, then f is strictly increasing if $x < e$ and it is strictly decreasing if $x > e$. Also we obtain: $r > s \geq e \Rightarrow r^s < s^r$ which generalizes the result of exercise 22 g chap. 3.

The number e is trascendent, i.e there is no non null polynomial function with coefficients in $\mathbb{Q}$ such that e is a root of it. That result was proved by Hermite in 1873. Here we will only prove that e is irrational.

Proposition: e es irracional.

Dem.: Let $S_{n-1}(x) = 1 + x + \cdots + \frac{x^{n-1}}{(n-1)!}$, according with (2):

$$0 < |e - S_{n-1}(1)| \leq \frac{e}{n!}$$

But as $(n-1)! \, S_{n-1}(1)$ is an integer, if e were rational, it would be enough to take n so big to have that $(n-1)! \, e$ be an integer and it would result:

$$0 < (n-1)! \, |e - S_{n-1}(1)| \leq \frac{e}{n}$$

That is, would obtain an integer that is > 0 and $< \frac{e}{n}$ which is a contradiction if $n > e$. ∎

The expansion in series of the exponential function, suggests to extend the definition of the function to complex exponent z as follows:

$$e^z = 1 + z + \frac{z^2}{2!} + \cdots + \frac{z^n}{n!} + \cdots$$

But that presupposes a knowledge of complex series and its convergence. Those and other prerrequisites will be developed in the next section.

In this section we wiil treat with two clases of complex valued functions, those with $\mathbb{N}$ as domain (that is sequences) and those of a real variable.

A sequence of complex numbers is a function $f: \mathbb{N} \to \mathbb{C}$, if we put $z_n = f(n)$, we will denote it by (z_n) or $(z_n)_n$ (the last notation in case that inside the brakets it would be more than one index). As in the real case, a sequence (z_n) of complex numbers is said to be convergent iff there exists $z \in \mathbb{C}$ such that $\forall \varepsilon$ real with $\varepsilon > 0$, there exists $n_0 \in \mathbb{N}$ such that:

$$n \geq n_0 \implies |z_n - z| < \varepsilon$$

And in such case we write $z_n \to z$ or $\lim_{n \to \infty} z_n = z$.

Defining as usual, the sum and product of two sequences (z_n), (w_n) of complex numbers, by:

$$(z_n) + (w_n) = (z_n + w_n)$$
$$(z_n) \cdot (w_n) = (z_n w_n)$$

The usual results on uniqueness and limits of the sum and the product, are valid, repeating the proofs for the real case, or using the following result, the proof of which is left an exercise.

Proposition: Let $z_n = a_n + b_n i$; $z = a + bi$ with a_n, b_n, a, b real numbers. We have:

$$z_n \to z \iff (a_n \to a \ y \ b_n \to b). \blacksquare$$

Also it is valid Cauchy's convergent criteria:

Proposition: A sequence (z_n) of complex numbers, is convergent if, and only if, it is a Cauchy sequence, that is, for each $\varepsilon > 0$ there exists n_0 such that:

$$n, m \geq n_0 \implies |z_m - z_n| < \varepsilon$$

Proof: Putting $z_n = a_n + b_n i$ with $a_n, b_n \in \mathbb{R}$, it follows readily that (z_n) is Cauchy if, Propositionand only if, (a_n) and (b_n) are both Cauchy, and the result follows from the correspondent result for real sequences. $\blacksquare$

As in the real case, given a complex sequence (z_n), the correspondent series, is defined as the sequence (S_n) of the partial sums $S_n = z_1 + \cdots + z_n$.

Proposition: Any absolutely convergent series $\sum z_i$ of complex numbers is convergent.

Proof:That the series converges absolutely means that the series $\sum|z_i|$ converges, so it is Cauchy, that is given any $\varepsilon > 0$, there is an n_0 such that:

$$m > n \geq n_0 \implies \left|\sum_{i=n}^{m}|z_n|\right| < \varepsilon$$

But by the triangle inequality $|\sum_{i=n}^{m} z_i| \leq \sum_{i=n}^{m}|z_i|$, and this implies that the series $\sum z_i$ is a Cauchy series, so that it is convergent. ∎

Consider now a function f of a real variable with values in $\mathbb{C}$, the definition of limit when $x \to a$ $(a \in \mathbb{R})$, is extended to this case, and we have the basic results on limits repeating verbatim the proofs of the real case. Putting:

$$f(x) = g(x) + i\, h(x)$$

with $g(x), h(x) \in \mathbb{R}$, g and h are then real funtions and we have: $l = \lim_{x \to a} f(x)$ exists, and only if, $l_1 = \lim_{x \to a} g(x)$ and $l_2 = \lim_{x \to a} h(x)$ exist, and in such case we have: $l = l_1 + i\, l_2$.

6 - COMPLEX EXPONENTIAL FUNCTION

Proposition: The series $1 + z + \dfrac{z^2}{2!} + \cdots + \dfrac{z^n}{n!} + \cdots$ is convergent $\forall z \in \mathbb{C}$.

Proof: As any absolutely convergent series is convergent, it will be enough to prove that the series $1 + |z| + \cdots + \dfrac{|z|^n}{n!} \ldots$ is convergent, but that follows for the considerations on the exponential real function that we have seen above. ∎

In view of the last proposition, we define: $e^z = 1 + z + \dfrac{z^2}{2!} + \cdots + \dfrac{z^n}{n!} + \cdots$ for any $z \in \mathbb{C}$ and we have:

Theorem: $z, w \in \mathbb{C} \implies e^z e^w = e^{z+w}$.

Dem.: Putting $S_n(z) = 1 + z + \dfrac{z^2}{2!} + \cdots + \dfrac{z^n}{n!}$, we have:

$$S_n(z)S_n(w) = \left(\sum_{i=0}^{n}\frac{z^i}{i!}\right)\left(\sum_{j=0}^{n}\frac{w^j}{j!}\right) = \sum_{k=0}^{n}\sum_{i+j=k}\frac{z^i w^j}{i!\,j!} = \sum_{k=0}^{n}\frac{(z+w)^k}{k!}$$
$$= S_n(z+w)$$

then:

$$e^z e^w = \lim_{n \to \infty} S_n(z) \lim_{n \to \infty} S_n(w) = \lim_{n \to \infty} \big(S_n(z) S_n(w)\big)$$
$$= \lim_{n \to \infty} S_n(z + w) = e^{z+w} \; \blacksquare$$

In particular if $z = x + iy$, we have $e^z = e^x e^{iy}$.

7 - IMAGINARY EXPONENTIAL FUNCTION

If $\alpha \in \mathbb{R}$, we have by definition:

$$e^{i\alpha} = 1 + i\alpha + \frac{(i\alpha)^2}{2!} + \cdots + \frac{(i\alpha)^n}{n!} + \cdots$$

The function that it defines, is called *imaginary exponential function.*

Proposition: The function defined by: $f(\alpha) = e^{i\alpha}$ is derivable $\forall\, \alpha \in \mathbb{R}$ and its derivative is $f'(\propto) = i e^{i\alpha}$. Moreover we have the following expression of $e^{i\alpha}$ in terms of its real an imaginary parts:

$$e^{i\alpha} = \left(1 - \frac{\alpha^2}{2!} + \cdots + (-1)^n \frac{\alpha^{2n}}{(2n)!} + \cdots\right)$$
$$+ i \left(\alpha - \frac{\alpha^3}{3!} + \cdots + (-1)^n \frac{\alpha^{2n+1}}{(2n+1)!} + \cdots\right)$$

Proofs: We define for $\alpha \in \mathbb{R}$: $\cos \alpha = Re(e^{i\alpha})$, $\operatorname{sen} \alpha = Im(e^{i\alpha})$, that is:

$$\cos \alpha = 1 - \frac{\alpha^2}{2!} + \cdots + (-1)^n \frac{\alpha^{2n}}{(2n)!} + \cdots; \qquad \operatorname{sen} \alpha$$
$$= \alpha - \frac{\alpha^3}{3!} + \cdots + (-1)^n \frac{\alpha^{2n+1}}{(2n+1)!} + \cdots$$

From where we obtain: $\cos(-\alpha) = \cos \alpha$ and $\operatorname{sen}(-\alpha) = -\operatorname{sen} \alpha$, then $e^{-i\alpha} = \cos \alpha - i \operatorname{sen} \alpha$ is the conjugate of $e^{i\alpha}$, and we have:

$$1 = e^{i\alpha} e^{-i\alpha} = \cos^2\alpha + \operatorname{sen}^2\alpha = |e^{i\alpha}| \quad (1)$$

On the other hand, from $e^{i(\alpha+\beta)} = e^{i\alpha} e^{i\beta}$ follows:

$$\cos(\alpha + \beta) = \cos\alpha \, \cos\beta - \operatorname{sen}\alpha \, \operatorname{sen}\beta \quad (2)$$
$$\operatorname{sen}(\alpha + \beta) = \cos\alpha \, \operatorname{sen}\beta + \cos\beta \, \operatorname{sen}\alpha \quad (3)$$

In a similar way all the relations of chap. 7 on trigonometric functions are obtained. From (1) follows:

$$-1 \le \cos\alpha \le 1 \;,\; -1 \le \operatorname{sen}\alpha \le 1$$

Moreover, from $D(e^{i\alpha}) = i e^{i\alpha}$ (where D means derivative with respect to α) results that $\cos\alpha$ and $\operatorname{sen}\alpha$ are derivable, and we have:

$$icos\alpha - sen\alpha = ie^{i\alpha} = D\left(e^{i\alpha}\right) = D(cos\alpha) + iD(sen\alpha),$$

then:

$$D(cos\alpha) = -sen\alpha \;\; ; \;\; D(sen\alpha) = cos\alpha.$$

From the series expansión of $cos\alpha$ follows, using results on alternate series, that:

$$cos\alpha \le 1 - \frac{\alpha^2}{2!} + \frac{\alpha^4}{4!}$$

We are going to prove that inequality in another way, a way from which we will obtain some useful results. Consider the function defined by:

$$f(\alpha) = 1 - \frac{\alpha^2}{2!} + \frac{\alpha^4}{4!} - cos\alpha$$

And let us find its first succesive derivatives:

$$f'(\alpha) = -\alpha + \frac{\alpha^3}{3!} + sen\alpha$$

$$f''(\alpha) = -1 + \frac{\alpha^2}{2!} + cos\alpha$$

$$f^{(3)}(\alpha) = \alpha - sen\alpha$$

$$f^{(4)}(\alpha) = 1 - cos\alpha$$

We have: $f^{(4)}(\alpha) \ge 0$ $\forall\alpha$, and as $f^{(3)}(0) = 0$ and since $f^{(4)}$ is the derivative of $f^{(3)}$, results $f^{(3)}(\alpha) \ge 0$ $\forall\alpha \ge 0$. Repeating the argument, we obtain that for $\alpha \ge 0$: $f''(\alpha), f'(\alpha)$ and $f(\alpha)$ are all ≥ 0. Since f is an even function, it follows that $\forall\alpha$:

$$cos\alpha \le 1 - \frac{\alpha^2}{2!} + \frac{\alpha^4}{4!}$$

Since the polynomial of the second member is null for $\alpha = \sqrt{6 - 2\sqrt{3}} < 1,6$ it must exists an α such that $0 < \alpha < 1,6$ and $cos\alpha = 0$. Call $\frac{\pi}{2}$ to the minimum value of α (it exists as cos is continuous) that satisfies those conditions (this is a definition of π). Furthermore, from $f_4(\alpha) \ge 0$ we obtain:

$$sen\alpha \ge \alpha - \frac{\alpha^3}{6} = \frac{\alpha}{6}\left(\sqrt{6} - \alpha\right)\left(\sqrt{6} + \alpha\right)$$

And as $\sqrt{6} > 1,6$, it follows that $sen\alpha > 0$ if $\alpha \in \left(0, \frac{\pi}{2}\right)$. Since $cos\alpha$ does not take the value 0 in $\left[0, \frac{\pi}{2}\right)$ and as its derivative $-sen\alpha$ is negative in that interval, we conclude that *cos* is strictly decreasing in it, then *cos* is a bijection between $\left[0, \frac{\pi}{2}\right]$ and $[0, 1]$. Moreover, since $sen^2(\alpha) + cos^2(\alpha) = 1$ it follows that $sen\frac{\pi}{2} = 1$. In a similar way, it is proved that sin   is strictly increasing in $\left[0, \frac{\pi}{2}\right]$ and it follows that *sen* is a bijection between $\left[0, \frac{\pi}{2}\right]$ and $[0, 1]$. From (2) and (3), follow:

$$sen\alpha = -cos\left(\alpha - \frac{\pi}{2}\right) \; ; \; cos\alpha = sen\left(\alpha - \frac{\pi}{2}\right)$$

which allow us to extend the study of that functions to the second quadrant, that is when $\alpha \in \left[\frac{\pi}{2}, \pi\right]$, obtaining that *cos* is a bijection between $[0, \pi]$ and $[-1, 1]$. The inverse function of that restriction of *cos* is denoted by *arcos* or cos^{-1}.

From the relations $cos\alpha = cos(-\alpha)$, $sen\alpha = -sen\alpha$, follow the extensión of the study of the trigonometric functions to the third and four quadrants, obtaining in in particular, that *sen* set up a bijection between $[-\pi, \pi]$ and $[-1, 1]$ and its inverse there is denoted by *arcsen*.

From the previous discussion it follows that the function $\alpha \rightarrow e^{i\alpha}$ is a bijection from $[0, 2\pi)$ on the unit circumference: $\{z = x + iy \in \mathbb{C}/x^2 + y^2 = 1\}$ and as $cos(2\pi) = cos0$, $sen(2\pi) = 1$, we have $e^{i2\pi} = e^{i0} = 1$. Then $\left(e^{i2\pi}\right)^n = e^{2\pi n} = 1 \; \forall n \in \mathbb{N}$ and taking inverses that relation is also valid for any integer n, so then:

$$e^{i(\alpha + 2k\pi)} = e^{i\alpha}, \quad \forall \alpha \in \mathbb{R}, \forall k \in \mathbb{Z}$$

Moreover, if $e^{i\alpha} = e^{i\beta}$, putting $\alpha = \alpha_1 + 2r\pi$, $\beta = \beta_1 + 2s\pi$ with $r, s \in \mathbb{Z}$ and $\alpha_1, \beta_1 \in [0, 2\pi)$, it follows that $e^{i\alpha_1} = e^{i\beta_1}$, and since in the interval $[0, 2\pi)$, $e^{i\alpha}$ is injective, we have $\alpha_1 = \beta_1$, so then $\beta = \alpha + 2k\pi$ with $k = s - r \in \mathbb{Z}$. Then the function $e^{i\alpha}$ is periodic with 2π as its minimum period. We conclude that the same is valid with the functions *sen* and *cos*.

To end this section we will prove:

$$cos\alpha = cos\beta \Longleftrightarrow \begin{cases} \beta = \alpha + 2k\pi \quad or \\ \beta = -\alpha + 2k\pi \end{cases}$$

For some $k \in \mathbb{Z}$. In fact, the implication $\Longleftarrow$ is clear, let us show the converse. Let $cos\alpha = cos\beta$, then $sen^2\alpha = sen^2\beta$, that is $sen\alpha = sen\beta$ or $sen\alpha = -sen\beta$. If $sen\alpha = sen\beta$, we have $e^{i\alpha} = e^{i\beta}$ so that $\beta = \alpha + 2k\pi$ for some $k \in \mathbb{Z}$. If $sen\alpha = -sen\beta$ we have $e^{i\alpha} = e^{-i\beta}$ and then $\beta = -\alpha + 2k\pi$ for some $k \in \mathbb{Z}$.

Exercises: 1) Prove that: $sen\alpha = sen\beta \Leftrightarrow \begin{cases} \beta = \alpha + 2k\pi & or \\ \beta = \pi - \alpha + 2k\pi \end{cases}$ for some $k \in \mathbb{Z}$.

2) Prove: $cos\dfrac{\pi}{6} = \dfrac{\sqrt{3}}{2}$.

8 - COMPLEX TRIGONOMETRIC, LOGARITMIC AND EXPONENTIAL FUNCTIONS

TRIGONOMETRIC FUNCTIONS: For $\alpha \in \mathbb{R}$, we have: $cos\alpha = \dfrac{e^{i\alpha} + e^{-i\alpha}}{2}$ and $sen\alpha = \dfrac{e^{i\alpha} - e^{-i\alpha}}{2i}$

So it is natural to extend the definition of the trigonometric functions to a complex argument z by:

$$cos\,z = \frac{e^{iz} + e^{-iz}}{2}\,; \sin z = \frac{e^{iz} - e^{-iz}}{2}$$

with these generalized definitions, the following relations remain valid for: $z, w \in \mathbb{C}$:

$$cos^2z + sen^2z = 1$$
$$cos(z + w) = cos\,z\,cos\,w - sen\,z\,sen\,w$$
$$sen(z + w) = sen\,z\,cos\,w + cos\,z\,sen\,w$$

Whose verification is left as an exercise.

Instead, this generalized functions are no longer bounded: $\nexists\,M \in \mathbb{R}$ such that $|cos\,z| < M$ for any $z \in \mathbb{C}$, for example, taking $z = ix$ with $x \in \mathbb{R}$, we have $cos(ix) = \dfrac{e^{-x} + e^{x}}{2}$ which is clearly not bounded. Incidentally $cos(ix)$ is the, so called, hiperbolic cosine, since the hiperbolic cosine and sine are defined by:

$$cosh\,x = \frac{e^{x} + e^{-x}}{2}\,; \quad senh\,x = \frac{e^{x} - e^{-x}}{2}$$

with $x \in \mathbb{R}$, but those definitions can be extended for $x \in \mathbb{C}$. The name hyperbolic comes from the relation: $cosh^2x - senh^2x = 1$.

LOGARITHMS: Given $w \in \mathbb{C}$ we will find all $z \in \mathbb{C}$ such that $e^z = w$. It must be $w \neq 0$ as $e^z e^{-z} = 1$. Putting $z = x + iy$, $w = |w|e^{i\alpha}$ with $x, y, \alpha \in \mathbb{R}$, we have $e^x = |w|$ and $y = \alpha + 2k\pi$ with $k \in \mathbb{Z}$, so that:

$$z = log|w| + i(\alpha + 2k\pi) \quad (*)$$

where log is the neperian logarithm defined in chap. 6. Then given $w \neq 0$, if we say that any $z \in \mathbb{C}$ such that $e^z = w$, is a logaritm of w, we have that each logarithm is given (∗) and then there are an infinity of such logarithms. Taking α such that $-\pi < \alpha \leq \pi$, then α is called the principal argument of w, denoted by $Arg\ w$, and taking $k = 0$ in (∗) it results the, so called, principal logarithm of w denoted by $Log\ w$:

$$Log\ w = \log|w| + iArg\ w$$

Note that for these principal logarithms, the relation is no longer valid valid the relation: $Log(ww') = Logw + Logw'$ for it is not valid the relation: $Arg(ww') = Arg\ w + Arg\ w'$. For example, taking $w = w' = -1$, we have tiene $Arg\ (-1)^2 = Arg\ 1 = 0$ but $Arg(-1) = \pi$ so that $Arg(-1) + Arg(-1) = 2\pi$. Note also that in case that w is positive real, we have $Arg\ w = 0$ and then $Log\ w = \log w$.

EXPONENTIALS: If $u, z \in \mathbb{C}$ if we define:

$$u^z = e^{z\ Log\ u}$$

With this definition, it is valid in general that $u^z u^v = u^{z+v}$, which is the characterictic property of the exponentials.

EJERCICIO: Fagnano, italian geometer, proved in 1719 that:

$$\frac{\pi}{4} = Log\left(\frac{1-i}{1+i}\right)^{\frac{i}{2}}$$

Verify it.

Solutions of starred exercises

Ex. E chap. 2: The implication $a + a = 0 \to a = 0$ can not be proved using only the axioms $S. 1, \ldots, D$.

The above property $(a + a = 0 \to a = 0)$ is equivalent, within the framework of the mentioned axioms, to say that $1 + 1 \neq 0$. In fact if the property is valid and $1 + 1 = 0$ it would result $1 = 0$ contradicting $P. 3$. Conversely, if $1 + 1 \neq 0$, from $a + a = 0$ follows $a(1 + 1) = 0$ and multiplying by the inverse of $1 + 1$ results $a = 0$.

To prove that $1 + 1 \neq 0$ can not be obtained from $S. 1, \ldots, D$, we must build a set with two operations, sum and product, such that it verifies $S. 1, \ldots, D$ and such that $1 + 1 = 0$. In order to do that, we take a set A with two elements, which we will denote 0 and 1. These may be arbitrary elements, we denote them 0 and 1 because we will define a sum and a product in A such that they will be neutral elements, respectively, of the sum and the product. For this to be so and for having $1 + 1 = 0$, the operations in A must be defined by the following tables:

+	0	1
0	0	1
1	1	0

·	0	1
0	0	0
1	0	1

In the set A with the operations just defined, the properties $S. 1, \ldots, D$ are valid and we have $1 + 1 = 0$, then it is impossible to derive the property $1 + 1 \neq 0$ from those.

Ex. H chap. 2: $x, y, z > 0$, prove:

$$x + y + z = 1 \implies \left(\frac{1}{x} - 1\right)\left(\frac{1}{y} - 1\right)\left(\frac{1}{z} - 1\right)$$

We have:

$$\left(\frac{1}{x} - 1\right)\left(\frac{1}{y} - 1\right)\left(\frac{1}{z} - 1\right) = \frac{1}{xyz} - \frac{1}{xy} - \frac{1}{xz} - \frac{1}{yz} + \frac{1}{x} + \frac{1}{y} + \frac{1}{z} - 1$$

but from $x + y + z = 1$ it follows, on one hand:

$$\frac{1}{xyz} = \frac{1}{x} + \frac{1}{y} + \frac{1}{z}$$

and on the other, according to exercise H.a:

$$\frac{1}{x} + \frac{1}{y} + \frac{1}{z} = \left(\frac{1}{x} + \frac{1}{y} + \frac{1}{z}\right)(x + y + z) \geq 9$$

Ex. 15.b, chap.3: If $a_1, ..., a_n$ are positive real numbers and $n \geq 2$, then:

$$a_1 + a_2 + \cdots + a_n = 1 \Longrightarrow \left(\frac{1}{a_1} - 1\right)\left(\frac{1}{a_2} - 1\right) \cdots \left(\frac{1}{a_n} - 1\right) \geq 2^{3(n-2)}$$

$$(1)$$

For $n = 2$ the result is clear and for $n = 3$ has already been proved. Assuming that (1) is valid, and that $n \geq 4$, $a_1 + a_2 + \cdots + a_n + a_{n+1} = 1$, $a_i > 0$, we must prove that:

$$\left(\frac{1}{a_1} - 1\right) \cdots \left(\frac{1}{a_n} - 1\right)\left(\frac{1}{a_{n+1}} - 1\right) \geq 2^{3(n-1)}$$

By (1) we have:

$$\left(\frac{1}{a_1} - 1\right) \cdots \left(\frac{1}{a_{n-1}} - 1\right) \cdots \left(\frac{1}{a_n + a_{n+1}} - 1\right) \geq 2^{3(n-2)}$$

so it will be enough to show that:

$$\left(\frac{1}{a_n} - 1\right)\left(\frac{1}{a_{n+1}} - 1\right) \cdots \left(\frac{a_n + a_{n+1}}{1 - a_n - a_{n+1}}\right) \geq 8$$

and this is clear by the case $n = 3$ since:

$$\left(\frac{a_n + a_{n+1}}{1 - a_n - a_{n+1}}\right) = \frac{1}{1 - a_n - a_{n+1}} - 1$$

Ex. 15.c, chap.3: If the a_i are positive real numbers and if $a_1 \cdot \ldots a_n = 1$ then:

$$a_1 + \cdots + a_n \geq n$$

For $n = 1$ it is trivial and for $n = 2$ was already seen. We assume the result valid for some $n \geq 2$ we will prove it for $n + 1$. Let then $a_1 \cdot \ldots a_{n+1} = 1$ and we have to prove that:

$$a_1 + \cdots + a_n + a_{n+1} \geq n + 1$$

By the inductive hypothesis, we have n inequalities:

$$a_1 + \cdots + a_{n-1} + a_n a_{n+1} \geq n$$
$$a_1 + \cdots + a_{n-1} a_{n+1} + a_n \geq n$$

$$\cdots\cdots\cdots$$

$$a_1 a_{n+1} + \cdots + a_{n-1} + a_n \geq n$$

and adding them up, we obtain:

$$(a_1 + \cdots + a_n)(n - 1 + a_{n+1}) \geq n^2$$

Then:

$$a_1 + \cdots + a_n + a_{n+1} \geq \frac{n^2}{n - 1 + a_{n+1}} + a_{n+1}$$

$$= \frac{n^2 + (n - 1)a_{n+1} + a_{n+1}^2}{n - 1 + a_{n+1}}$$

so that it will be enough to prove that:

$$n^2 + (n - 1)a_{n+1} + a_{n+1}^2 \geq (n + 1)(n - 1 + a_{n+1})$$
$$= n^2 - 1 + (n + 1)a_{n+1}$$

But this is equivalent to: $a_{n+1}^2 - 2a_{n+1} + 1 \geq 0$ and the result is clear.

The inductive step can also be obtained as follows: reordering, if necessary, the a_i we may assume that $a_1 \geq 1$ and $a_2 \leq 1$. By the inductive hypothesis, we have:

$$a_1 a_2 + a_3 + \cdots + a_{n+1} \geq n$$

then:

$$a_1 + a_2 + a_3 + \cdots + a_{n+1} \geq n + a_1 + a_2 - a_1 a_2 =$$
$$= n + 1 + (a_1 - 1)(1 - a_2) \geq n + 1$$

Ex. 22,e, chap. 3: $\left(1 + \frac{1}{n}\right)^n < 3$ for any $n \in \mathbb{N}$.

If $n = 1$ or $n = 2$ the result is clear. Let then $n \geq 2$, and we have:

$$\left(1 + \frac{1}{n}\right)^n = \sum_{i=0}^{n} \frac{n!}{(n - i)!\, i!\, n^i} = 2 + \sum_{i=2}^{n} \frac{n(n - 1)\dots(n - i + 1)}{n!} \frac{1}{i!} <$$

$$< 2 + \sum_{i=2}^{n} \frac{1}{i!} \leq 2 + \sum_{i=2}^{n} \frac{1}{2^i} = 2 + 1 = 3$$

Ex. 22,g, chap. 3: $r, s \in \mathbb{N}, r > s \geq 3 \implies r^s < s^r$.

Changing the notation, put $s = n, r = n + m$ with $n \geq 3$ and $m \in \mathbb{N}$. We will prove by induction on m t<many $n \in \mathbb{N}$, $(3 + \sqrt{5})^n + (3 - \sqrt{5})^n$ is natural and divisible by 2^n.

More generally, if a, b are real numbers such that $a + b$ and ab are integers, if c divides $a + b$ and c^2 divides ab, then $a^n + b^n$ is an integer and it is divisible by c^n.

for any natural n. In fact, the inductive step follows from the identity:

$$a^{n+1} + b^{n+1} = (a + b)(a^n + b^n) - ab(a^{n-1} + b^{n-1}).$$

Ex. 11.b, chap. 4: For any real a such that $a > 1$ and $m, n \in \mathbb{N}$, we have:

$$a^m - 1 \,/\, a^n - 1 \Leftrightarrow m \,/\, n$$

($\Leftarrow$): Is a particular case of ex. 11,a, chap 4.

($\Rightarrow$): Putting $n = mq + r$ with $0 \leq r < m$, we have:

$$a^n - 1 = a^r(a^{mq} - 1) + (a^r - 1)$$

Then, since by ($\Leftarrow$): $a^m - 1 \,/\, a^{mq} - 1$, and by hypothesis: $a^m - 1 \,/\, a^n - 1$, it results by the equation above: $a^m - 1 \,/\, a^r - 1$. Since $m < r$, it follows that $r = 0$, that is m/n.

Ex. 11.c, chap. 4: $a > 1$, $d = (n, m) \Rightarrow (a^n - 1, a^m - 1) = a^d - 1$.

We have $d = sn + tm$, where one and only one of the integers s, t is ≤ 0. Assume $t \leq 0$ then:

$$a^{-tm}(a^d - 1) = a^{sn} - a^{-tm} = (a^{sn} - 1) - (a^{-tm} - 1)$$

If c is a common divisor of $a^n - 1$ and $a^m - 1$, the above identity shows that c is a divisor of $a^{-tm}(a^d - 1)$ and as c must be coprime with a, it results $c \,/\, a^d - 1$. Since, moreover, by the above exercise $a^d - 1$ is a common divisor of $a^n - 1$ and $a^m - 1$, it follows that $a^d - 1$ is its greatest common divisor.

Ex 14, chap. 4: $\dfrac{1}{n+1}\dbinom{2n}{n}$ is an integer for any $n \in \mathbb{N}$.

We have:

$$(2n + 1)\dbinom{2n}{n} = (n + 1)\dbinom{2n + 1}{n}$$

and as $n + 1$ and $2n + 1$ are coprime, it follows that $n + 1$ divides $\dbinom{2n}{n}$.

Ex. 13, chap. 5: $11 \,/\, a^3 - b^3 \Leftrightarrow 11 \,/\, a - b$.

288

The statement can be translated in $\mathbb{Z}_{11}$ like follows:
$$\bar{a}^3 = \bar{b}^3 \Leftrightarrow \bar{a} = \bar{b}$$
this makes evident the implication $\Leftarrow$, while the implication $\Rightarrow$ states that the funcion of raising to the third power in $\mathbb{Z}_{11}$ is injective, which results clear from the following table:

$\bar{x}$	$\bar{0}$	$\bar{1}$	$\bar{2}$	$\bar{3}$	$\bar{4}$	$\bar{5}$	$\bar{6}$	7	$\bar{8}$	9	10
$\bar{x}^3$	$\bar{0}$	$\bar{1}$	$\bar{8}$	$\bar{5}$	$\bar{9}$	$\bar{4}$	$\bar{7}$	$\bar{2}$	6	$\bar{3}$	10

Ex. 10, chap. 6: For $a > 0$ we define recursively:
$$x_1 = \sqrt{a}, \quad x_{n+1} = \sqrt{a + x_n}$$
Assuming provisionally that there exist $s = supA$, we must have: $x_{n+1} = \sqrt{a + x_n}$ then $x_n \le s^2 - a$ that is $s^2 - a$ is an upper bound of A, hence $s \le s^2 - a$. Since:

$$s^2 - s - a \ge 0 \Leftrightarrow \left(s - \frac{1}{2}\right)^2 \ge \frac{1}{4} + a \Leftrightarrow s \ge \frac{1}{2} + \sqrt{\frac{1}{4} + a}$$

it will be enough to prove that $\frac{1}{2} + \sqrt{\frac{1}{4} + a}$ is an upper bound of A to obtain that the supremum exists and that $s = \frac{1}{2} + \sqrt{\frac{1}{4} + a}$.

Proceeding by induction, we have: $x_1 = \sqrt{a} \le \frac{1}{2} + \sqrt{\frac{1}{4} + a}$. Assuming that $x_n \le \frac{1}{2} + \sqrt{\frac{1}{4} + a}$ we have:

$$x_{n+1} = \sqrt{a + x_n} \le \sqrt{a + \frac{1}{2} + \sqrt{\frac{1}{4} + a}} =$$

$$= \sqrt{\left(\frac{1}{2} + \sqrt{\frac{1}{4} + a}\right)^2} = \frac{1}{2} + \sqrt{\frac{1}{4} + a}$$

Ex. 13, chap. 6: Given $a, b \in \mathbb{R}$ with $0 < a < b$, we define inductively:

$$a_1 = a; \quad b_1 = b; \quad a_{n+1} = \sqrt{a_n b_n}; \quad b_{n+1} = \frac{a_n + b_n}{2}$$

Prove:

1) $a_n < a_{n+1}$ and $b_{n+1} < b_n$ for any $n \in \mathbb{N}$.
2) $A = \{a_n | n \in \mathbb{N}\}$ is bounded from above, and $B = \{b_n | n \in \mathbb{N}\}$ is bounded from bellow.
3) $\sup A = \inf B$.

Note that if x, y are distint positive real numbers, their *geometric mean*: $\sqrt{xy}$ is strictly less than their *arithmetic mean*: $\frac{x+y}{2}$, then:

$$a_n < b_n \quad \forall n \in \mathbb{N} \quad (1)$$

Hence, $a_n^2 < a_n b_n$ and $a_n + b_n$, so that $a_n < \sqrt{a_n b_n} = a_{n+1}$ and $b_{n+1} = \frac{a_n + b_n}{2} < b_n$, that is:

$$a_n < a_{n+1} \quad \text{and} \quad b_{n+1} < b_n \quad (2)$$

From (1) and (2) follows that a is a lower bound of B and that b is an upper bound of A, so that there exist $s = \sup A$ and $t = \inf B$. Clearly $s \leq t$ and we will prove that the relation: $s < t$ leads to contadiction. We have:

$$b_{n+1} - a_{n+1} < \frac{b_n - a_n}{2} \quad (3)$$

as $b_{n+1} - a_{n+1} = \frac{a_n + b_n}{2} - a_{n+1} < \frac{a_n + b_n}{2} - a_n = \frac{b_n - a_n}{2}$. From (3) we obtain inductively that:

$$b_{n+1} - a_{n+1} < \frac{b - a}{2^n}$$

If it were $s < t$, by archimedianity there would exist $n \in \mathbb{N}$ such that $n(t - s) > b - a$, then $2^n(t - s) > b - a$ and so we have:

$$t - s > b_{n+1} - a_{n+1}$$

which is a contradiction, since $s > a_{n+1}$ and $t < b_{n+1}$.

Ex. 17, chap 6: There is one and only one function $f: \mathbb{N} \to \mathbb{N}$ such that for any $m, n \in \mathbb{N}$ it verifies:

1) $f(mn) = f(m)f(n)$.
2) $m \neq n$ and $m^n = m^n \Rightarrow f(m) = n$ or $f(n) = m$.
3) $m, n \geq 3$ and $m^n < n^m \Rightarrow f(n) < f(m)$

By ex.22g, chap.3, we have:

$$3 \leq n < m \Rightarrow m^n < n^m$$

from where it follows that, if $n < m$ the only solution of $m^n = n^m$ is given by $n = 2$ and $m = 4$. Hence by condition 2) follows: $f(2) = 4$

or $f(4) = 2$. But, by condition 1): $f(4) = f(2)f(2)$, so that we must have:

4) $f(2) = 4$.

Moreover, from 3) follows that: $3 \le n < m \Rightarrow f(n) < f(m)$, but $f(1) = 1, f(4) = 16$ and $f(6) = 4f(3) > f(4)$, then we have:

$$f(1) < f(2) = 4 < f(3)$$

hence:

5) f is strictly increasing,i.e.: $n < m \Rightarrow f(n) < f(m)$.

So far we have proved that, assuming 1), the conditions 2) and 3) impliy 4) and 5).

From the above discusion, follows that converse is also valid and we have that the conditions \ 1), 2) and 3) are equivalent to the conditions 1), 4) and 5).

The function "squaring" obviously satisfies 1), 4) and 5). We will see that, if a function $f: \mathbb{N} \to \mathbb{N}$ satisfies 1), 4) y 5), we must have $f(n) = n^2$ for any natural n.

In fact, given $n \in \mathbb{N}$ and taking $h, k, r, s \in \mathbb{N}$ (theorem 5.3, chap. 6) such that:

$$n^2 - 1 < 4^{\frac{h}{k}} < n^2 < 4^{\frac{r}{s}} < n^2 + 1$$

Since $2^h < n^k$ and $n^s < 2^r$ we have: $f(2^h) = 4^h < f(n)^k$ and $f(n)^s < f(2^r) = 4^r$, so that,

$$4^{\frac{h}{k}} < f(n) < 4^{\frac{r}{s}}$$

and hence: $f(n) = n^2$.

Ex. 18, chap. 6: If $a_1, \ldots, a_n$ are positive real numbers, then the following inequality is valid:

$$\sqrt[n]{a_1 \ldots a_n} < \frac{a_1 + \cdots + a_n}{n}$$

This can be obtained from ex. 15,c, chap.3. In fact, putting $P = a_1 \ldots a_n$, we have:

$$\frac{a_1}{P} \frac{a_2}{P} \ldots \frac{a_n}{P} = 1$$

and by the above mentioned exercise, follows:

$$\frac{a_1}{\sqrt[n]{P}} \frac{a_2}{\sqrt[n]{P}} \ldots \frac{a_n}{\sqrt[n]{P}} \ge n$$

which is equivalent to the inequality between the geometric and arithmetic means.

Ex, 19, chap. 6: Find all x such that:

$$log_x 2 + 3 \, log_{2x} 2 - 10 \, log_{4x} 2 = 0 \quad (1)$$

Putting $y = log_2 x$, we have:

$$2^y = x; \; 2^{y+1} = 2x; \; 2^{y+2} = 4x, (2)$$

then:

$$log_x 2 = \frac{1}{y}; \; log_{2x} 2 = \frac{1}{y+1}; \; log_{4x} 2 = \frac{1}{y+2}$$

and replacing in (1), we obtain:

$$\frac{1}{y} + \frac{3}{y+1} - \frac{10}{y+2} = 0$$

this is a second degree equation in y with roots: $y_1 = \frac{1}{2}, y_2 = \frac{2}{3}$, and by (2), we obtain two values of x: $x_1 = 2^{\frac{1}{2}} : x_2 = 2^{\frac{2}{3}}$, which satisfy equation (1).

Ex. 15, chap.7: Let u, v, w be distinct complex numbers. They determine an equilateral triangle if, and only if:

$$u^2 + v^2 + w^2 - uv - uw - vw = 0 \quad (1)$$

If we put $x = u - v, y = v - w, z = u - w$ (so then $z = x + y$), the conditions to be an equilateral triangle are:

$$\frac{x}{y} \notin \mathbb{R} \text{ (to be a triangle) and: } |x| = |y| = |z| \quad (2)$$

We have the following equivalences:

$$(1) \Leftrightarrow (u - v)^2 + (v - w)^2 + (w - u)^2 = 0$$
$$\Leftrightarrow x^2 + y^2 + (x + y)^2 = 0 \Leftrightarrow$$
$$\Leftrightarrow x^2 + y^2 + xy = 0 \Leftrightarrow \frac{x^2}{y^2} + 1 + \frac{x}{y} = 0 \Leftrightarrow \frac{x}{y} = \frac{-1 \pm \sqrt{3}i}{2} = e^{\pm i\frac{2\pi}{3}} \quad (3)$$

If we assume (3) then $\frac{x}{y} \notin \mathbb{R}$ and, as $\left|\frac{x}{y}\right| = 1$ we have $|x| = |y|$ and by symmetry we obtain (2). Conversely, assuming (2), from $|x| = |y|$ follows $x = ye^{i\theta}$ for some θ. We have:

$$z = x + y = y(e^{i\theta} + 1)$$

then $\left|e^{i\theta} + 1\right| = 1$ or $2\cos\theta = -1$. It follows that $\theta = \pm\frac{2\pi}{3} + 2k\pi$ and thus we obtain (3).

BIBLIOGRAPHY

1. AHLFORS L.: Complex Analysis. McGraw-Hill, New York (1966).
2. ANDREESCU,A. ANDRICA, D.: Complex Numbers from A to...Z. Birkäuser (2005).
3. BERLEKAMP E.: Algebraic Coding Theory. McGraw-Hill, New York (1968).
4. BIRKHFF S.- MAC LANE S.: Álgebra Moderna. Teide, Barcelona (1960).
5. BOURBAKI: Elements of History of Mathematics. Springer (1994).
6. BOYER, C.: A History of Mathematics. Merzbach, U. Wiley (2010).
7. CARDANO, G.: Ars Magna. Dover (2007).
8. CARTAN H.: Teoría elemental de funciones analíticas de una y varias variables complejas. Selecciones científicas, Madrid (1968).
9. CHILDS L.: A Concrete Introduction to Higher Algebra. Springer-Verlag, New York (1979).
10. DICKSON L.: History of the Theory of Numbers (3 vols.). Chelsea, New York (1971).
11. DUNHAM: Euler: The Master of Us All. M.A.A. (1994).
12. EDWARDS H.: Fermat's Last Theorem. Springer-Verlag, New York (1977).
13. GARLAND, T.H. Fascinating Fibonaccis. Dale Seymour pub. (1987).
14. GENTILE E.: Notas de Algebra 1. EUDEBA (1976).
15. GOLDHABER J. - ERLICH G.: Algebra. Macmillan, Toronto (1971).
16. GRIMALDI R.: Matemáticas Discreta y Combinatoria. Addison-Wesley, Washington (1989).
17. HALMOS P.: Teoría Intuitiva de los Conjuntos. Continental, Mexico (1966).
18. HARDY G.: Curso de Análisis Matemático (trad.de A Course of Pure Mathematics). Nigar, Buenos Aires (1962).
19. HARDY G.-WRIGHT E.: An Introduction to the Theory of Numbers. Oxford, New York (1960).
20. HICKS, S.-KELLEY, D.: Readings for Logical Analysis. W.W.Norton (1999).

21. HILBERT D.- ACKERMANN W.: Elementos de lógica teórica. Tecnos, Madrid (1975).
22. HOFMANN J.: Historia de la matemática (3 vols.). UTEHA, (1960).
23. IFRAH, G.: The Universal History of Numbers. Willey (2000).
24. KELLEY, D.: The Art of Reasoning: An Introduction to Logic and Critical Thinking. W.W.Norton & Co (2013).
25. KNEALE W.- KNEALE M.: El desarrollo de la lógica. Tecnos, Madrid (1972).
26. LANG S.: Algebra. Aguilar, Madrid (1971).
27. LE VEQUE W.: Topics in Number Theory. Addison-Wesley, Reading (1965).
28. LENTIN A.- RIVAUD J.: Algebra Moderna. Aguilar, Buenos Aires (1969).
29. Mc INERNY, D.C.: Being Logical: A Guide to Good Thinking. Random House (2005).
30. MORDELL L.: Diophantine Equations. Academic Press, London (1969).
31. MOSTOW G.- SAMPSON J.-MEYER J.: Fundamental Structures of Algebra. McGraw-Hill, New York (1963).
32. NEUGEBAUER, O.: The Exact Sciences in Antiquity. Dover (1969).
33. NEWMAN, J. R.: The World of Mathematics. (4 vol) Dover (2003).
34. RIBENBOIM P.: The book of Prime Number Records. Springer-Verlag, New York (1989).
35. RIBENBOIM P.: Thirteen Lessons on Fermat's Last Theorem. Springer-Verlag, New York (1979).
36. SCHAEFFER: An Introduction to Nonassociative Algebra. Dover (2017).
37. SHANKS D.: Solved and Unsolved Problems in Number Theory. Spartan, NewYork (1971).
38. TARSKY A.: Introduction a la logique. Gauthier Villars, Paris (1969).
39. VAN DER WAERDEN,: A History of Algebra. Springer (1985).

www.ingramcontent.com/pod-product-compliance
Lightning Source LLC
Chambersburg PA
CBHW071251220526
45468CB00001B/85